HIGHWAY ENGINEERING

HIGHWAY ENGINEERING

Professor T.D. Ahuja
(Director)
Institute of Engineering and Rural Technology,
Allahabad

STANDARD BOOK HOUSE

unit of: **RAJSONS PUBLICATIONS PVT. LTD.**

1705-A, Nai Sarak, PB.No. 1074, Delhi-110006 Ph.: +91-(011)-23265506
Show Room: 4262/3, First Lane, G-Floor, Gali Punjabian, Ansari Road, Darya Ganj,
New Delhi-110002 Ph.: +91-(011)43551085 Tel Fax : +91-(011)43551185
Fax: +91-(011)-23250212
E-mail: sbh10@hotmail.com www.standardbookhouse.com

Published by:
RAJINDER KUMAR JAIN
Standard Book House
Unit of: Rajsons Publications Pvt. Ltd.
1705-A, Nai Sarak, Delhi - 110006
Post Box: 1074
Ph.: +91-(011)-23265506 Fax: +91-(011)-23250212

Showroom:
4262/3. First Lane, G-Floor, Gali Punjabian
Ansari Road, Darya Ganj
New Delhi-110002
Ph.: +91-(011)-43551085, +91-(011)-43551185
E-mail: sbhl0@ hotmail.com
Web: www.standardbookhouse.com

Third Edition : 2015

Price: **₹ 280.00**

ISBN: 978-81-89401-36-8

Typeset by:
C.S.M.S. Computers, Delhi.

Printed by:
R.K. Print Media Company, New Delhi–110039

Preface

After the First World War the importance of highways was felt and realized. The concept of highway engineering has changed during the last two decades. The thumb rule concept has become a thing of the past. With the increasing importance of highways for the prosperity and integrity of the country and with the increasing cost of construction and maintenance of highways, the trend of construction, planning and designing has also changed. The Central Road Research Institute and P.W.D. research centres all over the country have contributed a lot in the design, planning road user safety, construction and economy etc.

The present work is the outcome of author's long association with the subject as a teacher and as a student. Efforts have been made to present the subject matter in a very lucid and comprehensive manner. The author does not claim any originality but sufficient pains have been taken in compiling the work by consulting important works and Road Research Journals. The subject matter is presented from the introduction so that the book may prove useful to diploma and degree students as well as practising engineers. The book presents acceptable theory and construction practices. Important topics such as bituminous roads, stabilized earth roads, traffic engineering, pavement design and highway planning and economics have been compre hensively dealt. Hill Roads including construction and layout of tunnels have been given special emphasis. Airport engineering, though it is not a part of highway engineering, has also been touched so as to introduce the subject matter.

I take this opportunity to express my gratitude to Padamshri R.S. Gahlowt, Chairman and Managing Director (Retd). Hindustan Steel Co. Ltd. for his valuable guidance, help and blessings and my friend and colleague Shri G.S. Birdie, Consulting Engineer for the preparation of a large number of drawings and consultations.

Any suggestion for the improvement of the book in the forthcoming editions will be thankfully acknowledged and welcomed. For errors or omissions and constructive criticism from the readers and users are welcome.

Allahabad **T.D. AHUJA**
2011

Foreword

Roads serve the same purpose in developing the economy of a country as the veins of human body which are the lifelines of human existance. With the Technological and Scientific developments, there has been immense growth in Mechanisation in all the construction activities, more particularly the roads. Not only there has been phenominal growth in the traffic intensity, the speed and axle loads have also undergone unprecedented increase. All these factors call for a rational and scientific approach to the design and upkeep/maintenance of roads.

Shri T.D. Ahuja, Director Institute of Engineering and Rural Technology Allahabad, who has a long and commendable record in the history of this one of the leading institutes, has set himself to the task of compiling book on "Highway Engineering" which is bound to be of consider use not only as Textbook but also a reference book to the Post-Diploma students specialising in Highway Engineering.

I have no doubt that this book on "Highway Engineering" would benefit all those who are engaged in the pursuit of their educational achievement or on designing and construction/maintenance of Highways and Rural Roads. The construction of roads is no more limited to Highways and Rural Roads only but their importance has gained considerable urgency in setting up of Heavy Industrial and Metallurgical Plants and Projects, where the roads have to cater to the needs of heavy axle loads and fast moving traffic, as observed during my association with Govt, of India Public Sectors Organisations for the last three decades in various positions while in service and the post retirement period as Adviser (Engg.) U.P. State Planning Commission (1974–82) and as consultant to various Public Sector Organisations engaged on heavy-construction projects.

R.S.GAHLOWT
F1533/LM1543IRC
Ex-Chairman & Managing Director
Hindustan Steel Works Constn. Ltd (G.I.)
and Ex-Director Steel Authority of India
Awarded **"PADMA SHREE".**

Contents

1

Introduction

In this Chapter you will study,

• Definitions • Means of Communications • History of Highway Development • Development of Roads in India • Role of Transportation in Rural Development • Rural Roads Development Plan • Fearures of Ancient Roads • Indian Road Congress • Nagpur Road Conference • Salient Features of Nagpur Plan • Role of Indian Road Congress • Types of Roads • Third Road Plan (1981–2001) Classification • Classification of Roads • Road Financing in India • Advantages of Roads • Improtance of Roads in India • Problems

GENERAL

Road Engineering is one of the important branches of engineering. It deals with the construction, design and maintenance of roads of different kinds. The first chapter has been particularly devoted to the definition of certain terms used in road engineering, types of roads and importance of roads for a country. The importance of roads in a country are comparable to the veins in the human body. Just as the veins supply the blood to the different parts of a body, so are the roads in a country, they convey men, material and information to the different parts of a country. By going through this chapter the reader will have a fair idea about the various technical terms used in road engineering and the various types of roads existing in our country.

1.1 DEFINITIONS

1. Road. A way for vehicles and for other types of traffic over which they may lawfully pass. It includes the entire area comprising the roadway and all structures pertaining to the road within the limits of the defined boundary or 'right-of-way'.

2. Highway. It is main and an important road in a road system by all means.

3. Roadway. It is the portion of the road (included within the construction limits) usually used for traffic. It includes carriage way and the shoulders.

4. Carriageway. It is the portion of the roadway designed and constructed for vehicular traffic.

5. Right-of-way. (i) The land secured and reserved for development of a road and all structures pertaining to the road, (ii) The privilege of use of a way, acquired by law for the traffic, custom or usage.

6. National Highway. These are the most important roads connecting capital cities of different states and territories.

7. State Highway. These are the main roads within a state connecting important towns of the state.

8. District Roads. These are less important roads within the district boundaries, connecting its various towns, tehsils etc.

9. By-pass Road. The road constructed around the city or town in such a way, to avoid congested areas or other obstructions in the movement of thorough traffic.

10. Loop Road. It is a route formed by a road or a series of roads to avoid an obstruction or provide an alternative way for traffic.

11. Ring Road. It is the circumferential road constructed around an urban area to enable free flow of traffic.

12. Radial Road. It is a road providing direct link between the centre and outskirts of an urban area.

13. Drive way. It is the way to secure access from a road to private property.

14. Service-Road. It is a subsidiary road constructed between a road and buildings or properties facing thereon and connected only at the selected points with the main road. Usually water, sewer and gas pipes are laid through this road and connections given to the buildings.

15. Fair-weather Road. It is a road which can be used for the traffic during dry weather only. In monsoons this road is closed for traffic.

16. Island. It is a central or subsidiary area in a carriage way, generally at road junctions, shaped and placed so as to constrain and conuol the movement of the traffic.

17. Transition length. It is the length of the transition curve connecting a straight-length of a road with another main curve which may be circular or transitional.

18. Sub-way. It is the underground passage or tunnel to permit the movement of traffic, or to accommodate service pipes, sewers, cables etc.

19. Fly-over. It is the road junction so designed so that traffic streams are divided to enable them to pass over or under each other.

20. Traffic density. It is the number of vehicles using the road per hour during peak periods and is the average of several peak days. The daily traffic is approximately ten times the maximum hourly traffic.

21. Formation. It is the final ground surface after completion of earthwork.

22. Formation width. It is the finished top width of earth work in fill or cut to lay the road structure. It is also known as 'roadway'.

23. Base course. It is the first layer of road structure laid over the soil formation.

24. Base-coat. It is the intermediate course between the base course and the wearing coat.

25. Sub-crust. It is an intermediate layer which acts as a cushion between the foundation and the pavement.

26. Wearing coat. It is the topmost coat of the road, over which the traffic moves.

27. Carpet. It is the wearing coat of bitumen or tar concrete having thickness more than 25 mm.

28. Pavement. It is the hard crust placed on the soil formation.

29. Creteways. It is a carriageway in which a cement concrete wearing surface is provided for the wheel tracks only. Usually sugar mills construct the creteways for collection of sugarcane from the interior of rural areas.

30. Edging. It is the bricks or blocks of stones embedded along the edges of a pavement for its protection from damage by traffic.

31. Berm, Haunch or Shoulder. This is the portion of the roadway just beyond the edges of a carriageway, which is used by the traffic for occasionally passing, during overtaking etc. The strip of the roadway between side drain and the lower edge of the bank is also included in it.

32. Camber. It is also known as transverse slope. It is the convexity given to the curved cross-section of the carriage way for draining water. It may be defined as crossfall of the road.

33. Crown. The cross-sectional highest point of the road.

34. Super-elevation, Banking or Cant. It is the inward tilt or transverse inclination given to the cross-section of the road at the curve positions, to reduce the effect of the centrifugal force on the moving vehicles.

35. Footpath. Footpaths are particularly provided in the case of urban roads, and are 15 cm to 20 cm higher than the road surface. Footpaths are generally made of bricks or cement concrete. Shoulders are generally in level with the road surface, having a slope towards the drain side. Shoulders and footpaths provide lateral strength to the road and prevent the edges of the road from wear and tear.

36. Cycle track. In some urban roads, running through congested areas, some portion of the road-way is reserved for cycle traffic and is called *cycle track,* while some portion, generally in the centre, for high speed vehicles such as motor cars, trucks etc. and is called *motor way* or *express way.* Footpaths are reserved for pedestrians only.

1.2 MEANS OF COMMUNICATIONS

Communication is defined as the means of moving men and material (or postal information) from one place to another. There are four means of communication :

1. Roadways
2. Railways
3. Waterways
4. Airways

1. Roadways. These are the means of communication on land. Roadways help in carrying large number of passengers and goods from one place to another. They play a very vital role in the development of a country. Road transport is a means of detailed distribution between homes, shops, factories etc. Roads thus form an integral part of the country to facilitate transport from one place to another.

2. Railways. Railways are a very good means of communication for men, material and information. Railways are steel tracks laid on the ground, over which the train moves. In the case of railways the tractive resistance between the steel rail and steel tyre is reduced to $\frac{1}{6}$th or $\frac{1}{5}$th of the pneumatic tyre on a modern highway. The steel track can take heavy axle loads nearly three times as heavier as the road and the trains can run at a higher speed.

3. Waterways. Waterways are known to men for a very long time. The existence of large sheets and depths of water are taken advantage of, for transport of men and material by means of boats and ships. The cost of transport is very moderate as compared with the cost of mechanical transport by road or rail. But these means of transport are limited or restricted to only those places, where there is enough water. This method of transport is very slow as compared to other methods.

4. Airways. This method of communication is the quickest but it is very costly. Hence the method is confined to the rich and the aristocracy. Elaborate arrangements are being made for this sort of communications. Roads and railways work as feeders without the help of which this means of communication is not possible.

Hence we see from the above explanations that roads are of utmost importance for all other types of communication. Roadways are inter-communicating links. The entire development of a country depends on the efficient and widespread network of roads throughout the country.

1.3 HISTORY OF HIGHWAY DEVELOPMENT

Roads have been put to use from a very early time. The first hard surface was constructed in 3500 B.C. in Mesopotamia. A stone paved road was located as early as 1500 B.C. on the Island of Crete in Mediterranean Sea. A hard surface road was constructed between Babylon and Egypt in 539 B.C. Actual development of the roads started in the 18th century in France and England. Mc Adam in England used crushed stone as road surfacing material (1576–1836) which method is still used, though in a modified form. The road development was very slow in USA till nineteenth century. Road development got a momentum in the first two decades of the 20th century with the improvement of motor vehicles which proved to be easiest method of transporting men and material from one place to another.

1.4 DEVELOPMENT OF ROADS IN INDIA

History tells us that Indians were adapt in the science of Road Construction long-long ago. Excavation of Mohanjodaro and Harappa has revealed that even in 3500 B.C., Indians knew this science. In about 600 B.C., a metallic road 6 m to 7.5 m wide existed in Rajgir near Patna. The road was made of stone. In about 300 B.C. Kautilya got constructed a National Highway connecting North West Frontier Province and Patna. Chandra Gupta Maurya had opened a special department for looking after the construction and maintenance of roads. In about 269 B.C. during the regime of Ashoka, there was a good network of roads in India. In the days of Ashoka, trees were planted along the sides of the roads and rest houses were constructed at a distance of 6 to 10 km along the road side.

Muslim ruler Mohamed Tughlaq constructed a road connecting Delhi to Daultabad. Shershah became famous for the construction of several roads in India. The longest road constructed in his time was from Lahore to Sonargaon (Bengal) at present this road is known Grand Trunck Road (GT).

During the Mughal period about 24 long roads connecting different parts of the country were constructed.

In the British period Lord William Bentinck was the first who revived the idea of road construction. It was only during the time of Lord Dalhousie that a Central P.W.D. was formed to look after the work of road construction. Lord Mayo and Lord Ripon contributed a lot in the construction of roads.

Wilh the development of railways, the road construction received a serious setback. Road construction was given a secondary importance. But the circumstances changed after World War I. Motor transport came to the forefront. By the time the existing roads were badly deteriorated and they could not keep pace with the increased tyre traffic. The Central Government became conscious of this. In 1930, a Central Road Organisation was set up and in 1935 a Transport Advisory Committee was formed. In 1961, the Road Congress met to discuss the development of roads in India. In 1931, a semi-official body called the Indian Road Congress was set up to give recommendations for the development of roads.

After the Second World War I, on the recommendations of I.R.C. the conference of all Chief Engineers was held in December, 1943 in Nagpur. It drew a 10-year plan for the construction and development of nearly 5,29,600 km of all kinds of roads and bridges at an estimated cost of Rs. 448 crores. This plan is known as Nagpur Plan.

In September 1950, the Central Road Research Institute was started in Delhi. This institute is financed and controlled by Central Transport Ministry. It gives technical advice to the State and Central Governments on various aspects of road construction and development. The Government of India through its five-year plans is taking keen interest in the development of all kinds of roads.

1.5 ROLE OF TRANSPORTATION IN RURAL DEVELOPMENT

India lives in villages. More than 73% of the Indian population lives in villages where there is no proper transportation or means of communication. There are villages where it is impossible to reach during rainy season. They are completely cut off from the mainstream. Development of such remote areas is only possible with the development of roads and communication links. With the development of roads in rural areas many problems of rural India can be solved such as marketing and transportation of goods, medical facilities, law and order, schooling for children, exportation etc. With the development of roads in rural area the urge for the migration to urban centres will decrease, thus helping in balanced development of the country as a whole.

1.6 RURAL ROADS DEVELOPMENT PLAN

The number of big and small villages in our country is nearly 5,75,856. 57% of the villages have population more than 1500. 36.3% villages are with population between 1000 and 1500 and the rest of villages with population less than 1000. Less than half of the villages are connected with all weather roads. Government have since realised the importance of village roads and are actively engaged in planning minimum need based road development in rural and semi-urban areas for an overall development of the country. With rural development and rural construction departments the rural road development will get boost up.

It is estimated that about Rs. 11,000 crore will be required for providing all weather roads to nearly 90% of the villages. In the seventh five year plan out of Rs. 3,800 crore for road development nearly Rs. 1,200 crore was earmarked for rural roads only. Various states are actively involved in planning rural roads in the eighth plan.

1.7 FEATURES OF ANCIENT ROADS

(i) *Roman Roads.* During the period of Roman Civilisation many roads were built of thick stone blocks. Roman roads were constructed without the consideration of gradients. Total thickness of these roads used to be between 0.75 m to 1.2 m. For the construction of these roads an open trench was excavated by removing the loose soil. One or two layers of boulders with or without lime were laid. Over the boulders, a layer of lime concrete was laid. Over one or two layers of lime concrete dressed stones were laid for the wearing surface.

(ii) *Metcalf Road.* John Metcalf constructed a road in England in the year 1780–85. He was responsible for the construction of 290 km of roads in the northern region of England. Metcalf used stone boulders or big stones and laid them on edge. The cavities or voids in between the boulders were filled with smaller stones and spalls. The surface was then compacted by bigger pieces of stone to make the weaving surface.

(iii) *Tresaguets Road.* Piere Tresaguets of France developed a new technique of construction of road in the eighteenth century B.C. Tresaguets gave due consideration for sub-soil moisture, sub-grade strength and the drainage conditions. During the period of Nepoleon major development of road system in France took place.

In Tresaguets system of road construction, sub-grade was properly prepared. Large foundation stones were packed tightly on the sub-grade, laid on edge. The voids on interstices of these stones were filled with smaller stones. For the first time camber was provided in the road for drainage. The shoulders were also made sloping towards outside to drain off rain water.

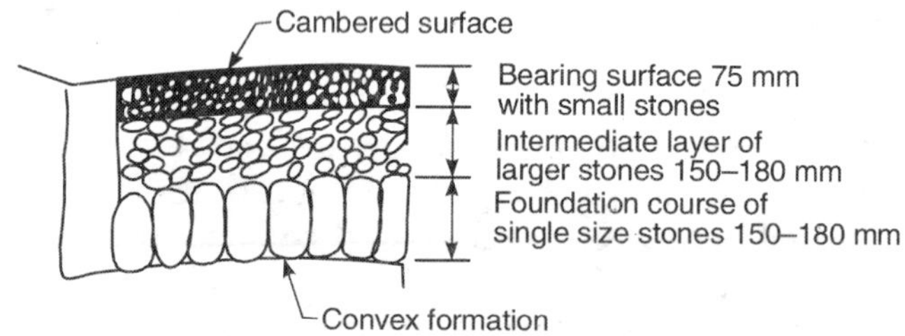

Figure 1.1 *Tresaguets road structure*

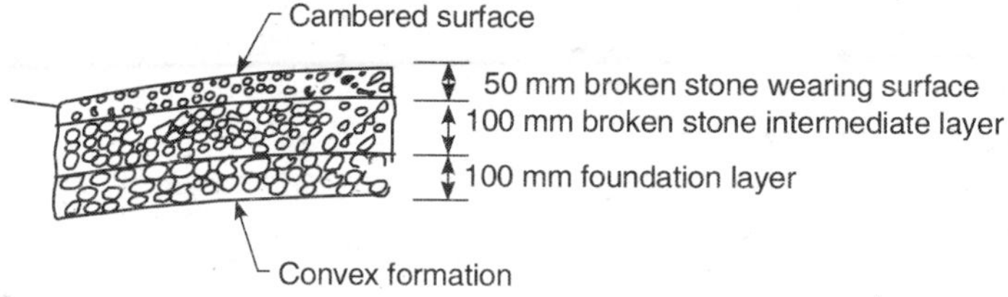

Figure 1.2 *Macadam road structure*

(iv) *Telford Road.* In England in the year 1786 or so, Thomas Telford started the work on road construction. He emphasized the work on road construction. He also pleaded for definite cross slope or camber in the surface of road, depending on the surface condition and intensity of rainfall. The constructional features of a Telford road are :

(a) The sub-grade was kept horizontal.

(b) Heavy foundation stones of size 17 cm were kept on the edges and bigger pieces of 22 cm size were kept in the middle of road.

(c) Two layers of broken stones were kept on the foundation layer. These layers of stones were properly compacted.

(d) A wearing course of nearly 4 cm thickness was then laid and completed.

(e) The total thickness of road varies from 35 cm at the edges to 42 cm at the centre.

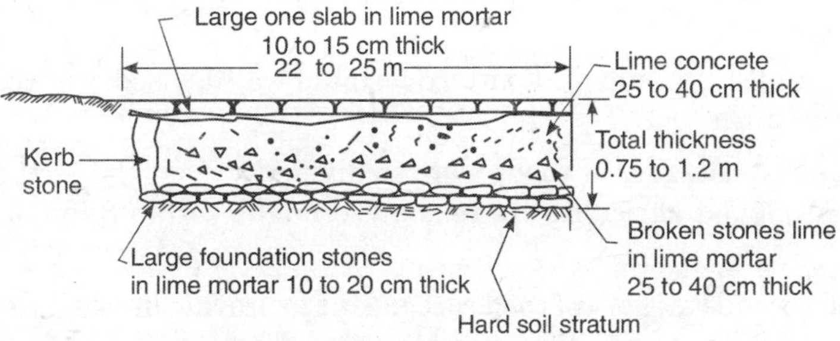

Figure 1.3 *Roman road structure*

1.8 INDIAN ROAD CONGRESS

In 1934 Indian Road Congress (I.R.C.) a semi-official technical body was formed by the Govt. of India to pool in experience, expertise and ideas for the planning and development of roads in the country to meet the demand of vehicular traffic. I.R.C. was also entrusted with the responsibility of formulating standards and specifications of different type of roads for their construction and maintenance. The I.R.C. has since become the main forum for exchange of ideas on the development and planning of road network. Now the Indian Road Congress has become an active body of Govt. of India controlling standardization of specifications and recommendations as regard materials, design and construction of roads and bridges. The I.R.C. publishes journals, research papers, standard specifications for different roads. I.R.C. has become the father of Highway Engineering.

1.9 NAGPUR ROAD CONFERENCE

During the Second World War II the importance of roads was realized. In 1943 a Conference of Chief Engineers of all states was convened at Nagpur at the initiative of Indian Road Congress to chalk out a road development plan for the country. A 20 year plan 1943–63 was finalised. It was the first attempt of co-ordinated planning for the development of roads in the country. During the first two five year plans, the Nagpur plan target was achieved.

1.10 SALIENT FEATURES OF NAGPUR PLAN

The following are the salient features of Nagpur Plan :

(i) The responsibility of construction and maintenance of various type of roads was assigned to central government, various state governments, departments, local bodies and other departments specifically.

(ii) It was a 20 year plan aimed at integrated road planning and development.

(iii) A definite star and grid pattern formula was devised over the existing irregular pattern.

(iv) The first category of roads were so divided so that the farthest points and the developed agricultural areas are brought within 8 km of metalled road.

(v) The second category of roads are meant to provide internal road system linking small villages with first category roads. The road length in the second category roads is worked out on the basis of villages of different population ranges.

(vi) Agricultural and industrial development allowance in the coming 20 years was taken into consideration.

(vii) The length of railway track in the adjoining area was also considered in deciding the length of first category roads.

1.11 ROLE OF INDIAN ROAD CONGRESS

History of Roads date back to Vedas. The Rig Veda speaks of mahapath on which royal chariots used to ride. Ramayan describes roads of varying nature used by Lord Rama in his journey to Lanka. There was a network of roads linking various parts of the country during the period of Ashoka. Upto Second World War II very little attention was paid to the construction of surface roads mainly due to paucity of funds. The Indian Road Congress (I.R.C) came into being with the principal objective of providing a national forum of pooling the experience and ideas of construction, maintenance and planning of the network of roads in the country. In more specific terms the following are the broad objectives of I.R.C :

1. To promote and encourage the science and practice of building and maintenance of roads.

2. To provide a channel to the expression of collective opinion on matters for the development of road in our country.

3. To promote use of standard specifications and to formulate newer specifications.

4. To advise regarding education, experience, research connected with roads.

5. To hold periodical meetings to discuss technical questions regarding roads.

6. To suggest legislation for the development improvement and protection of roads.

7. To suggest new methods of construction, administration, planning designing, testing of materials and maintenance of roads.

The vital role of road network in a country can be further emphasised, for the socio-economic development of a country. It is only roads which serve as a feeder for railway, airways and seaways and can penetrate deep into the rural areas. The importance of roads is increasing day by day and so its problems of construction and routine maintenance are to be taken care of. Due to financial constraints the problem of construction of new roads and their proper maintenance can be tackled through concerted efforts backed by experience and pooling of all available resources, knowledge and expertise. The I.R.C. has played a significant role in the promotion of these activities, through seminars, lectures and group discussions. Over the years the I.R.C. has touched upon and contributed to the improvement of the various facets of road development. Adaptation of standardized practice in design, construction and maintenance with regard to variabilities in terrain, soil and climate also comes under the purview of I.R.C. The Road Congress has been closely associated with road planning in the country which forms the forerunner to any development activity. Indian Road Congress also plays an important role for the development of rural roads. In 1958 the I.R.C. set up a committee which was then called the Community Project Road Maintenance Committee. The aim of this committee was to recommend measures and means for proper and adequate maintenance of rural roads which are otherwise fair weather roads. In 1969, the committee was renamed as Rural Roads Committee for bringing out a comprehensive report dealing with planning, specifications, financing, construction and maintenance of all rural roads.

The Indian Road Congress is also seized of the problems of road safety measures. It is a well-known fact that at present India has one of the highest accident rate in the world. The I.R.C. is constantly working on Road Safety Measures and Devices. The Govt. of India has already constituted a National Road Safety Council for the purpose. A National Transportation Safety Board was also constituted. To reduce the number of accidents and to ensure overall safety of the traffic, to reduce travel cost and time are some of the important objectives of National Transportation Safety Board.

The Nagpur Plan proposed a formula for calculating the length of National Highway, State Highway and District Roads.

Mileage of National and State Highway and Major District Roads

$$= \frac{A}{8} + \frac{B}{22} + 1.6N + 8T + D - R$$

where,

A = Agricultural area in sq. km of the concerned area

B = Area of non-agricultural land in sq. km

N = No. of towns having a population between 2001–5000.

T = No. of towns having a population of more than 5000

D = Development allowance at the rate of 15% of road length so calculated for industrial and agricultural development for the next 20 years

R = Existing length of railway track in kilometres.

1.12 TYPES OF ROADS

The roads in India are divided as (i) All weather roads and (ii) Fair weather roads.

All weather roads are those roads which are used throughout the year without any interruption. The fair weather roads as the name suggested, are those roads which can be used only during fair weather. During rains and floods these roads are not negotiable and the areas connected by these roads are cut-off from the main stream.

The roads can be paved or unpaved. The hard top road such as W.B.M. or bituminous roads may be called paved roads whereas earth or gravel roads may be called un-paved roads. These can also be called or unsurfaced roads and the paved roads may be called as surface roads.

1.13 THIRD ROAD PLAN (1981–2001) CLASSIFICATION

On the basis of third road plan, the roads in our country are classified as

(i) Primary roads

(ii) Secondary roads

(iii) Tertiary roads or system

Primary roads are the major roads of a country. They are of two types (a) National highways and (b) Express ways. The express ways are a superior class of roads having better facilities and design standards and are meant for high speed vehicles. They are for thorough routes having high volume of traffic. Express ways are provided with divided carriage ways, controlled access, grade separation etc. The express ways may be owned by central government or state government.

The secondary system may be a State Highway and Major District roads.

The Tertiary roads or system consist of two category of roads viz., other District Roads and Village Roads.

1.14 CLASSIFICATION OF ROADS

According to the Nagpur Plan, the roads in India are classified as under :

1. National Highways
2. State or Provincial Highways
3. District Roads
   (a) Major District Roads
   (b) Minor or other District Roads
4. Village Roads

1. National Highways. These are the important roads of the country connecting important towns and capital cities of different states and important cities to ports etc. They are running throughout the length and breadth of the country. Roads connecting neighbouring countries are also called National Highways.

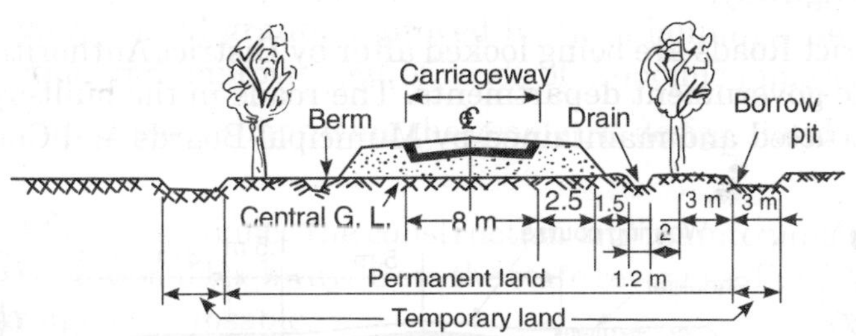

Figure 1.4 *National highway*

The National Highway should have two-lane traffic at least 8 m wide with at least 2 m wide shoulders on each side. However, this has been modified taking in view the present need and increased vehicles on the roads.

The construction and maintenance responsibility of the roads is of the Central Government Department such as Military Engineers Service (M.E.S.), Central Public Work Departments (C.P.W.D.) etc.

2. State Highways. These are the important roads of a particular state, which connect important cities and towns within a particular state. They also connect important cities with National Highways. The provincial or state highways should also have 8 m wide carriageway with 2 m wide shoulders on each side. The State Government Department of Public Works look after the construction and development of these roads. The central government give grants for the development of the state highways.

80 per cent is allotted to the state government on the basis of actual petrol consumption by the states.

A total of Rs. 600 million has so far been given by the central government.

State Governments also levy taxes under the State Motor Vehicles Act 1952, under the following heads. The rates are different in different states :

1. Registration fee on vehicles.
2. Issue and renewal of driving licence.
3. Passenger tax on vehicles carrying goods and passengers.
4. Sales tax on motor vehicles and their parts.

The local bodies like municipal boards, district boards, corporations etc. also levy some taxes in the form of Wheel tax, Toll tax etc. at various rates. These toll taxes are sometimes very irritating to the passengers particularly travelling in motor cars.

1.16 ADVANTAGES OF ROADS

Roads are arteries of a country. Roads are a pre-requisite to speed, and speed is essential for progress. It is difficult to enumerate the advantages of roads. Without roads nothing is possible in a country. However, following are the main advantages of roads :

(1) A network of roads is an asset to national defence, during war as well as peace time.

(2) Roads facilitate the movement of men and material from one place to another.

(3) Better law and order can be maintained if there is a good network of roads in a country.

(4) Educational and cultural contacts can be maintained with each other, with the help of good roads.

(5) Roads help in the growth of trade and other economic activities in and outside the villages and towns.

(6) Roads serve as a feeder for railways, airways and waterways.

(7) Natural resources of an area are easily tapped and improved with the help of good network of roads.

(8) They are essential for the economic prosperity and general development of a country. The country having comparatively more mileage of roads is said to be a more advanced and prosperous country.

1.17 IMPORTANCE OF ROADS IN INDIA

India is a vast country and to connect its different parts with a good network of roads is essential. India's deficiency in agriculture and economic progress is also due to the lack of good roads, especially in villages. The vast difference of culture between different parts of this country, for example North and South, is also due to the lack of good roads connecting the two major portions. Some of the interior portions are absolutely cut off from the remaining country due to lack of roads. In villages mostly fair weather roads are there i.e., those roads which can be used only in fair weather, and disconnect the villages from towns and railway stations during rainy season. Hence for the uplift of villages and economic development of this country, good and uptodate roads are very essential. India lags behind many other countries as far as the mileage of roads is concerned. This stresses the urgent and dire necessity of the planning and development of adequate road system in this country.

The importance of road transport can be better understood by the following analysis :

(i) Roads can be constructed at a comparatively lower cost than other mode of conveyance.

(ii) The road transport offers a quick and assured delivery of goods. Road transport due to its quick delivery reduces huge investments and large inventory as compared to rail network.

(iii) Roads offer a flexible service as the number of buses and trucks can be reduced or enhanced according to the need.

(iv) Road transport can provide service to remote corners and door to door from origin to destination. Rails on the other hand cannot provide door to door facility. Moreover the rails depend on the road network as the roads are a feeder to the railways.

(v) Road transport packaging is much simpler and cheaper than railways.

(vi) Road transport has a high potential for employment.

(vii) The road transport provides a personalised service and individual attention is being paid to each and every aspect of transportation of men or material.

(viii) For short distances road transport is much superior and easier than any other transport system including airways.

(ix) Road travel by personal cars or two wheelers gives and satisfies personal satisfaction and pleasure.

(x) Road transport is the only means of communication to remote villages, inaccessible localities and hilly terrain.

(xi) Roads have helped the effective administration of this vast country. Maintenance of law and order have been possible through the network of roads only. National integration and cohesion amongst the people of various states and linguistic diversities have been brought about by roads.

(xii) India having a rural based economy, agricultural output plays a very prominent role in our economy. Roads have fostered mobility of agricultural produce and modern inputs such as fertilizers, high yielding seeds, agricultural implements, pesticides etc.

PROBLEMS

1.1. What are the different means of communication? Which of them, in your opinion, is the most important for general development of a country. Give reasons for your answer.

1.2. How are roads classified in India? What is the necessity of classifying roads in this manner. Describe in brief.

1.3. Define the following terms :

Right of way, carriageway, footpath, express way, urban roads.

1.4. Draw a neat sketch showing cross-section of the following roads :

(i) National Highway, (ii) Urban Road, (iii) Village Road.

1.5. What are the advantage of roads to a country?

1.6. Write short notes on :

(a) Shoulders.

(b) Footpaths.

(c) Ribbon development.

(d) Major and minor District Roads.

Along the roadsides, construction of shops, hotels and other development of buildings, is known as *Ribbon Development*. The following are the disadvantages of ribbon development:

(a) It produces congestion on the road.

(b) Future widening of the road becomes very costly and sometimes impossible.

(c) Chances of accidents increase.

(d) It causes hindrance to free flow of fast moving vehicles.

1.7. Write an essay on the importance of roads in India.

1.8. Describe the importance of Nagpur Plan. Explain its salient features.

1.9. Compare the construction methods of Telford Roads and Macadam Road.

1.10. Write short notes on

(i) Expressway.

(ii) Radial system of road planning.

(iii) Role of transportation in Rural Development.

(iv) Nagpur Plan.

(v) Unpaved roads.

BIBLIOGRAPHY

1. History of Roads in India I.R.C.–1963.
2. Report of the Road Development Committee 1927–28, Government of India.
3. Chief Engineers Conference on Road Development in India (Nagpur Plan) 1943, I.R.C. Publication.
4. Kahyali–Highway Engineering–1984.
5. O'Flaherty–Highways–UK.
6. Road Facts India–1963, Ministry of Surface Transport (Roads wing) Govt. of India.
7. Sharma and Sharma–Highway Engineering, Asia Publication–1962.
8. History of Road Development in India, Central Road Research Institute–1963.
9. AA.S.H.O.A Policy on Arterial Roads in Urban Areas–Washington DC.
10. Planning Mannual—California Division of Highway Planning.

2

Geometrics of Roads

In this Chapter you will study,

• Requisites of a Good Road • Road Structure • Characteristics of Pavements • Road Chamber • Width of Carriageway • Medians • Road Kerbs • Ribbon Development • Right of Way • Super-Elevation • Method of Providing Super-Elevation • Relation between Camber and Gradient • Widening of Roads • Road Gradient • Sight Distance • Curves • Transition Curves • Planning of a Highway • Dimensions and Weight of Vehicles • Design Speeds • Carriagway • Traffic Lane Capacity • Traffic Requiement

GENERAL

In this chapter, we will study the technical aspects of roads. The design and standards of a road are the most important aspects of a road construction. For the success and failure of road all possible data regarding the type of traffic, its intensity etc., should be collected and sufficient thought should be given, to all these aspects, before finalising a particular road project.

2.1 REQUISITES OF A GOOD ROAD

In order that a road surface may give a satisfactory service throughout the year, it must satisfy the following conditions :

1. It should remain dry.
2. It should have a good carriageway.
3. It should have smooth gradients and large and smooth curves.
4. It should have a good wearing surface.
5. It should be easy in construction and cheap in maintenance.
6. It should have an impervious surface.

2.2 ROAD STRUCTURE

Like other engineering structures a road has also a foundation and a super-structure. The top of the grounds on which the foundation of the road rests, is called *Sub-Grade.* The top of the sub-grade should be 60 cm above the highest flood level (H.F.L.) of that area. The foundation of the road is also called *Soling* or *Base* and the super-structure of the road is called *wearing layer, wearing course,* or *road surfacing.* In those places, where the bearing capacity of the soil is poor and the intensity of traffic is high, an additional layer between the soling and sub-grade is provided. This additional layer is called *Sub-base.*

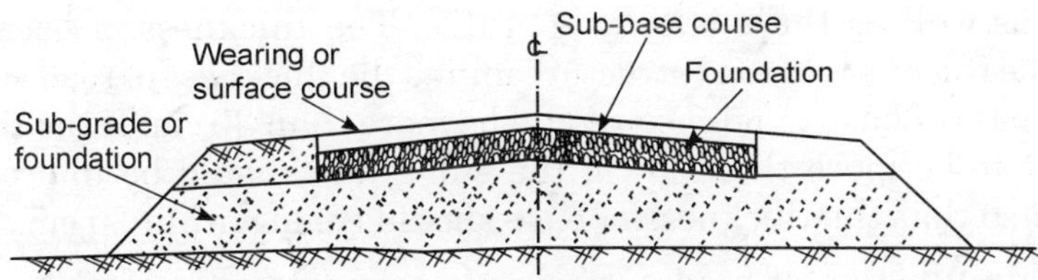

Figure 2.1 *Road structure*

The soling or the base is the most important part of a road structure. The strength and durability of a road depends on the type of soling provided. The base consists of a layer of hard murum, hand packed boulders or rubble, bricks etc. Lean cement concrete slab is also sometimes used as a soling. The type and thickness of soling depends on many factors, including bearing capacity of soil, intensity of traffic etc. The function of a road base or soling is to transmit load of the traffic, from the surfacing to the sub-grade. Hence the strength and durability of a road depends upon its sub-grade, on which the load of the traffic is coming ultimately, and the base.

The base or soling can be of a flexible type stone boulders, brick-bats or brick ballast or a rigid type such as cement concrete slab. In case of flexible soling, it is necessary that the bearing capacity of the sub-grade should be uniform and the soft patches etc. should be refilled with good earth. In case of rigid base like cement concrete slab, it is not necessary as the cement concrete slab will act as a beam over the weak spots.

The thickness of a road structure can be calculated by the following empirical formula:

$$d = \sqrt{\frac{23w}{70p} + 0.7T^2 - 0.84T}$$

where

d = thickness of road structure, including surfacing and soling in cm

w = maximum wheel load including 50% impact. Hence w is 1½ times the maximum static wheel load. In India the value of static wheel load is taken as 2268 kg for the purpose of design.

T = width of wheel in contact with road surfacing in cm. (The value of T is taken as 12 cm)

(For only bullock cart traffic, the value of static wheel load is taken as 1270 kg and las 4.5 cm)

p = Safe bearing capacity of soil in kg/cm^2

The thickness of surfacing depends on the type of traffic and the type of material as well as the intensity of traffic. The thickness of base will be equal to the thickness of road structure minus the thickness of road surfacing. The thickness of base in no case should be more than 30 cm. If the thickness comes out to be more than 30 cm the sub-grade should be improved and stabilized so that the thickness of base should come within 30 cm.

The main function of road surfacing is to provide a smooth and stable running surface suitable for the type and intensity of traffic anticipated. The surfacing should have the following requisite properties :

1. It should be impervious so that it should protect the base, the sub-base and sub-grade from the action of weather and rain water.

2. It should be durable, so that the maintenance charges should be the minimum.

3. It should be stable and should transmit the load of the traffic to the base of sub-grade without undue deformation and should be sufficiently flexible to adjust slight unequal settlements.

4. It should be non-slippery especially in rainy season.

5. It should be dustless.
6. It should be cheap in construction and maintenance.

2.3 CHARACTERISTICS OF PAVEMENTS

The important characteristics of road surfaces may be summarised as friction, drainage, light reflection, surface un-evenness etc.

Friction : It is one of the most important factors deciding the design speed, distance required for stopping or accelerating the vehicles. When a vehicle negotiates a horizontal curve, centrifugal force is developed which is countered by the lateral friction developed between the wheels of the vehicle and the road surface. Frictional force plays an important role in accelerating and retarding the speed of vehicles. The coefficient of friction or skid resistance is responsible for safe movement of vehicles. The coefficient of friction or skid resistance depends on the various surface conditions, weather conditions and wheel surface. The braking efficiency of the vehicle should be such so as to prevent rotation of wheels on applying brakes when the maximum coefficient of friction comes into play. Similarly on horizontal curves the skid resistance or coefficient of friction should be sufficient to counteract the centrifugal force on skidding or even over-turning may take place. Slip occurs when a wheel revolves more than the corresponding longitudinal movement along the roads. Slipping usually occurs in the driving wheels of a vehicle when a vehicle is abruptly stopped from a high speed or when a vehicle accelerates suddenly from a stationary position. Slipping occurs when the surface is wet, muddy or loose.

Unevenness : Pavement surfaces with less undulations permit higher operating speeds. The pavement surface should be even and smooth but not slippery i.e., should have nominal skid resistance or coefficient of friction. Pavement un-evenness affects operating cost, comfort and safety. The operating cost consists of fuel, wear and tear of tyres and the vehicles. Loose surfaces increase tractive resistance and increase fuel consumption. The un-evenness of the surface is measured by measuring the commulative un-evenness or undulation recorded per unit length of horizontal surface. Unevenness index is cm per kilometre. The standard unevenness index should vary between 150 cm/km to 250 cm/km.

Undulation or unevenness indicator has been designed and patented by Central Road Research Institute, New Delhi for instantly measuring the un-evenness index.

Reflection : The visibility, on the road particularly for the fast moving vehicles during night and on bright sunny days, very much depends on the light reflected by the road. Cement concrete surfaces produce too much of glare during day time. The glare due to head lights on wet rainy days, is very alarming and dangerous. Black top roads reflect very little light and provide very poor visibility during night particularly when the surface is wet, but during day time the surface do not produce any glare.

Camber : Camber is provided for draining of rain water from the surface of road as quickly as it is possible so that the water does not enter the road surface. The camber is provided on the straight reach of the road, by raising the middle portion of the road above the edges. On curvatures instead of providing camber, super elevation is provided. The amount of camber depends upon the surface material and its finish, type of traffic and the intensity of rainfall. More the impervious surface such as cement concrete or bitumen lesser or flatter the camber. Too steen cambers will have the following disadvantages :

(i) Lateral or transverse tilt of vehicles will cause discomfort and a drag on the steering of vehicles. It will also cause unequal wear and tear on the tyres of the vehicles.

(ii) It can cause topping or overturning of highly laden vehicles.

(iii) Wearing of central portion of the road as all the vehicles will try to negotiate on the central portion of road.

TABLE 2.1 *Camber prescribed by various authorities*

Authority	*Type of surface*	*Camber*
ASSHO	High	1–2%
	Intermediate	15–3%
	Low	2–4%
U.K.	All surfaces	2.5%
I.R.C.	(i) High type bituminous surfacing or cement	2.0 - 2.5% (1 in 60 to 1 in 50)
	(ii) Thin bitumen surfacing	2.0 - 2.5% (1 in 50 to 1 in 40)
	(iii) W.B.M. or gravel road	2.5 - 3.0% (1 in 40 to 1 in 33)
	(v) Earth and Moorum roads	3.0 - 4.0% (1 in 33 to 1 in 25)

2.4 ROAD CAMBER

The road surface has convexity upwards, with its highest point in the centre, in the straight portion of the road. The highest point on the surface is called *crown*. The word 'camber' is defined as the slope of the line joining the crown and the edge of the road surface. Thus a camber of 1 in 30 means that for a

30 metre wide road the crown of the road will be $\frac{1}{2}$ metre above the edge of the road or for a 60 metre wide road the crown will be 1 metre above the edge of the road. The camber is also sometimes referred to as *cross-fall* or *cross-slope*.

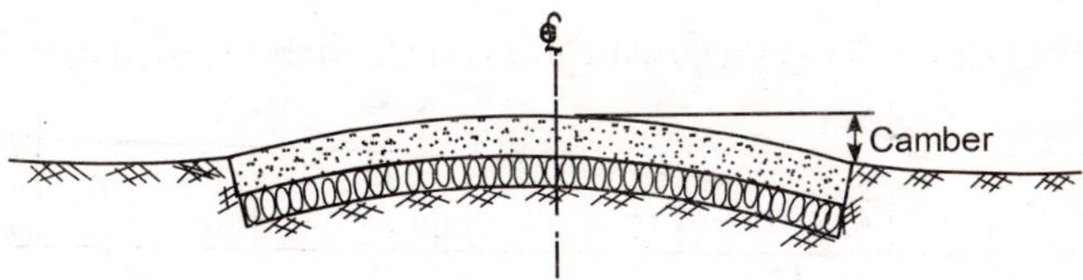

Figure 2.2 *Road camber*

The main object of providing a camber is to drain off rain water from the surface of the road, as quickly as possible. Hence a hard smooth surface will require less camber than a soft and rough surface. The amount of cross-fall or camber depends on the rainfall of that particular locality in which the road is to be constructed and the permeability of the road surfacing material. The steeper the camber, the more inconvenient it is for the traffic. In roads having steeper camber, the central portion of the road surface will deteriorate quickly as the traffic will tend to run on the central half portion. The following cambers have been recommended for different road surfaces:

TABLE 2.1 *Road cambers*

S. No.	*Type of Road*	*Recommended Camber*
1.	Earth roads and footpath	1 in 20 to 1 in 24
2.	Gravel road	1 in 24 to 1 in 30
3.	Murum road or Kankar Road	1 in 24 to 1 in 30
4.	Water Bound Macadam Road	1 in 30 to 1 in 48
5.	Bituminous High Speed Road	1 in 48 to 1 in 60
6.	Bituminous Concrete or Sheet Asphalt with speed limit	1 in 36 to 1 in 40
7.	Cement concrete roads	1 in 60 to 1 in 72
8.	Pavings etc.	1 in 48 to 1 in 60

Types of cambers. In general three types of cambers are provided for road surfaces:

1. Barrel camber.
2. Sloped camber.
3. Composite camber.

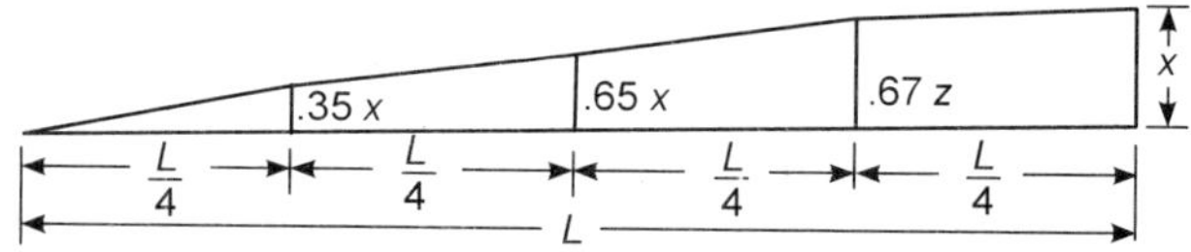

Figure 2.3 (a) *Elliptical*

Barrel camber consists of a continuous curve either parabolic or elliptical.

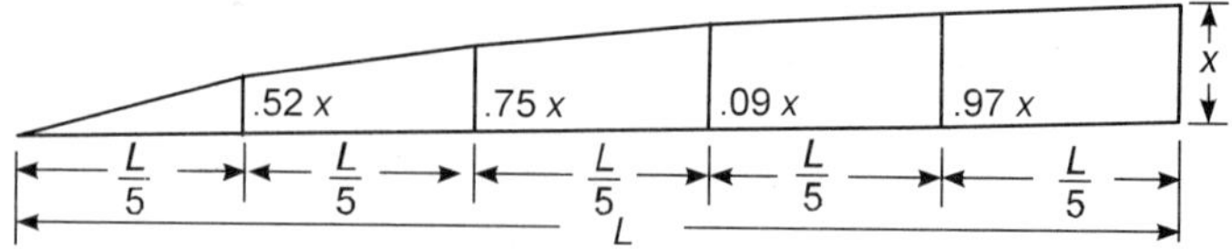

Figure 2.3 (b) *Parabolic*

Sloped camber consists of two straight slopes joining at the centre.

Figure 2.3 (c) *Sloped camber*

Composite camber consists of two straight slopes with a parabolic crown in the centre.

Figure 2.3 (d) *Composite camber*

2.5 WIDTH OF CARRIAGEWAY

Carriageway is the width of roadway meant for the movement of vehicular traffic. Traffic lane is defined as the portion of road meant for single line movement of vehicles. The width of carriageway will depend on the number of traffic lanes provided. As per I.R.C recommendations the maximum width of a vehicle is 2.44 m and the minimum width of a carriageway for a single traffic lane is 3.8 m leaving a clearance of 0.68 m on either side. For a two lane pavement a carriageway of 7.0 m is recommended. The number of traffic lanes will depend on the predicted volume of traffic. In some highways traffic separators or medians are provided between the lanes for traffic moving in opposite direction.

2.6 MEDIANS

Medians or separators are provided between the two traffic lanes meant for traffic coming in opposite directions to prevent head-on collision. Medians also help in channelising the traffic into their respective lanes. Medians or separators can be provided by lane markings, mechanical separators embedded in the road surface, raised platforms etc.

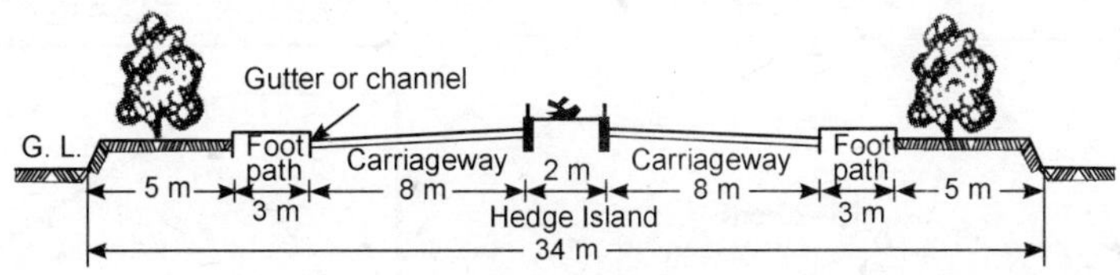

Figure 2.4 *Medians*

Pavement marking is the simplest of all. Ideally speaking the two lanes of opposite movement of traffic, should be separated by 6.0 m wide medians so that the glare of head-lights of two vehicles coming in opposite direction is minimised. The medians should be of uniform width on a particular road but if reduction in width is unavoidable a transition of 1 in 15 to 1 in 20 should be provided. For multilane highways medians must be provided. The width of a medians should not be less than 1.2 m.

2.7 ROAD KERBS

Road kerbs are indicators between the edge of a carriageway and the footpath, road islands, refuge islands, medians etc. The road kerbs may be of three types :

(i) Low kerbs (ii) Barrier type kerbs (iii) Semi-barrier type kerbs.

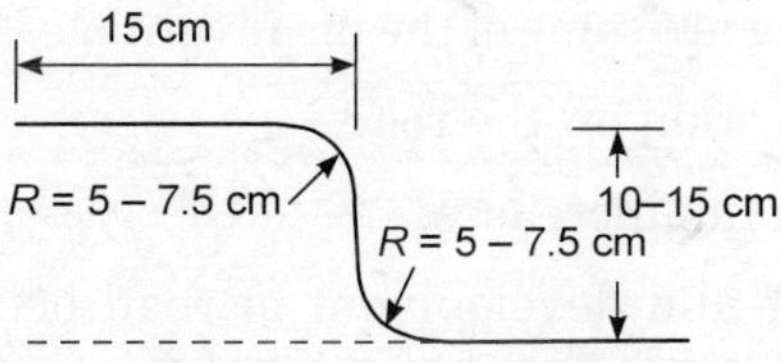

Figure 2.5 *Low kerb*

(i) *Low Kerbs* : They are also called mountable kerbs. These kerbs are indicators between the boundary of a road and the shoulders. The height of the kerb is such that the drivers find no difficulty in crossing these kerbs and use the shoulders in case of emergency. Normally the height of low road kerbs is only 10 cm above the payment edge. These kerbs channelise the traffic and also allow longitudinal drainage of road.

(ii) *Barrier type kerbs* : These type of road kerbs are provided in urban roads or city roads where the road passes through built up areas. The height of these kerbs is generally kept at 20 cm above the pavement surface.

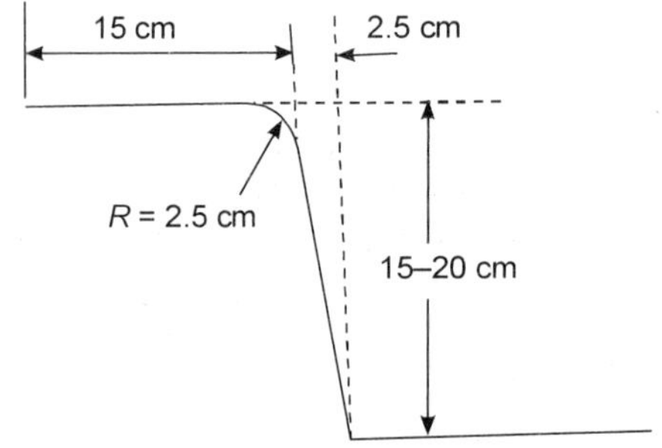

Figure 2.6 *Mountable kerb*

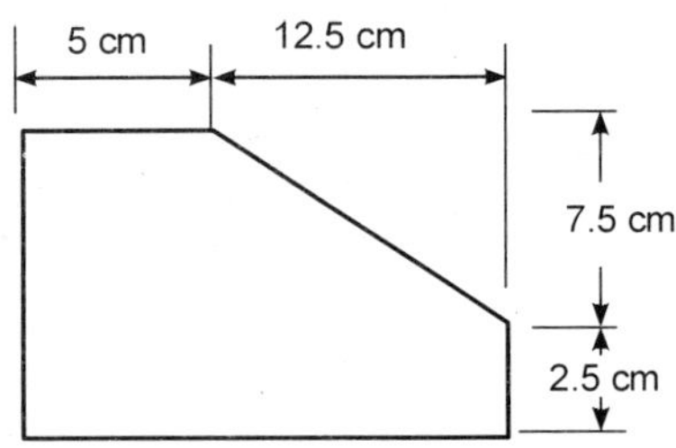

Figure 2.7 *Semi barrier type kerb*

(iii) *Semi-barrier type kerbs* : These type of road kerbs are used on the two ends of the highway separating the footpath and the pavement where pedestrian traffic is high. The kerb is generally 15 cm high with a batter of 1:1. This type of road kerbs prevent encroachment of parking spaces, footpaths, etc.

2.8 RIBBON DEVELOPMENT

When a new road is constructed near a city or in the sub-urban, construction of buildings, such as shops, restaurants etc. starts. Sometimes the development of building is so haphazard that it becomes nuisanse for the traffic. Development of buildings in an un-planned manner is called *Ribbon Development.* Following are some of the disadvantages of ribbon development.

1. It produces congestion on the road.
2. Chances of accidents increases.
3. Future widening and development of road becomes very costly and sometimes very difficult.
4. Approach to cities and market places becomes hazardous and unattractive.
5. It causes hinderance to the free flow of the thorough traffic.
6. It reduces the sight distances on crossings.

7. Due to ribbon development, average travel speed decreases which causes inconvenience to the traffic.

2.8.1 Prevention of Ribbon Development

The following precautions should be taken for the prevention of ribbon development.

1. No construction should be allowed along the road side atleast 50 metres from the centre of the road.

2. No building should be allowed to be constructed on the bye-passes and near the crossings at 100 metres from the centre of the lane.

3. Road arbouculture should be effectively planned for preventing ribbon development.

For easy, comfortable and safe flow of traffic the following bold features should be adopted for modern highways :

1. Access to the main road should be completely controlled.
2. Sufficient number of traffic lanes should be provided.
3. Fast and slow moving traffic should be segregated.
4. Parking spaces should be clearly indicated and provided.
5. Cross and opposite traffic should be separated by providing fly-overs, over and under bridges at convenient points.
6. Pedestrians should be segregated.
7. Comfortable and easy sight distances should be provided.
8. Separate by-passes should be provided for loading and un-loading of goods and passengers.
9. Modern street lighting should be provided in urban roads.

2.9 RIGHT OF WAY

It is defined as the land width acquired along the alignment of road. The *land width* of road will depend on its importance and future development. A minimum land width for different category of roads and different locations have been fixed by I.R.C. While acquiring land for the construction of a new road, care should be taken to see that sufficient land width is acquired for future development also as the cost of land invariably increases soon after the road is constructed particularly in sub-urban areas. The land width is governed by the following factors :

1. It primarily depends on the category of roads, e.g., National Highway, State Highway, District road and Village road.

2. Height of embankment or depth of cutting will also matter in deciding the land width.

3. Side slopes and soil type.

4. Sight distance and nature of curves.

5. Drainage system and catchment areas.

6. Nature of surroundings.

2.10 SUPER-ELEVATION

When a fast moving vehicle negotiates a horizontal curve, the centrifugal force acts on the vehicle and the stability of the vehicle is disturbed. This force is experienced by the wheels at right angle to the direction of motion. The frictional resistance between the wheel and road surface will act in the opposite direction. If the value of the centrifugal force exceeds the frictional resistance, the vehicle will have side slip and the outer wheels of the vehicle will be raised up from the road surface thereby causing instability to the vehicle. In passing from a straight to a curved path, a vehicle is under the influence of two forces namely, (i) the weight of the vehicle, and (ii) the centrifugal force, both of them acting through its centre of gravity.

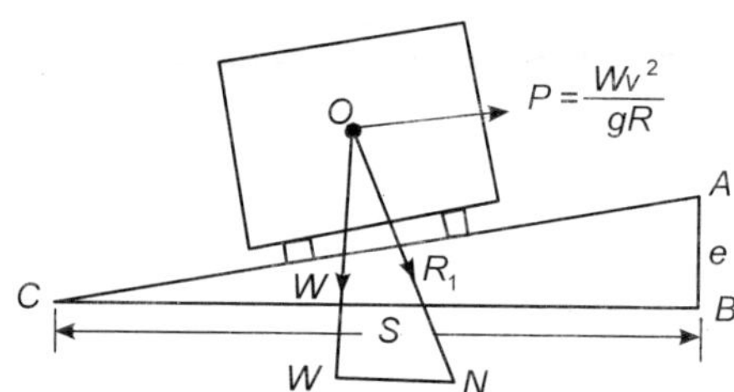

Figure 2.8 *Centrifugal force*

From Fig. 2.8,

$$\frac{AB}{CB} = \frac{WN}{QW}$$

or

$$\frac{e}{s} = \frac{p}{w} = \frac{Wv^2}{gR}/W$$

$\therefore$

$$e = \frac{v^2}{gR}$$

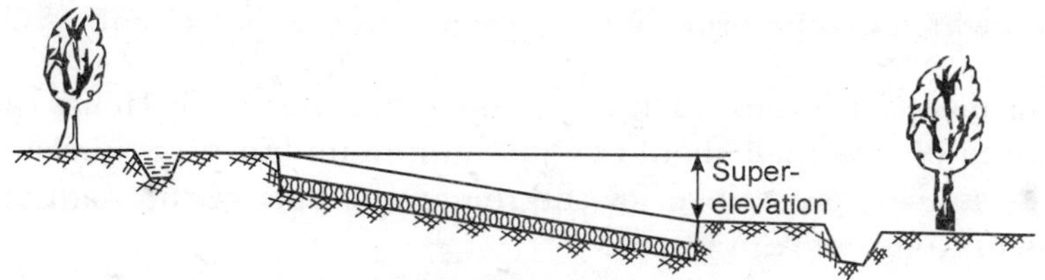

Figure 2.9 *Super-elevation*

The centrifugal force always acts in the horizontal direction and its effect is to push the vehicle off the track. To balance this, it is necessary to make the road surface perpendicular to the resultant of the above two forces i.e., the outer edge of the road is raised above the inner edge. To avoid this, the outer edge of road at the horizontal curves is raised above the inner edge. *Super-elevation* is defined as the inward tilt or transverse inclination given to the cross-section of the road surface, throughout the length of the horizontal curve to reduce the effect of centrifugal force on the running wheels. It is also sometimes termed as *Cant* or *Banking.* It is expressed as the difference of heights of two edges of the carriageway to the width of the carriageway. Thus a super-elevation of the 1 in 20 means that for a 20 metre wide road the outer edge of the carriageway is 1 metre above the inner edge at the vertex of the horizontal curve.

The amount of super-elevation can be calculated by the following formula :

$$e = \frac{V^2}{126R}$$

where

e = amount of super-elevation.

V = speed of vehicle in km p/h.

R = radius of curve in metres.

Example. Find the amount of super-elevation on a horizontal curve, having a radius of curvature 300 m. The average speed of vehicle can be taken as 50 km p/h.

$$e = \frac{50 \times 50}{126 \times 300} = \frac{1}{15.12}$$

The super-elevation varies from $\frac{1}{14}$ to $\frac{1}{16}$ but the maximum super-elevation which can be provided in special cases is $\frac{1}{10}$. Greater the super-elevation, more the inconvenience to the slow moving traffic. Hence the super-elevation to be provided should be just minimum but it should not be less than the camber, prescribed for the remaining part of the road to ensure effective drainage of rain water.

Table 2.2 gives the recommended minimum radius of horizontal curves on the flat country at various speeds of the vehicles.

TABLE 2.2 *Minimum radii for horizontal curves for flate country*

S. No.	*Design speed of vehicle*	*Minimum Recommended Radii*		*Absolute minimum radii*
		For flat country	*For urban areas*	
	km/hr	*m*	*m*	*m*
1	100	500	—	370
2	80	300	—	250
3	60	250	—	155
4	50	170	125	100
5	40	130	100	60
6	30	100	60	50
7	25	—	50	30

Table 2.3 gives the minimum recommended radii for horizontal curves of hill roads on various types of roads.

TABLE 2.3 *Recommended minimum curve radii for hill roads*

S.No.	*Classification road*	*Recommended minimum radius of curves*			
		In steep terrain		*In mountainous terrain*	
		Snow bounded area	*Areas unaffected by snow*	*Snow bounded areas*	*Areas unaffected by snow*
		m	*m*	*m*	*m*
1.	State and National Highways	33	30	60	50
2.	Major District Roads	15	14	33	30
3.	Other District Road	15	14	23	20
4.	Village Roads	15	14	15	14

2.11 METHOD OF PROVIDING SUPER-ELEVATION

The changing of the cambered surface of a road to a one way slope on a curved length is done progressively along a certain length of the road. This length of the road is equal to the *transition curve.* A transition curve is a curve so shaped as to follow the natural path and commences to move in a segment of a circle. A transition curve may be defined as *A curve whose radius*

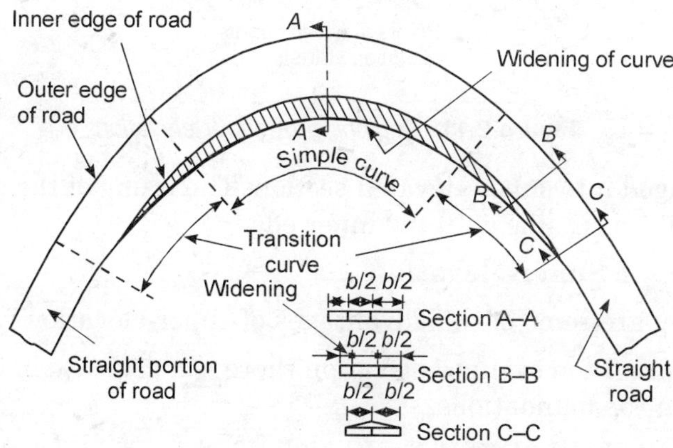

Figure 2.10 *Simple curve with transition curve and widening of road*

gradually changes from infinity to a selected minimum or the converse for the purpose of easy changing of direction of a road. Hence a transition curve joins the straight portion of a road, with one end of a circular curve.

In cases when the transition curve is not provided, the length along with the super-elevation is provided, is equal to half the length of the circular curve, and the change from the cambered section to the required super-elevation starts at $\frac{1}{4}$th the length of the curve measured along the straight portion of the road from the tangent point and the full super-elevation is reached at $\frac{1}{4}$th the length of the curve measured along the curve from the tangent point.

Now the super-elevation is introduced gradually. The surface is assumed to be rotated about the crown that is the outer edge is raised at a uniform rate. The rate of rise of outer edge over the inner edge will be equal to thetotal super-elevation divided by the length in which the cambered section

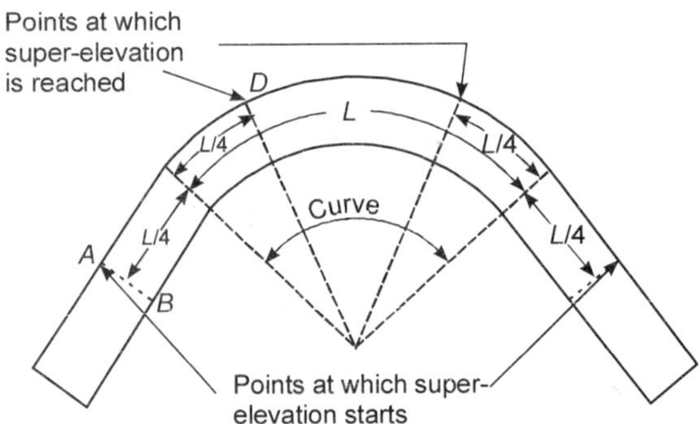

Figure 2.11 *Beginning of super-elevation*

is to be changed into super-elevated section. The rising of the outer edge is continued till it is in line with the inner edge.

Advantages of Super-elevation

The following are some of the advantages of super-elevation :

1. By the provision of super-elevation there is a decrease in the intensity of stresses on the foundations.

2. As the whole width of the road is drained to one side where there is super-elevation, no gulleys are formed on the outer edge of the road.

3. Super-elevation increases the stability of the fast moving vehicles when they negotiate a horizontal curve.

4. On super-elevated curves the vehicles are not necessarily slowed down.

2.12 RELATION BETWEEN CAMBER AND GRADIENT

Camber is provided to drain off the rain water from the surface of road as quickly as possible. Steeper cambers help in the draining off rain water quickly whereas a flatter camber helps in the smooth flow of the traffic. So we have to strike a balance between the two. Gradients also help in draining the rain water. Water should not travel a longer distance on the road as it causes erosions on the surface. It is sometimes misconceived that on steeper gradients the camber can be reduced. But the fact is that on steeper gradients, the camber should be increased as otherwise the rain water will travel almost parallel to the centre line or crown of the road.

Let AB be the half width of the road and equal to $b/2$. Let A be the crown and B be edge of the road. AD is perpendicular to BE. From the triangle ABD and triangle EAB

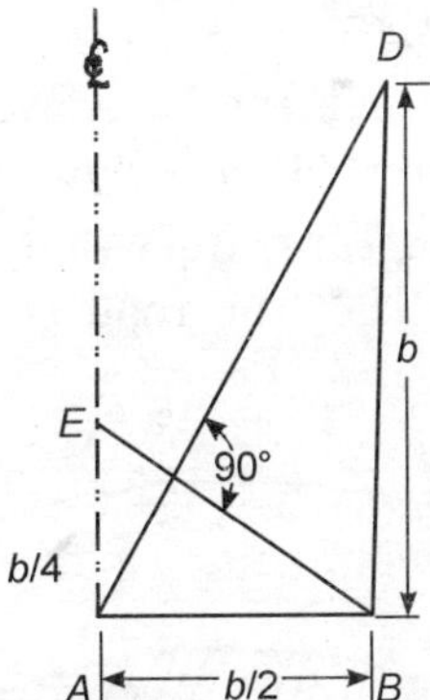

Figure 2.12 *Relation between camber and super-elevation*

$$\frac{DB}{AB} = \frac{AB}{EA}$$

$$EA = \frac{AB^2}{DB} = \frac{(b/2)^2}{b} = \frac{b}{4}$$

Also $$AB \times C = AE \times p$$

where c is percent grade of camber and p is percent grade of longitudinal gradient.

or $$\frac{b}{2} \times c = \frac{b}{4} \times p$$

or $$c = \frac{p}{2}$$

The point E and B are at the same level.

Thus the camber of a road should be double the gradient or gradient should be equal to half of the camber for an efficient drainage of rain water and smooth traffic flow.

2.13 WIDENING OF ROADS

When a vehicle negotiates a horizontal curve, the steering wheel turn sideways and occupy more width of the carriageway than on straight portion of the road. Hence the carriageway is increased on the entire portion of the curve on the inside. The widening is effected gradually, with a maximum on the central portion. When the radius of the curve is more than 460 m, this widening is not necessary. The following are the other *main reasons* of widening of carriageway on the curves :

1. The sight distance or vision is increased while driving on the outer half of the curve, only when the pavement is widened.

2. Most drivers use only the central portion of the road while moving on the curves. Thus, if the width of the road is increased this will prevent accidents to a good extent.

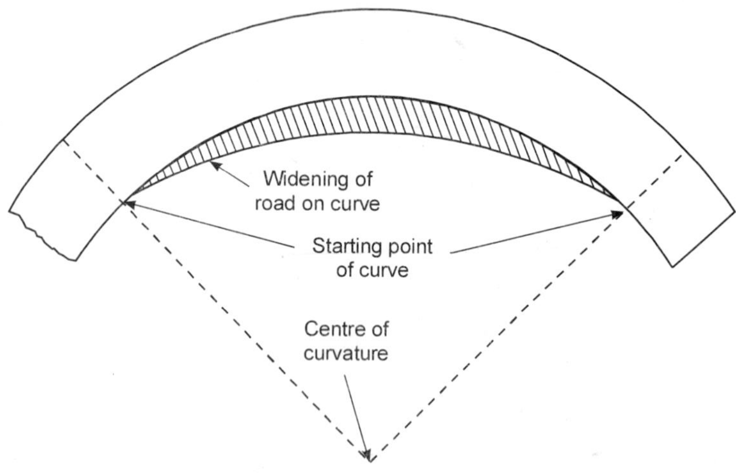

Figure 2.13 *Widening of road*

The value of uniform width added to the inside of the entire circular curve corresponding to the radii of curves is as follows :

2.13.1 Method of Calculating Widening of Road

Consider Fig. 2.14, in the triangle

$$R_2^2 = R_1 + l^2$$

$$R_2 = R_1 + l_x$$

where l_x = extra width required and l is the length of the vehicle between the axles.

Substituting the value of R_2, we have

$$(R_1 + l_x)^2 = R_1^2 + l^2$$

$$R_1^2 + l_x^2 + 2Rl_x = R_1^2 + l^2$$

Ignoring l_x^2, we have $2R_1^2 l_x = l^2$

or $$l_x = \frac{l^2}{2R_1}$$

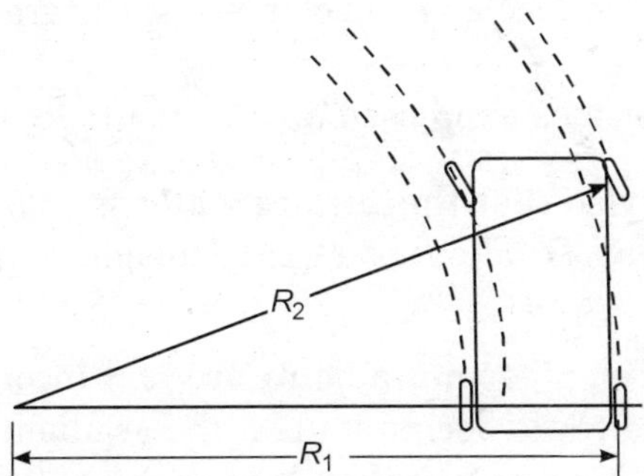

Figure 2.14 *Movement of a vehicle on a curve*

The widening of road is further increased due to psychological reasons. The I.R.C. has recommended the following formula for additional widening:

$$l_{x_2} = 0.1\times \frac{V}{\sqrt{R}}$$

where

V = designed speed of vehicles in km/hr

Hence total widening is

$$0.1 \times \frac{V}{\sqrt{R}} + \frac{l^2}{2R} \text{ or } 0.1 \times \frac{V}{\sqrt{R}} + \frac{18n}{R}$$

where n = no. of lanes and l has been assumed as equal to 6 m.

TABLE 2.4 *Additional width of roads at curves*

Radius of circular curve	*305 m to 460 m*	*150 to 305 m*	*60 m to 150 m*	*less than 60 m*
Value of uniform width for each traffic lane of carriage way having one or two traffic lanes	0.3 m	0.6 m	0.9 m	1.25 m

2.14 ROAD GRADIENT

The ground is never dead flat, and hence the road to be provided will also have rises and falls along the length of the road. The rate of rise or fall of the road surface along its length is called gradient. It is expressed as the ratio of the difference of heights between the two points and the distance between them. Thus, if the difference of levels between two points A and B is 1 metre and their distance apart is 50 metres, the gradient is said to be 1 in 50. It is also sometimes expressed as a percentage, i.e., $\frac{1}{50} \times 100 = 2\%$. Gradient is necessary so that the surface water may be drained off easily through the side drains. Gradient of a road depends on the following factors :

1. Nature of traffic. Steeper gradients are very inconvenient to the slow moving traffic. Hence while deciding upon the gradient to be provided the nature of traffic will have to be taken into account. Roads only meant for slow moving traffic such as bullock carts, etc., must not have very steep gradients.

2. Nature of ground. The amount of gradient to be provided is directly related to the nature of the ground. Ground with steep undulations will have roads with steep gradient as it is sometimes not possible to provide unnecessary deep excavations or cuttings.

3. Rainfall of the locality. The gradients in a road are mainly provided to drain off rain water from the side drains as quickly as possible. Hence more the rainfall, steeper gradients will have to be provided, in the side drains and the road.

Type of gradients. The gradients which are provided in different portions of a road length are as follows :

1. Maximum Gradient.
2. Ruling Gradient.
3. Minimum Gradient.
4. Average Gradient.
5. Exceptional Gradient.
6. Floating Gradient.

1. Maximum Gardient. It is the maximum or steepest gradient which is to be permitted on the road, which on no account is to be exceeded, as steeper gradients are very incovenient to the traffic, specially to the slow moving traffic. This is also called *Limiting Gradient.* Its value has been fixed as 1 in 15 for *Hill Roads* and 1 in 20 for other roads. These steeper gradients are sometimes to be provided to avoid deep excavations and avoid long detours.

2. Ruling Gradient. It is the permissible gradient in the alignment of a road. This gradient is such that vehicles, whether they are animal driven or power driven, can overcome long distances of this gradient, without much fatigue or uneconomical fuel consumption. *Ruling Gradient,* is therefore, defined as the suitable gradient within which the Engineer must endeavour to design the road. The Indian Road Congress has recommended the value of ruling gradient as 1 in 20 in hills and 1 in 30 in plains.

3. Minimum Gradient. It has also been found that for efficient drainage of water from the road surface, a certain minimum gradient is essential. Such an essential gradient which has to be provided for the purpose of road drainage, is called *Minimum Gradient* and its value is usually fixed as 1 in 200. However, the minimum gradient will depend upon the nature of the ground as well as on the rainfall of that particular area. For cement concrete roads, a minimum gradient of 1 in 330 can be provided.

4. Average Gradient. It is defined as the total rise or fall between any two points chosen on the alignment divided by the horizontal distance between the two points. The determination of the average grade is useful in carrying out the first paper location or preliminary survey.

5. Exceptional Gradient. During the alignment of a road, there may come certain patches of length, where a gradient may have to be provided which may be either less than the minimum known as *Exceptional Gradient.* Such a gradient becomes necessary to avoid deep cuttings or excavations. Exceptional gradient should not be provided in a length more than 100 metres, in any case.

6. Floating Gradient. When a motor vehicle descends a gradient, some tractive effect is required to maintain it at uniform speed, but to the descent, there is the negative tractive force or resistance. This resistance will be equal to the tractive effort required to maintain the vehicle at a uniform speed with a particular gradient, such a gradient is called *Floating Gradient.*

2.15 SIGHT DISTANCE

When a fast-moving vehicle negotiates a horizontal or vertical curve, a certain distance of visibility, through which a driver can see the opposite vehicle, a pedestrian or some fixed object, is essential so that the driver may react and avoid any collision or accident. *Sight distance* or *visibility* is defined as the distance measured along the centre line of a road, over which a driver can see the opposite object on the road surface and the provision of this distance is necessary to avoid any accident. The distance should be such that drivers and pedestrians should be given sufficient time to react to an emergency and not only to avoid accident but extend road courtesy to each other. In short, a sight distance is the length of the road, which a driver can see, especially on the curves. Sight distance can be:

1. Crossing sight distance or safe sloping sight distance.
2. Non-passing or non-overtaking sight distance.
3. Passing or overtaking sight distance.
4. Lateral sight distance.
5. Stopping sight distance.

1. Crossing sight distance. On roads and highways, two vehicles coming in opposite direction, on seeing each other have to reduce their speed to enable each other to use the pavement edges or shoulders, as the case may be. This distance in such a case is taken as twice the distance required for a vehicle to come to stop, and is called crossing sight distance. So on horizontal and vertical curves, this minimum sight distance must be provided to avoid any collision of two vehicles coming from opposite directions.

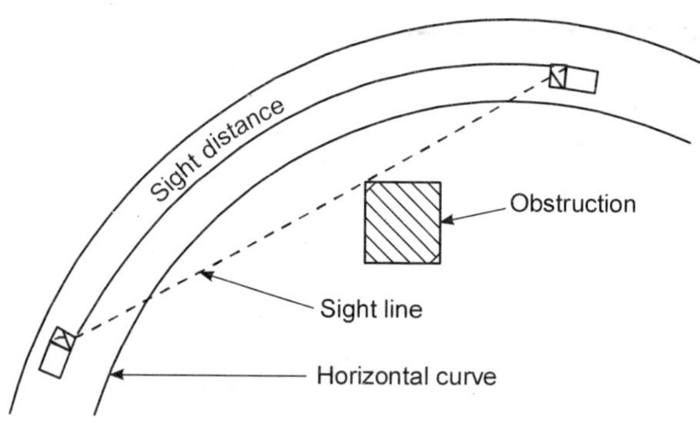

Figure 2.15 *Crossing sight distance*

Tables 2.5 and 2.6 give the minimum setback distance *s* required at horizontal curves on two lane roads for safe stopping sight distance.

2. Non-passing or non-overtaking sight distance. It is defined as the longest distance at which a driver whose line of sight is 1.2 m above the road surface can see the top of an object 10 cm high on the surface of the road.

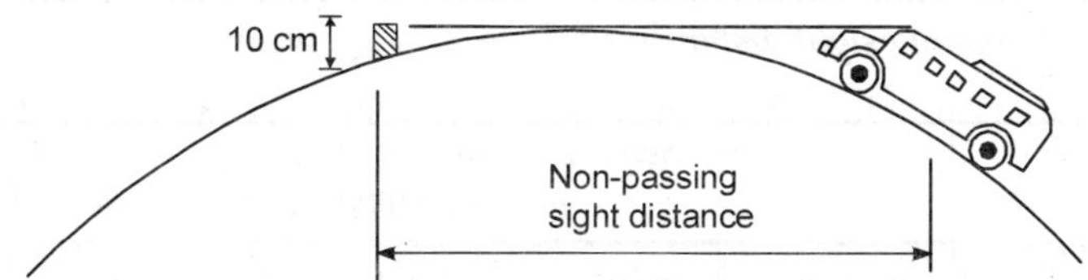

Figure 2.16 *Non-passing sight distance*

TABLE 2.5 *Minimum setback distance s at horizontal curves on two lane roads on flat country*

S. No.	Radius of curve along the centre line of the road	Setback distance 's' for various designed speed V and safe stopping sight distance D					
		D = 20 m V = 20 km/hr	D = 45 m V = 20 km/hr	D = 60 m V = 50 km/hr	D = 90 m V = 65 km/hr	D = 120 m V = 80 km/hr	D = 190 m V = 100 km/hr
	m	m	m	m	m	m	m
1	50	3.2	7.0	—	—	—	—
2	75	2.8	5.5	13.0	—	—	—
3	100	2.5	5.0	11.0	—	—	—
4	125	2.5	3.7	9.0	11.8	—	—
5	150	2.5	3.6	7.0	2.0	—	—
6	175	2.5	3.3	6.5	8.0	13.0	—
7	200	2.5	3.0	6.0	7.5	11.0	—
8	250	2.5	3.0	5.0	6.5	8.5	17.5
9	300	2.5	3.0	4.5	6.0	7.5	15.5
10	350	2.4	2.8	4.0	5.0	7.0	13.0
11	400	2.4	2.7	3.5	4.5	6.5	12.0
12	450	2.4	2.6	3.0	4.3	6.0	11.0
13	500	2.4	2.5	3.0	4.0	5.5	10.5
14	550	2.4	2.5	3.0	4.0	5.3	9.5
15	600	2.3	2.5	3.0	4.0	5.0	9.0

practice the perception time for a normal driver is assumed to be 2.5 seconds.

If the speed of the vehicle is V km/hr or $\frac{1000\ V}{60 \times 60} = 0.278V$ metre/sec. then distance travelled in 2.5 seconds will be $0.278 \times 2\ 5\ V = 0.695\ V$ metres.

Similarly the distance travelled during the time the brakes are applied and the vehicle comes to a halt is given by

$$d = \frac{V^2}{254f}$$

where

f = coefficient of fraction between the road surface and tyres of the vehicle and its value varies from 0.35 to 0.40.

2.16 CURVES

Curves are provided on the highways in order that the change of direction at the intersection of straight alignments either in horizontal or vertical plane, shall be gradual. The necessity of providing curves arises due to the following reasons :

(i) Topography of the country.

(ii) To provide access to a particular locality.

(iii) Restriction imposed by some unavoidable reasons of land etc.

(iv) Preservation of existing amenities.

(v) Avoidance of certain religious, monumental or some other structures.

(vi) Making use of existing sight of ways.

Factors affecting the design of curves. The following factors will influence the design of curves :

(i) Design speed of the vehicle.

(ii) Allowable friction.

(iii) Maximum permissible super-elevation.

(iv) Permissible centrifugal ratio.

Types of Curves. Curves are of two types viz., horizontal and vertical. The horizontal curves allow change in direction of the road while the vertical curves change in gradient. The curves used in the design of highways are :

(i) Circular curves ; and (ii) Transition.

Circular curves are of three types (i) Simple, (ii) Compound, and (iii) Reverse.

Transition curves can be divided into four groups: (i) True spiral or clothoid, (ii) Cubic spiral, (iii) Cubic parabola, and (iv) Lamniscate.

Circular Curves. *Simple curve.* A simple circular curve consists of a single arc connecting two straights. In this country particularly a curve is expressed in terms of degrees subtended at the centre by an arc of 30 m radius.

Compound curve. A compound curve consits of a series of two or more simple curves that turn in the same direction, and join at common tangent points. At each common tangent point, the adjacent curves have a common tangent and their centres are on the same side of the curve.

Reverse curve. A reverse curve consists of two simple curves of opposite direction that join at the common tangent point called the point of reverse curve. Their centres are on opposite sides of the curves.

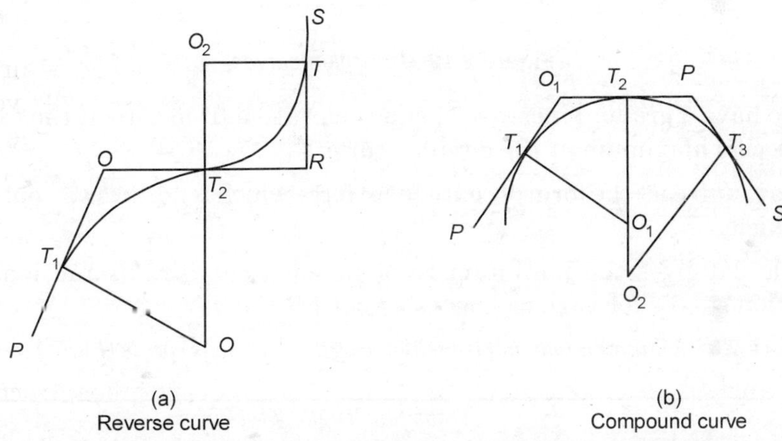

(a)
Reverse curve

(b)
Compound curve

Figure 2.18

2.17 TRANSITION CURVES

On a highway or railway, curves of varying radii are introduced. A transition or spiral curve or an easement curve as it is sometimes called, is one whose radius changes gradually from an infinite value to a finite value or vice versa for the purpose of giving easy change of direction. This tends to counteract the swaying outwards of a vehicle when subjected to sudden application of a centrifugal force at the instant of its entering or leaving the curve.

Advantages of Providing Transition Curves

1. To obtain *transition* from the tangent to the circular curve and from the circular to the tangent.

2. To obtain a gradual increase of curvature from a value of zero at the tangent to a maximum at the circular curve.

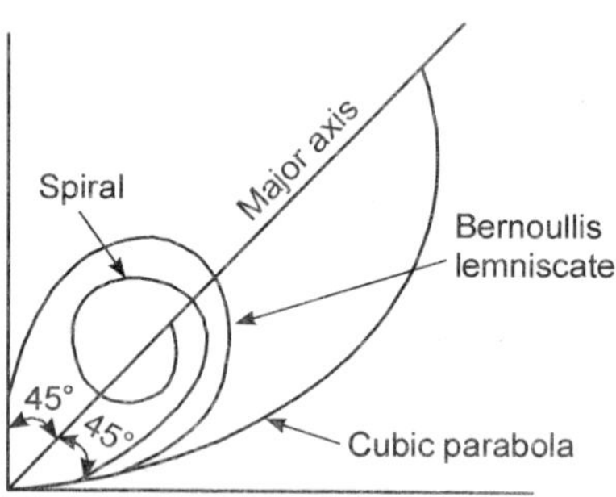

Figure 2.19 *Transition curves*

3. To have a gradual increase of super-elevation from zero at the tangent to a specific maximum at the circular curve.

4. To counrteact the form of centrifugal force which tends to sway outwards the vehicle.

Table 2.8 gives the minimum recommended curve radii and length of transition curves for various speeds as per I.R.C.

TABLE 2.8 *Minimum radi and transition length for various speeds (I.R.C.)*

S.No.	Curve radii	Transition length of the curves								
		Hill roads				Plain and rolling country roads				
		Speed of vehicle in km/hr				Speed of vehicle in km/hr				
		25	30	40	50	30	50	65	80	100
1	45	30	30	–	–	75	–	–	–	–
2	60	15	15	30	–	60	–	–	–	–
3	75	15	15	30	–	45	–	–	–	–
4	90	15	15	30	45	30	45	–	–	–
5	120	15	15	15	30	25	60	–	–	–
6	150	15	15	15	15	25	45	75	–	–

Contd.

Table 2.8 Contd.

7	200	15	15	15	15	15	45	60	–	–
8	250	–	–	–	–	15	30	45	90	–
9	300	–	–	–	–	15	25	26	75	–
10	400	–	–	–	–	–	15	25	45	100
11	600	–	–	–	–	–	–	25	45	80
12	750	–	–	–	–	–	–	25	30	65
13	900	–	–	–	–	–	–	25	30	50
14	1000	–	–	–	–	–	–	–	–	45

Note. It is usually not necessary to provide transition curve on the horizontal curves having radius greater than 1000 m.

2.17.1 Method of Calculating the Length of Transition Curve

The length of transitions curve should be such so as to fulfill the following requirements of providing the transition curves.

1. To introduce gradually the centrifugal force between the tangent point and the beginning of the circular curve avoiding a sudden jerk on the vehicles.

2. To enable the driver to turn his vehicle gradually according to his convenience and comfort.

3. To gradually introduce the super-elevation designed and the widening of curve so as to improve the aesthetic appearance of the road

At the tangent point the centrifugal acceleration V^2/R is zero as R the radius is infinity. At the end of the transition curve the value of R is minimum. The centrifugal acceleration is to be distributed over the length of the transition curve to be provided. The length of the transition curve should be such that the centrifugal acceleration should be developed at such a low rate that it does not cause any discomfort to the passangers of the vehicles. It is now clear that longer the length of the transition curve lesser will be the change in the centrifugal acceleration. Let the length of the transition curve be L and t is the time taken by a vehicle moving with a speed v m/sec, to traverse the curve of length L. The rate of change of centrifugal acceleration will be given by

$$c = \frac{v^2}{R.t} = \frac{v^2}{\dfrac{R.L}{v}}$$

$$= \frac{v^3}{R.L} \text{ metres/ sec}^2$$

According to I.R.C. the rate of change of centrifugal acceleration can also be calculated by the following formula :

$$c = \frac{80}{75+V} \text{ m/sec}^3$$

where V is the designed speed of the vehicle in km/h.

The value of c varies from 0.5 to 0.8.

Now the length of the transition curve will be

$$L = \frac{v^3}{R.C}, \qquad R = \text{radius of circular curve}$$

or

$$L = \frac{v^3}{(3.6)^3 C.R.}$$

The value of C will vary from 0.5 to 0.8 for maximum comfort.

In an open country road the amount of super-elevation should not conform to the raising of the outer edge of the basement to that extreme. It is not desirable to raise the outer edge of the pavement at a larger rate than 1 in 150 relative to the grade of the centre line of the pavement. The length of the transition curve should be atleast 150 times the total amount by which the outer edge of the pavement is to be raised with respect to the centre line. The length of the transition curve in urban/built up areas is taken as 100 time and in hills as 60 times the height through which the outer edge is raised.

If e be rate of the super-elevation designed for a pavement W and w be the maximum widening of pavement at the curve, the total raising of the outer edge will be $(W + w)\, e$.

For a rate of change of super-elevation N varying from 150 to 60 (for country road to hill roads)

$$L = \frac{EN}{2} = \frac{eN}{2}(W + w)$$

According to I.R.C. standards, the length of the horizontal transition curve should not be less than the value given by the following formulae :

For plain tension

$$L = \frac{2.7V^2}{R}$$

For hilly terrain

$$L = \frac{V^2}{R}$$

Hence from the above it is clear that the length of the transition curve depends on

(i) Radius of the circular curve.

(ii) Design speed and width of pavement.

(iii) Amount of super-elevation.

(iv) Rate of providing super-elevation.

2.18 PLANNING OF A HIGHWAY

Before planning and designing a highway the engineer must keep in view the following basic considerations :

1. The design of the highway should be safe and efficient during the day and night as well as during fair and bad weather.

2. The design should be suitable for the normal volume of traffic as well as for peak hours. It should have scope for future expansion.

3. The design should conform to the characteristics of the vehicles their design, speed, expected life, future traffic trends and speeds.

4. The design of the highway should be consistent and abrupt changes should be avoided so that the drivers are not confronted with difficult and serious situations.

5. The planner of the highway should have complete knowledge of the movement of traffic and the hazards, weight and size of vehicles which are going to operate on the highway.

6. The design should incorporate the aesthetics of the surrounding area so as to make the journey more pleasing.

7. The design should be as simple as possible from the construction as well as user's point of view.

8. The design should include all aspects of road such as crossings, signal, road signs, packings, kerbs, road markings, danger zones, speed limits etc.

2.19 DIMENSIONS AND WEIGHT OF VEHICLES

The dimensions and weights of vehicle have considerable influence on the design of highways. The dimensions of the vehicles have the following effects :

(i) The width of the vehicle influence the width of carriageway, width of shoulders.

(ii) The maximum height of vehicles passing over a roadway will decide the clearance to be provided under the bridges, underways or underpasses, electrical service lines etc.

(iii) The maximum length of vehicles will influence the curvatures horizontal as well as vertical, sight distances etc.

(iv) Weight of vehicles influence the structural design of pavements and bridges, gradients etc.

Dimension of road vehicles

Maximum width	2500 mm
Height	
(i) Single decker	3840 mm
(ii) Double decker	4800 mm
Length	
(i) Single unit with low axles	4600 mm
(ii) Single unit with more than two axle	12000 mm
(iii) Tractor with semi Trailor combination	15000 mm
(iv) Tractor with full trailor combination	18000 mm

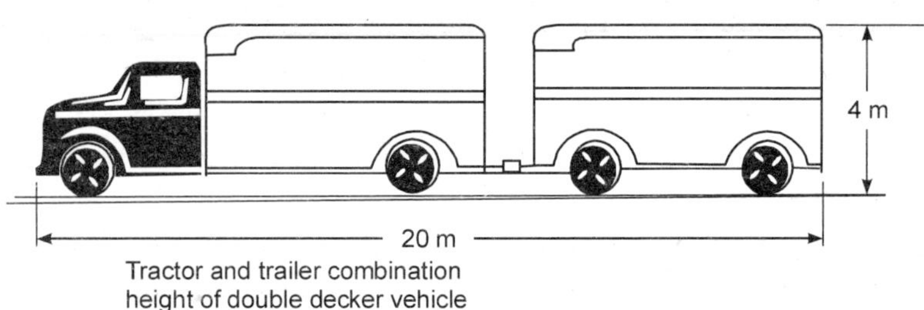

Figure 2.20 *Tractor and trailor*

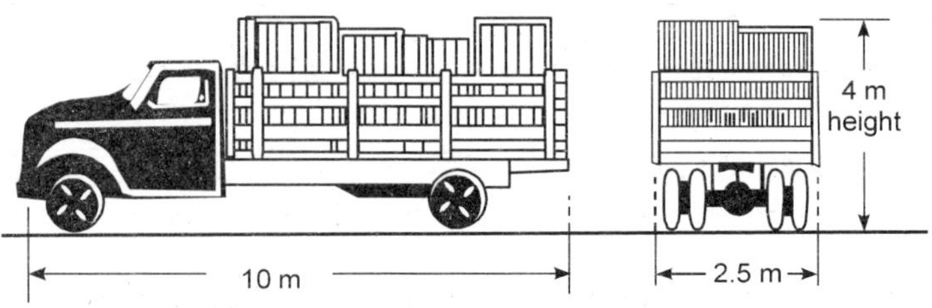

Figure 2.21 *Single unit*

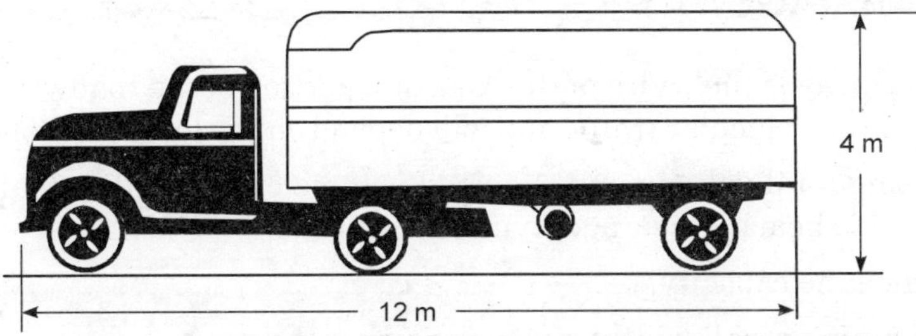

Figure 2.22 *Semi-trailor tractor*

Weight of Vehicles

The weight of a vehicle is expressed in terms of its axle load which is transmitted by its wheel to the road pavement. According to I.R.C. no axle load according to Indian conditions should exceed 6550 kg. The gross load of any combination of vehicles shall not exceed as calculated by the following formula:

$$W = 1025\,(I + 24) - 3L^2$$

where

W = gross weight of the vehicle

L = distance between the extreme axles.

2.20 DESIGN SPEEDS

All the roads are not meant for all type of traffic and hence it is not advisable and economical too, to design all the roads for many speeds. A road has to be desgined for a specific speed known as 'Design Speed'. The design speed is defined as the approximately maximum speed on which most of the vehicles will move. The design speed depends on the following :

(i) Type and conditions of the pavement surface.

(ii) Nature, type and intensity of traffic.

(iii) Nature of the terrain.

(iv) Alignment of road.

The Indian Road Congress has recommended the following standard design speeds for various category of roads in the country :

	Plains	*Hills*
National Highways	80 km/hr	50 km/hr
State highway	60 km/hr	40 km/hr
District Roads	40 km/hr	30 km/hr
Village Road	30 km/hr	20 km/hr

2.21 CARRIAGEWAY

The carriageway is the width of the road or a portion of the roadwidth which is meant for the vehicular traffic and will depend on the following conditions :

1. Existing total volume and expected volume of the traffic, peak hourly intensity and the daily average, type of traffic etc.

2. Traffic lane capacity.

3. Maximum overall width of vehicles using the road.

4. Clearance between the body of the vehicle and the edge of road and between the two vehicles on the road.

2.22 TRAFFIC LANE CAPACITY

Number of lanes to be provided in a carriageway will depend on the capacity of lane to accommodate vehicles. This capacity will be maximum under ideal conditions and will be called as Basic Capacity so to define, basic capacity of a lane is defined as the capacity to accommodate number of vehicles which pass a definite point on the lane during an hour under the most ideal roadway and traffic conditions that can possibly be achieved. The basic capacity of a lane is calculated taking into consideration the following points :

(i) Only one type of vehicle uses the lane (this is only a hypothetical proposition).

(ii) All vehicles are running at the same speed and maintain equal distances.

(iii) There are atleast two lanes for the exclusive use of traffic moving in one direction.

(iv) Adequate lane width, clearance and shoulders should be provided.

(v) There should be complete segregation of traffic without any interference from the cross traffic.

(vi) There should be no restrictive sight distances, improperly designed super-elevations or super elevated curves intersections and grades.

2.23 TRAFFIC REQUIREMENT

Width of a carriageway is influenced by the type of traffic. For only bullock cart traffic 2.5 m to 3.0 m width of carriageway is sufficient, but for fast moving traffic greater width will be required. Roads should not be designed for the present day requirement but for the anticipated future requirement also. Future requirement are worked out on the basis of the statistical data. A 20 year plan should be envisaged for the highway. The number of traffic lanes will depend upon the traffic density both in character and volume. With increased density wider lanes will have to be provided.

The Indian Road Congress has accepted the following Clayton's formula for calculating the capacity of a roadway for commercial and light vehicles :

$$\text{Ligth vehicles} = \frac{5280V}{15+1.47V+\dfrac{V^2}{30}}$$

$$\text{Commercial vehicle} = \frac{5280V}{15+1.47V+\dfrac{V^2}{30}}$$

The capacity of a roadway is reduced on account of the following factors:

(i) *Mixed traffic.* On almost all the roads the problem of mixed traffic exists. There is hardly any road which has only one type of traffic. On account of one large size slow moving vehicle such as a ladden bullock cart, on a single lane roadway, the whole lot of vehicles behind it will have to slow down, thus reducing the capacity of the road. This sometimes creates traffic jams also.

(ii) *Delays at the intersections.* Inordinate delay in the movement of traffic is caused due to intersections which reduces the actual capacity of a road.

(iii) *Gradient and camber.* Sometimes due to improper gradient and cross slopes, the rain water is not properly drained off and due to mud etc. the road surface becomes slippery and then naturally the speed is reduced and capacity is reduced.

(iv) *Alignment.* When the alignment is not proper due to which the sight distance is reduced and causes hindrance in the uniform flow of the traffic and thereby reduces the capacity of the road.

(v) *Grades.* Due to steep grades on a hilly terrain the spacing between the vehicles is automatically reduced both on up-hill and down-hill, which reduces the capacity of the road.

Due to all these problems the practical capacity of a road is reduced to a great extent then its theoretical capacity.

PROBLEMS

2.1 (a) What should be the qualities of a good surface?

(b) Draw a neat sketch, showing the different parts of a road structure.

2.2. What do you understand by the term 'camber'. Why camber is provided and on what factors the camber of a road surface depends. What are the disadvantages of a high camber. How much camber will you recommend for the following roads :

(a) Cement Concrete.

(b) Earth Road.

(c) Water-bound Macadam Road.

2.3. What is super-elevation, when and why it is provided? On what factors the super-elevation depends. Describe in brief the change over of cambered surface to that of super-elevated surface.

2.4. What is the maximum gradient and ruling gradient? Why the value of a gradient is different for roads in plain and roads in hilly places?

2.5. Describe in brief the meaning of 'Sight Distance'.

2.6. Describe the necessity of super-elevation, transition curve and widening on curves.

2.7. Write short notes on :

Transition curve, non-passing sight distance, introducing of super-elevation, composite camber, floating gradient, average gradient.

2.8. What are the requirements of a good road surface? How you will achieve to construct an ideal road ?

2.9. What do you understand by the term alignment of a road? What are the ideals which are kept in mind while aligning a road ?

2.10. Write short notes on :

(a) Super-elevation.

(b) Maximum gradient.

(c) Widening of road.

(d) Dual carriageway.

(e) Soling.

(f) Barrel camber.

2.11. What is super-elevation ? On what factors the super-elevation of a road depends ? What are the disadvantages of super-elevation ?

2.12. State the method of calculating the length of transition curves. What are the objectives of providing transition curves.

2.13. State in brief the important factors of highway surfaces which improve their efficiency and serviceability.

2.14. Establish a relation between Camber and gradient of a road. What should be the minimum gradient to be provided in a road?

2.15. State the important features responsible for smooth and efficient flow of traffic.

BIBLIOGRAPHY

1. Geometric Design Standards for Rural Roads I.R.C. – 1980.
2. Manula for Planning Urban Roads, U.S. Department of Surface Transport– 1964.
3. Guidelines for the regulation of mixed traffic in urban areas. Traffic Engineering Committee –I.R.C. : 36 – 3 –1975.
4. Lateral and Vertical Clearances of Under-passes of vehicular traffic I.R.C– 54 –1974.
5. Continuous Traffic Count on N.H. 45 in Tamil Nadu. Highway Research Station – 1980.
6. Tentative Guidelines on Capacity of Roads in Rural Areas – I.R.C. : 64 – 1976.
7. Design Tables for Horizontal Curves I.R.C. : 35 –1970.
8. Vertical Curve Design for Highways I.R.C. Journal Volume XVI–1–1981.
9. O'Flahacty – Highway – McGraw Hills, 1980.
10. Roads in Urban Areas, Department of Environment (U.K.) 1956.
11. Road User Cost Study in India I.R.C. : 1982.
12. Ribbon Development along Highways and Its Prevention. Special Report No. 15 I.R.C – 1974.
13. Standard Specifications and Code of Practice for Rural Roads I.R.C. : 5 – 1970.
14. Post War Development of Roads in India – Conference Report of Chief Engineers I.R.C. Publication –1961.
15. Date on the Growth of Traffic in Maharashtra Highway Research Institute, Nasik, Maharashtra –1980.
16. Dimensions and Heights of Road Design Vehicles I.R.C. : 3 –1954.
17. Highway Capacity Manual, Highway Research Board, U.S.A–1965.
18. Horizontal and Transition Curves Paper No. 119 I.R.C. Journal Vol. XI – 3, 1947.
19. Black More F.C. Capacity on Single Lane Intersection, Ministry of Transport R.R.I. Report No. I.R. 356.

20. Recommended Practices for the Construction of Embankments of Rural Roads I.R.C – 65 –1976.
21. Report on Research Project R –3 Scheme P.W.D. Rajasthan, Jodhpur 1979.
22. Highway Capacity Report– Highway Research Boards –1980.
23. Recommended Practice for Sight distance on Rural Roads I.R.C.– 1980.
24. Vertical Curves for Highways–Paper No. 136, Journal of I.R.C. Vol. XVI–1, 1951.
25. Report on Research Project R-3 Scheme U.P. P.W.D. Research Institute, Lucknow.
26. Pragya A., U.P. P.W.D. Research Institute Publication, 1983.

3

Road Project

In this Chapter you will study,

• Survey • Preparation of Maps • Map Study • Acquisition of Land • Alignment • Factors Controlling Alignment of Roads • Re-Allgnment • Earth Work • Side Slopes • Some Technical terms in Earth Work • Mass Diagram • Traffic Surveys • Estimates • Project Report

GENERAL

Road project is a comprehensive term. Road project means planning, designing and construction of a new road, connecting two terminus points or a town with an existing road. The road project will consist of the following works :

1. Survey.
2. Preparation of maps.
3. Acquisition of land.
4. Road alignment.
5. Earth work.
6. Traffic surveys.

7. Estimates.
8. Project report.

In this chapter, we will underline the various stages of road project in brief.

3.1 SURVEY

To determine the location of a proposed road and to collect the necessary data, the following road surveys are undertaken :

1. Reconnaissance survey.
2. Preliminary location survey.
3. Final location survey.
4. Construction survey.

1. Reconnaissance Survey. This is nothing but a rapid examination of the ground and its adjacent natural features. This preliminary survey is done generally without the help of any instrument. The reconnaissance or *Recci* as it is sometimes called, is to determine the suitability of each alignment marked on the map during map study. *Recci* also helps in conducting the preliminary survey. During the *Recci* one has to examine a very wide belt on either side of the general location of the route; and thus covering all possible alignments drawn during map study, has to improve upon the alignments, if possible. It is during the *Recci* that the visual inspection of the proposed alignment and the surrounding country is carried out. Special attention is paid to :

(i) River crossing and determination of suitable site for bridge or culvert construction.

(ii) Alignment which will entrail heavy earth work or steep gradients. (Which should be avoided, if possible).

(iii) Sources of supply of water, and other materials required for the construction of road.

2. Preliminary Location Survey. It is a rough type of survey, which is conducted to have a fair idea of the surrounding areas. The preliminary survey is done with the help of a plane-table, compass, chain or steel tape. Rough levels are also taken at different points so as to have an idea of the earth work.

After the preliminary survey, the necessary plans, i.e., drawings are prepared corresponding to the survey work and a rough estimate of the cost of the road and its ancillary works for each proposed route is prepared. The most economical and the best of these alignments is selected.

3. Final Location Survey. Before the final location survey is started, the centie line of the finally selected route is marked on the ground. New leves are taken along this centre line. Cross-sectional levelling is also done at the required points of centre line. Actually speaking the final location survey is the detailed levelling along the centre line and at right angles to it. From this survey work, final plans, designs and estimates are prepared.

4. Construction Survey. The construction survey is done for the setting out of the road. It consists of the following steps :

(i) Clearing of jungles, bushes, grass and any other objectionable materials.

(ii) Setting out the centre line and edges of the road. The centre line is defined by the alignment pegs which are driven at regular distances. On the curves pegs will be driven close together. The line between these pegs is clearly demarcated by cutting a narrow V-shaped cut in the ground along this central line. This cut is called *lockspit* or *dagbel.*

(iii) Setting out areas for *barrow pits* and *spoil banks.*

(iv) Fixing bamboo and string profiles at regular distances for earth work.

3.2 PREPARATION OF MAPS

The following drawings are generally required :

(i) *Topographical maps.* These maps show the general details of the area such as existing metalled and unmetalled road, cart tracks, position of wells or other sources of water, hospitals, schools, police stations or outposts etc. Contours are also shown. The topographical maps also show villages and towns having a population more than 500. These maps are generally prepared by the Survey of India to a scale of 1′ = 1 mile or 1 cm = 0.66 km.

(ii) *Population maps.* These maps show the distribution of population in a particular area. These maps show the important topographical features of villages, towns etc. and their population.

(iii) *Agricultural and industrial maps.* These maps are prepared on the basis of agricultural and industrial productivity. The maps are divided into three groups following :

(a) Those villages and towns which are highly developed agriculturally or industrially. Sometimes industries are also divided according to their importance, (b) those which are particularly developed, and (c) those which can be developed.

(iv) *Proposed plans.* These maps show the proposed alignment, and detailed sections of the proposed road, along with the earth work details such as cuttings and embankments. The details of ancillary works of roads, such as bridges, culverts, and railway crossings etc., are also drawn on separate sheets. The alignment maps should clearly show the position of these ancillary works etc.

3.3 MAP STUDY

In the topographic map of the area, all possible alignments can be marked. Survey of India department supply maps with 15 or 30 metre contour intervals. On these topographical maps, important features of the area such as ridges.valleys, rivers, ponds, constructed buildings, religious monuments etc. are shown. Various possible alignments are then marked on the map and advantages and disadvantages of various possibilities are worked out.

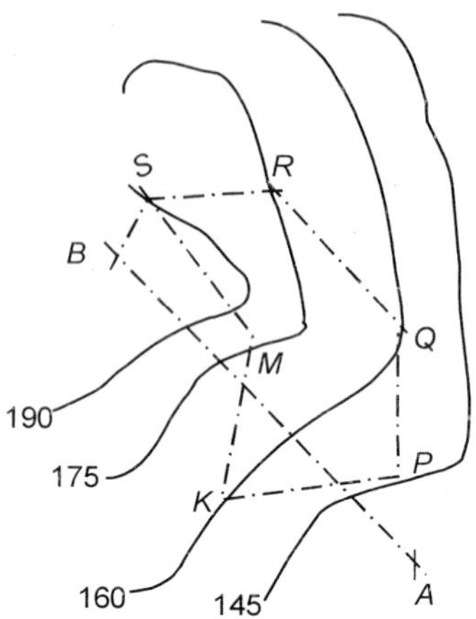

Figure 3.1 *Alignment*

Suppose the contour interval is 15 m. *A* and *B* are the two stations to be connected by a road. *AB* can be the shortest alignment. The minimum road length between *A* and *B*, if a ruling gradient of 1 in 20 is to be restricted, will be (190 – 145) × 20 = 900 metres. Now if the straight line *AB* does not measure 900 metres, it means the alignment is not possible. For locating the possible alignment within the ruling gradient of 1 in 20, the method is that two points on two consecutive contours should be so located that their distance is 300 metres (20 × 15 = 300 m) by an arc. So two possible routes *PQRS* and *PKMS* are located keeping in view the ruling gradient. Now the two alternative routes are studied separately and the final alignment is decided.

3.4 ACQUISITION OF LAND

For the construction of a road and its ancillary works, some land is to be acquired. It is always advisable not to pass the alignment through those lands which are agriculturally developed or through built-up areas, because it becomes difficult to acquire the required land. There are two types of lands viz., permanent land and temporary land. Permanent land is the land-width which has permanently been acquired for the development of the road and its ancillary works. The following are permanent land-widths recommended by the Indian Road Congress: (Table 3.1).

TABLE 3.1 *Permanent land widths*

Class of road	*Normal width*	*Minimum width*
National Highway	60 m	45 m
State Highway	45 m	30 m
Major District Roads	30 m	20 m
Minor District Roads	25 m	15 m
Village Roads	20 m	15 m

Apart from this permanent land, some more land is also acquired for barrow-pits and spoil-banks etc. This land, is called *'Temporary Land'*. This temporary land, after the construction of road is given to the farmers on lease, only for agricultural purposes.

3.5 ALIGNMENT

The course or route along which the centre line of a road is located in the plan, is called *Road Alignment*. Before starting the actual construction, the centre line of the road is first marked on the plan and then on the site. The following points should be kept in mind while aligning a particular road :

(i) The alignment should be as short and straight as possible.

(ii) The straight alignment should be deviated when it is required to give the benefit of the road to an intermediate town, an important village, a railway station, a market place or some other important highway.

(iii) The alignment of a road should cross another road, a railway line or a stream, preferably at right angles.

(iv) The alignment should cross a river or a stream at such points where the width of river or stream is minimum and where good and durable foundations are possible for the construction of a bridge or culvert.

(v) There should be minimum number of crossings, bridges or culverts in the alignment of the proposed road.

(vi) As far as possible, alignment should not pass through thick forests, built up areas, or cultivated land.

(vii) There should be minimum of cutting and minimum of banking. Practically speaking the amount of cutting and of banking i.e., filling should be equal.

(viii) The alignment should ensure easy gradients and smooth curves.

(ix) The alignment should pass through such points where the material for the construction and maintenance of road is easily available.

(x) Unnecessary zig-zags in the alignment should be avoided.

3.6 FACTORS CONTROLLING ALIGNMENT OF ROADS

The various factors which govern the alignment of roads are :

(i) *Obligatory points.* For an alignment to be shortest, it must be straight. For connecting the obligatory points which the road alignment has to pass, the alignment has to be deviated. The various obligatory points are bridge, railway track, intermediate town, high ridges, mountain pass, tunnel etc. Various alternatives are to be worked out for a suitable alternative. For a crossing a hillock or a ridge, sometimes a tunnel is more suitable than going around a ridge. Sometimes some important points are to be touched and the alignment is deviated. The various obligatory points and the deviation in the alignment are shown in Fig. 3.2.

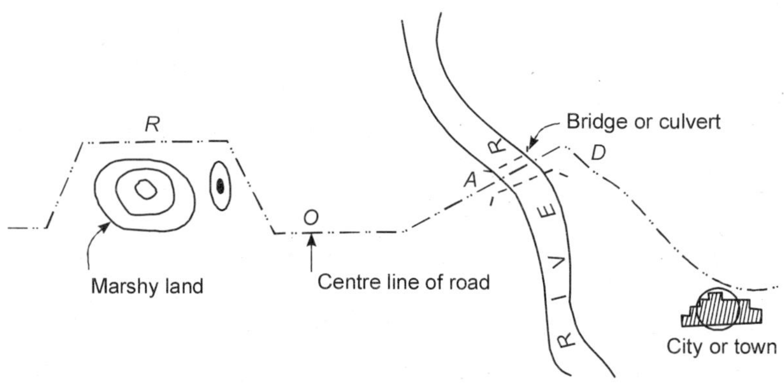

Figure 3.2 *Obligatory points*

There are certain obligatory points which should be avoided such as religious places, costly structures, thick jungles, marshy lands, grave yards, military protected areas, etc. Alignment of such areas will have to be avoided otherwise either the alignment will be too costly in construction and maintenance or it will create other problems of law and order, acquisition of land etc.

(ii) *Traffic.* The volume and trend of traffic also affect the alignment of road, so while aligning a road desire line should be drawn before taking a final decision.

(iii) *Geometric design.* Road curves, gradients, sight distance and such other factors also affect the alignment of roads. If the straight alignment causes steep gradients, then it will have to be avoided. Similarly if straight alignment does not provide passing or crossing sight distance, then too the alignment will have to be deviated. Similarly care will have to be taken so as to avoid steep curves even if we have to deviate from the shortest alignment. The gradient should be within the ruling gradient.

(iv) *Economic Factors.* Economic considerations play a very important role in the alignment of road. The economic considerations are easy and cheap acquisition of land, travel cost or running cost. Longer the route, more the construction, maintenance and running cost. Deep cuttings high embankments, tunnelling, railway crossings etc. are the other economic considerations of alignment. Drainage, hydrological factors, political considerations, monotomy, military strategic points also demand considerations in the alignment of roads.

3.7 RE-ALIGNMENT

Most of the Indian roads of pre-independence era had been constructed to meet the demand of the traffic which was much slower than the present day traffic. Most of these roads were either constructed in stages or were upgraded in phases due to which the originality of the road have been diminished. They are now found to be very much deficient in their characteristics of design, alignment, drainage conditions etc. There are many National Highways having very narrow bridges and culverts that too unsuitable for heavy traffic, with sharp bends, steep gradients, inadequate sight distances and poor lane width. These defects of highways are to be rectified by re-alignment of the present road. It will be worthwhile to adopt more liberal values of geometric design parameters where excessive costs are not involved. It may not be always possible to re-align the road for the sake of making improvements and rectifying defects. At the national level it has been decided that National Highways should as far as possible, be able to cater to the traffic demands moving at the design speed, fulfilling the safety requirements and comforts, both for the present and future needs. For achieving this object, it is imperative on the part of the traffic engineer to plan improvements and remove deficiency of geometries etc. within the means. The following situations may warrant the necessity of re-aligning a road :

1. For improving geometric designs such as horizontal curves, super elevation, sight distance etc.

2. For improving design elements like steep gradients, undulations, hair curves or zig-zag curves etc.

3. For raising the level of the road on account of change in H.F.L. and water logging.

4. For re-constructing weak culverts, narrow bridges etc.

5. For constructing over bridges or railway crossings, grade separation etc.

6. For constructing a bye-pass so as to avoid congested localities.

7. For connecting some strategic point such as a new railway station, some tourist spot or some defence requirements.

During the process of re-alignment, special care should be taken to improve the geometric design of the highway such as smooth curve, super-elevation, sight distance etc. This is possible only when the alignment is considered as a whole and not in piece meals. Improvement in transition curves is not a very costly process and hence can be improved only by improving the horizontal curves. Sight distances will automatically improve. While alignment attempts are made to improve overtaking sight distances on summit curves. Stopping sight distances should also be improved on strategic points. Water logged area and sub-merging of road or a portion of road should be avoided.

The following procedure should be adopted for the re-alignment of a project :

1. Extensive survey should be conducted for collecting various datas like traffic, soil conditions, drainage and drainage structures, right of way and its cost etc.

2. Detailed drawings such as topographical plans, detailed cross-sections at various points, longitudinal section, contour maps etc. should be prepared for the existing and the proposed alignment/alignments for the purpose of comparison.

3. Compare the economics of the proposed alignment to see whether the project is worthwhile (see chapter on Highway Economics).

4. Preparation of design of pavement, detailed working drawings of culverts, bridges, fly-overs etc. and their estimates.

5. The construction work along the new alignment should be started by first marking the centre-line, dog bevelling, earth work etc.

3.8 EARTH WORK

Earth work in its widest sense includes excavation in rock as well as ordinary clayey soil and formation of banking.

(a) *Cutting.* It denotes excavation, under the surface of the earth and the removal of the material. When the depth of cutting is less than 3 m it is called shallow-cutting and when it is more than 3 m, it is called deep-cutting. Cutting more than 16 m should be avoided.

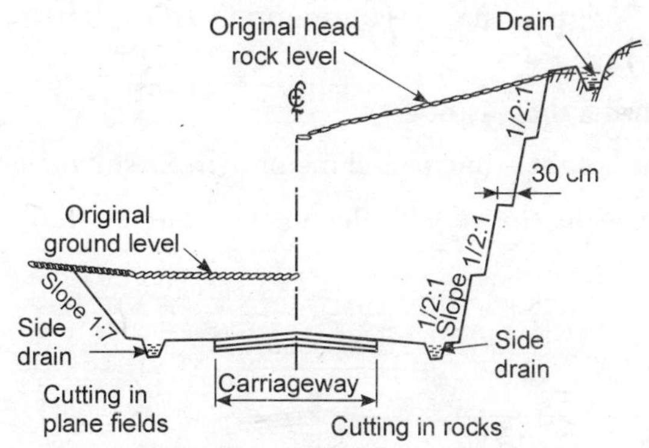

Figure 3.3 *Cutting*

(b) *Embankment.* It denotes lump of earth, collected together from the excavation and built to a definite shape to carry a roadway (or a railway) over it, above the ground level. In a low lying area embankment more than 3 m is called high embankment.

Excavated earth, when broken down, increases in volume. This loose earth when deposited in banks will settle down under its own weight and through the action of atmosphere. The decrease of height of embankment is called settlement and the amount by which the new embankment is to be made higher than the formation level is called *allowance for settlement.*

3.9 SIDE SLOPES

Earth embankment of usual height will stand safely with side slopes of 1 in $1\frac{1}{2}$. In cutting a side slope of 1 : 1 will serve the purpose well. Rocks when cut are even stable at 1 in $\frac{1}{2}$ or 1 in $\frac{1}{4}$. Such slopes were adopted in as they involve minimum of earth work, but the tendency these days is to provide flatter slopes as they result in the safe operation of vehicles and the maintenance cost is also reduced. Steep slopes in embankments have the following disadvantages:

1. Steep slopes result in accident hazards. The vehicle can overturn if the wheel goes over the edge.

2. Steep slopes on gutter ditches may also result in accidents.

3. Steep slopes erode badly, hence the maintenance cost is increased much.

4. Tree plantation or grass growing is difficult.

Indian Road Congress has recommended the following side slopes in embankments:

(a) Upto 60 cm height—1 in 4.

(b) Over 60 cm height—natural slope or 1 in 2 whichever is flatter.

In cutting, the side slopes, should not be steeper than 1 in 2 except in solid rocks.

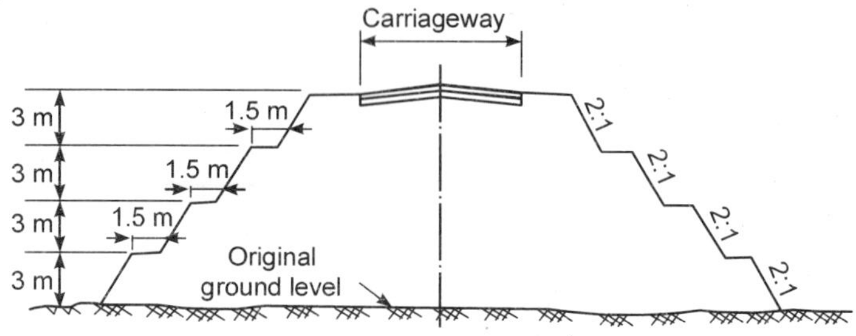

Figure 3.4 *Embankment*

3.10 SOME TECHNICAL TERMS IN EARTH WORK

(1) *Balancing of earth work.* The excavated earth from cutting should be utilized in forming embankments, when the cutting precedes or follows a fill so long as the *lead* is not very great. This cutting and formation of banks simultaneously along the alignment of a road is generally known as *balancing of earth work.*

(2) *Berms.* A strip of land is always left in between the toe of the embankment and the inner edge of the barrow-pit and the boundary line to prevent damage of banks by erosion. In case of high embankment and deep cuttings, the long side slopes are stepped to avoid speedy flow of rain water and thereby causing *rain-cuts.* These horizontal strips are called *berms.*

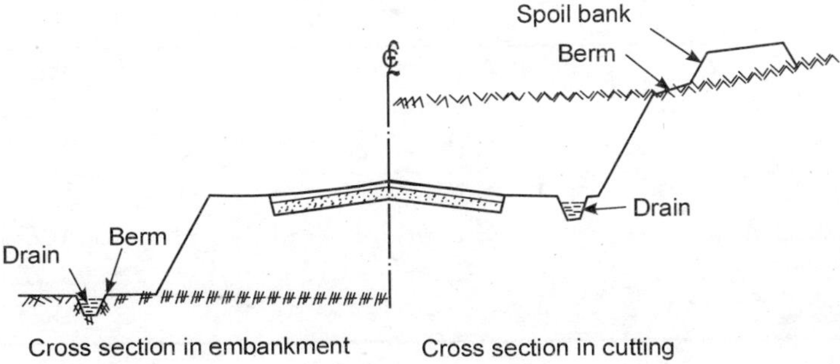

Figure 3.5 *Berms*

(3) *Spoil banks.* Earth from excavation, if not utilized in banking or to fill up the low lying areas nearby, is deposited in heaps, parallel to the length of the cutting in the form of a bank at some distance away from the top edge of the cutting. These banks are called *spoil banks.* The height of spoil banks should be nearly 1.5 m. Between the toe of the spoil bank and top edge of cutting a small drain should be provided for draining off rain water.

(4) *Borrow pits.* It is a pit usually prismoidal in shape dug out at some distance from the toe of the embankment from which earth is borrowed for the purpose of constructing banks.

(5) *Lead and lift.* The horizontal distance through which excavated earth is to be taken and carried for banking or spoil banks etc. is called *lead,* and the vertical height through which earth is lifted is called *lift.* Lead is measured from the centre line of the barrow pit to the centre line of the bank. The initial lead and lifts are taken as 100 ft (30 m) and 5 ft (1.5 m) respectively.

(6) *Angle of repose.* Loose earth or any granular solid when heaped together, slides down and finally takes up a shape of equilibrium. The maximum angle, at which the surface of the earth stands up by itself, with the horizontal, is called the natural angle or *angle of repose.*

(7) *Benching.* Embankments on ground which have a considerable slope or cross fall, say of more than 1 in 10, the ground is benched to prevent slipping down of earth. *Benching* means cutting of series of steps at right angles to the line of slope along the length of the bank.

(8) *Formation.* It is the finished surface of sub-grade of a roadway (or railway) in cutting, or embankments. The formation width is defined as the finished top width of an embankment or a cutting. The portion of the formation width which is left unmetalled on either side is called *shoulder.* The shoulders should be nearly 1.5 m wide. The following are the formation widths recommended for various classes of roads by the Indian Road Congress (I.R.C.) :

Table 3.2 *Road formation widths*

Class of Road	*Minimum formation width in plane country*	*Minimum formation width in hilly country*
National Highways	12 m	8 m
State Highways	9.75 m	8 m
Major District Roads	7.25 m	6.75 m
Minor District Roads	7.25 m	6.75 m
Village Roads	5.5 m	4.25 m

Procedure of earth work in embankment. For a roadway (or railway) construction, earth work is essential throughout the length of the road. It may be either in the form of embankment or cutting. The following procedure should be adopted for earth work in embankment:

(1) If suitable earth is available from cutting, banking should be done by the excavated earth from the cutting portion only.

(2) In case of high embankments, earth should be taken from cuttings and barrow pits.

(3) Banks should not be constructed entirely with sand but if it is unavoidable they should be covered with a layer of good earth. That is, sand layers should be covered with a layer of good quality of earth. Otherwise, if possible the earth to be used for banking should contain two parts of sand and one part of clay.

(4) When the ground has a considerable cross fall, it should be benched to prevent slipping down of soil.

(5) After dogbelling, the usual demarcations of pegs line, on both sides of the central line, should be made. Bamboo profiles are set up at an interval of 80 m.

(6) Embankments are raised in regular layers of 25 cm to 30 cm thickness. The layers should be horizontal having a uniform slope. Each layer should be compacted and consolidated properly.

(7) Allowance for shrinkage is to be given for each particular class of soil.

(8) The formation should be finished only after two successive monsoons. The gradients and camber should then be checked.

3.10.1 Procedure for Earth Work in Cutting

1. The cutting should be started from the extreme dogbelled edges and from cross-section to cross-section.

2. In shallow cuttings the central portion is first excavated with vertical faces but deep cuttings are taken in steps, then the slopes are finally formed.

3. During cutting, check measurements should be taken at intervals from the guard pillars on each side of cutting, so that excavation should not be carried below the formation level.

4.Suitable catch-water drains are to be provided on either side or on one side of the cutting.

5. The earth obtained from cutting should be utilized for making embankments adjacent to it, otherwise it should be deposited in the form of spoil banks on one or both sides as may be necessary. The spoil banks are formed in the same way as the embankment.

3.11 MASS DIAGRAM

This is also called mass-haul diagram. It is proposed to enable the engineer to have a knowledge of the amount and extent of free-haul and overhaul on a project. This helps the construction engineer and the contractor about the extent of earth work, so that necessary earth moving machinery may be required. The mass haul diagram is a graphical representation of the amount of earth work involved in the scheme and the manner in which it can be handled most economically. It shows accumulation of earth along the proposed alignment and the economical direction of haulage and positioning of borrow pits and spoil banks or heaps can be planned and estimated.

Before drawing a mass haul diagram it is worthwhile considering how the accumulated volumes of cut and fill materials at each station are tabulated. The characteristics of a mass haul diagram are:

1. The ordinate at any point along the curve represents the earthwork accumulated to that point.

2. The maximum ordinate (+) indicates a change from cut to fill as one proceeds along the centre line from an arbitrary assumed station. The minimum ordinate (–) shows a change from fill to cut. These maximum and minimum points may not necessarily coincide with the apparent points of transition.

3. A rising curve at any point indicates as excess of excavation over embankment material at this point. A falling curve indicates the reverse.

4. A steeply raising or falling curve indicates heavy cuts or fills. Flat curves show that the earth quantities on a particular station are small.

5. The slope of loops indicate the direction of haul. A convex loop, for example, shows the haul from cut to fill is to be left to right, while a concave loop indicates that the haul to be from right to left.

6. Since the ordinates of a curve are plotted from cut volumes and adjusted fill volumes, then any line parallel to the baseline which cuts off a loop

intersects the curve at two points between which the amount of cut is equal to the fill. Such a line is called a balancing line and the intersecting points are called balancing points.

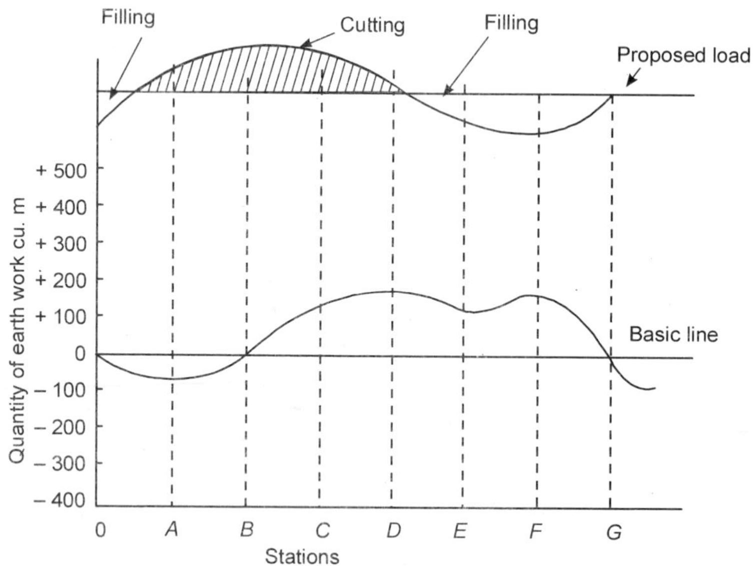

Figure 3.6 *Mass diagram*

7. The area between the balancing line and mass-haul curve is a measure of the haul between the balancing points. If this area is divided by the maximum ordinate between the balance line and curve, the value obtained is the average distance that the cut material must be hauled in order to make the fill. This distance can also be estimated by drawing a horizontal line through the mid-points of this maximum ordinate until it intersects the loop at two points, the length of this line is very close to the average haul distance when the shape of loop is smooth.

8. Balancing lines need not be continuous the vertical break between any two balancing lines merely indicates unbalanced earthwork between adjacent points of termination of the lines. Adjacent balance lines should never overlap, as this will mean using the same part of the mass diagram twice.

3.12 TRAFFIC SURVEYS

Before deciding width of formation and the type of surfacing to be provided, it is essential to have a detailed study of the nature of the traffic anticipated over the proposed road. The following information should be collected in addition to data relating to existing trades, industries and agricultural products etc.:

(1) Nature and intensity of traffic anticipated.

(2) Possibilities of trade and industrial development resulting from the proposed road.

(3) General resources of the locality.

From the above information the width of formation and the type of surfacing to be provided is decided. Traffic survey also helps in the proper alignment of the proposed road.

3.13 ESTIMATES

The following items of estimates are to be prepared. Actually speaking the items of estimates depend upon the nature of the project. In some cases, where the project is minor, a few items are required :

(i) Acquisition of land.

(ii) Clearing of jungles from the site.

(iii) Earth work.

(iv) Materials to be used for the construction.

(v) Estimates (which are prepared separately) for ancillary works such as bridge, culvert, railway crossing, etc.

(vi) Supplying and fixing of mile, furlong and boundary stones.

(vii) Arboriculture.

(viii) Construction of temporary site buildings etc.

3.14 PROJECT REPORT

Before the construction is actually started, a project report is prepared and submitted for approval. A project report is a concise statement, containing important information regarding the project. It should also include the possibilities of future developments and the benefits to the general masses. The following detailed information should be included in the project report.

(1) Details of rainfall, floods and other climatic conditions.

(2) Justifications for the selection of the particular alignment of the proposed road.

(3) History of the existing road system and other means of communications.

(4) Earth work details.

(5) Justification for the type of surfacings, you have proposed.

(6) Benefits served to the people in general, by the construction of this road.

(7) Details of bridges, culverts, railway crossings, etc.

(8) Availability of constructional materials and labour near the alignment.

PROBLEMS

3.1. What are the essential constituents parts of a project ? Describe in brief how you will proceed to start a road project.

3.2. What is meant by earth work, what is the procedure for earth work in embankment ?

3.3. For a road project, how many types of engineering survey are there. What is the difference between traffic survey and engineering survey ?

3.4. Write short notes on :

(i) Reconnaissance (ii) Topographical maps

(iii) Construction survey (iv) Spoil banks

(v) Barrow pits (vi) Dagbell.

3.5. Write down specifications for earth work in cutting.

3.6. What is meant by alignment of roads ? What are the points to be kept in mind while aligning a road ?

BIBLIOGRAPHY

1. Report on Research Project R-3 U.P. P.W.D. Research Institute, 1980.
2. I.R.C. Traffic Census on Non-urban Roads–I.R.C. : 9–1970.
3. Kahyali L.R.–Highway Engg.–1984.
4. Date on Growth of Highway Traffic in Maharashtra Engineering Research Institute, Nasik.
5. Report on the National Transport Policy Committee (Pande Committee), Planning Commission–1980.
6. Report of the Committee of Transport Planning and Coordination Committee (Tirlok Singh Committee) Planning Commission, 1966.
7. Road User Studies, Central Road Research Institute–1952.

4

Highway Drainage

In this Chapter you will study,

Importance of Drainage • Surface Drainage • Surface Water • Design of Side Drains • Run off for Highway Drainage • Highway Drainage Design Procedure • Sun-Soil or Sub-Surface Drainage • Methods of Providing Sub-Soil or Sub-Surface Drainage • Water Table • Seepage • Capillary Control • Cross Drainage Works • Drainage in City or Urban Roads • Drainage in Cross-Roads

GENERAL

Adequate and efficient drainage is an essential factor in the design and maintenance of a road. With efficient drainage system the efficiency and life of a road can be considerably increased. The water enters the road structure from the following sources:

(i) From the top of pavement.

(ii) Surface water from the sides of pavements.

(iii) From the under-side of the pavements by sub-soil water.

Road or highway *drainage* means removal of surface water and sub-soil water. The removal of rain water from the surface of road is called *surface drainage* and the removal of sub-soil water from the sub-grade is called *subsurface-drainage.*

4.1 IMPORTANCE OF DRAINAGE

The following points will justify the necessity and importance of highway drainage:

1. Due to improper drainage, the rain water will percolate deep into the sub-soil and will cause instability of sub-grade.

2. Variation of moisture causes variation of volume thereby causing failure of sub-soil and the road structure.

3. Prolonged contact of water with bitumen cause failure due to stripping of bitumen from the aggregate which will result pot holes.

4. In cold climate when the ground temperature reaches freezing point the excess sub-soil water may cause frost action and damage the road structure.

5. Due to poor drainage waves and corrugation are developed which is the main cause of failure of flexible pavements.

6. The rigid pavements fail due to mud pumping due to the pressure of water in the sub-grade.

7. Increased moisture content of sub-grade will cause lowering of its strength and bearing capacity.

4.2 SURFACE DRAINAGE

During the rainy season, water is collected at the top of the road surface and percolates into the road surface and the sub-grade and thereby weakens the sub-grade. To prevent this percolation of water into the sub-grade from the surface of the pavement, the following remedies have been suggested:

(1) Providing a water-tight or impervious type of road surfacing.

(2) Providing sufficient cross and longitudinal slopes in the road.

(3) The top surface of the berms should be of an impervious material and should have proper slopes towards side drains.

(4) Providing side drains on both sides of the road. The drains should have a proper slope or gradient so that the water from the road surface should be drained off easily. The side drains should not be less than 1.85 m or 6 ft from the edge of the formation of the road. In case of cutting, the surface drains should be just after the edge of formation.

(5) In those places where the rainfall is heavy, the side drains should be trapezoidal in section and in those places where there is less rainfall the section of the drain is triangular [(Figs 4.1 (a) and (b)].

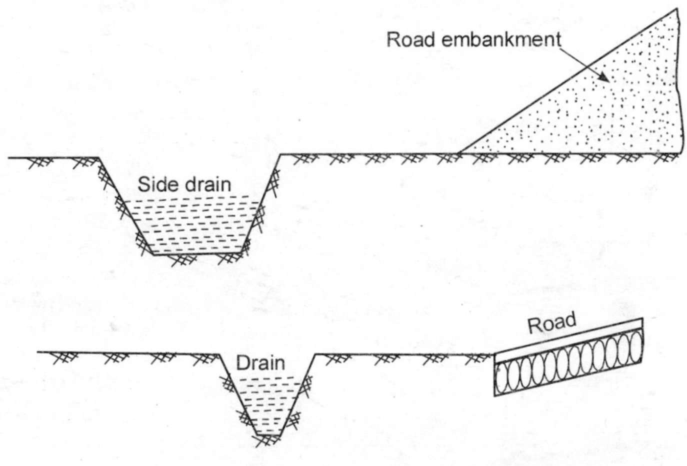

Figure 4.1 *Surface drainage*

(6) The height of the embankment should be nearly 60 cm or 2 ft. above the highest flood level (H.F.L.) of the area.

4.3 SURFACE WATER

The disposal of water in country or rural roads will depend on the condition of formation i.e., whether the road is in cutting or in embankment. When the road is in bank it is the usual practice to allow the flow of surface water to continue across the shoulders down to the side slopes to the natural ground or side drains. In case of high embankments the side slopes arc properly protected by shrubs, turf or grass. The flow of water from the road surface is fairly distributed evenly, when the slopes are unprotected and the sheet flow of water is not maintained, erosion of soil takes place. The water from the surface of road is collected in the side drains on the outer edge and then drained off. In another method a kerb is provided at the outer edge of the pavement to carry the water to the catch basin or other collecting devices from which it is removed and allowed to flow through the side drains or slope to the natural ground. This practice is not favourable as the water collected near the kerb will cause hazard to traffic.

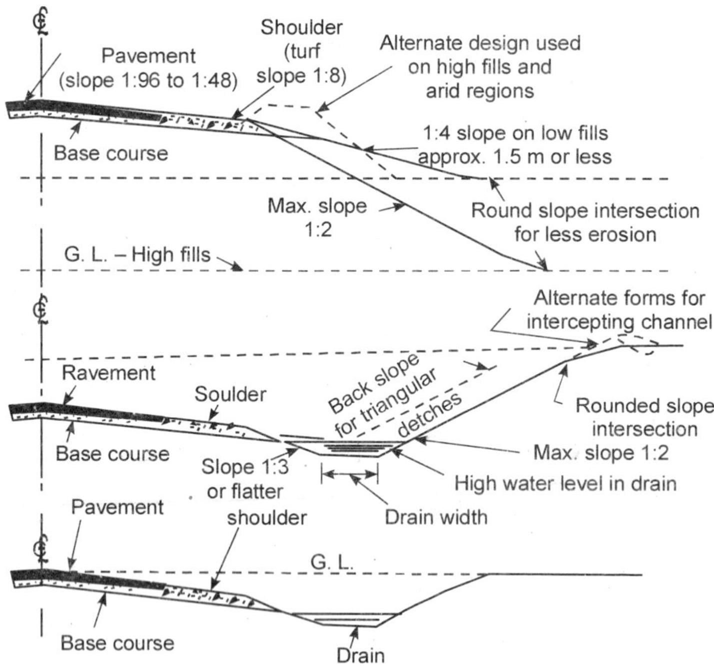

Figure 4.2 *Surface water features*

When the road is in cutting or on the ground line, water is disposed off to the side drains which should slope down towards the nearest water course to flow away from the road. Side drains are required where the ground by the side of the road is at a higher level because if the ground is lower, the water will flow away by natural drainage. In case of a long high embankment of poor soil as in the case of approaches to bridges the surface water can be collected through V-shaped drains near the edge of formation and then allowed to flow the slope by means of paved drains. Aprons are also provided at the foot of the slopes to prevent penetration of water. On low embankments the slopes can be protected by turfing and by providing flat slopes. Convex corners at the edge of the embankments should be avoided. The corners should be rounded off. On side long ground, it is seldom necessary to have a drain at the toe of the bank.

The water which enters the pavements or shoulders or berms is intercepted by providing a granular base below the surface. Usually the granular base is omitted and instead a turf is provided on the surface to maintain flow and penetrate seepage.

The side drains in country or rural roads are unlined or kutcha drains of trapezoidal shape of suitable cross section and longitudinal slope. The side drains are provided parallel to the alignment of the road and are called longitudinal drains.

4.4 DESIGN OF SIDE DRAINS

Design of side drains is based on the principle of flow in open-channels. First of all the run off is calculated through which the discharge of the drain. Run off is the quantity of water that is supposed to flow through the drains. If Q is the quantity of water to be flown through the side drains and V the velocity, the area of cross section of the channel will be

$$A = \frac{Q}{V} \quad (Q \text{ in m}^3\text{/sec, } V \text{ in m/sec})$$

The velocity in open drains on the country side which are unlined should be such to prevent silting and erosion. The allowable velocity of flow depends on the soil conditions. Normally for sand and silly soil 0.3–0.5 m/sec velocity is permissible and for loamy soil 0.6–0.9, clay 0.9–1.5, gravel 1.2–1.5 m/sec are recommended velocities. Assuming uniform and steady flow through the drain of uniform cross section, Manning's formula is used for determining the velocity of flow

$$V = \frac{1}{n} R^{2/3} \cdot S^{1/2}$$

where

V = average velocity in m/sec

R = hydraulic radius (cross sectional area of flow divided by wetted parameter)

n = Manning's constant (roughness coefficient)

Types of soil	*Value of n*
Ordinary earth smooth grade	0.02
Rock or rough rubble	0.04
Earth with vegetation	0.08 – 0.1
Smooth rubble	0.03

S = longitudinal slope of drain

4.5 RUN OFF FOR HIGHWAY DRAINAGE

The run off for a highway drainage is calculated by the following formula

$$Q = C.i.A$$

where

Q = run off in m^3/sec

C = run off coefficients (it is the ratio of runoff and rate of rainfall)
i = intensity of rainfall mm/sec
A = drainage area in 1000 m^2.

The value of C for various pavement surfaces are

Bituminous road	0.8
Cement concrete road	0.9
Gravel road	0.35
W.B.M. road	0.7
Compacted and stabilized soil	0.33–0.55
Soil covered with turf	0.05–0.30

The design value of intensity of rainfall is to be determined for the expected duration of rainy season and frequency. The time for water to flow through

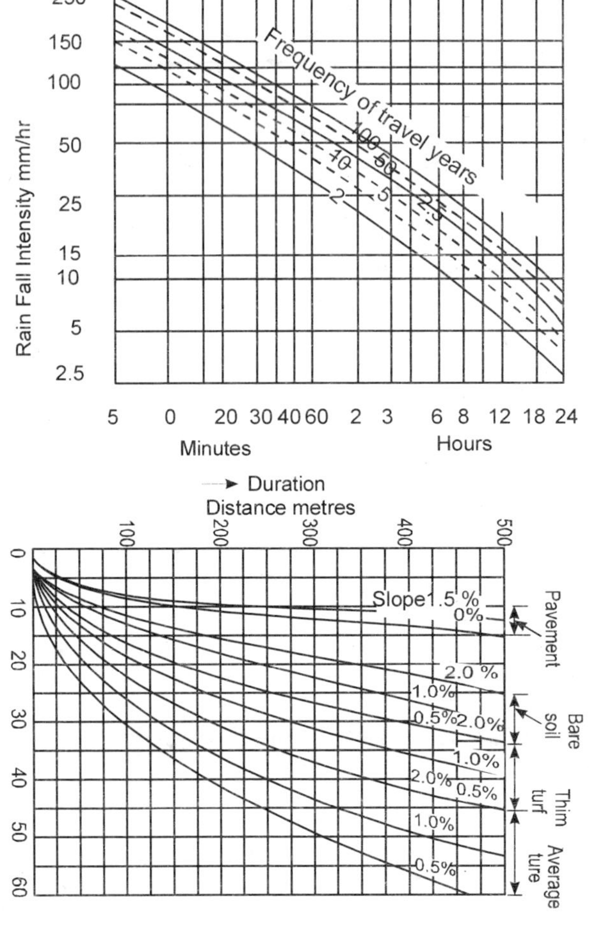

Figure 4.3 *Rainfall intensity curves*

the drain from the inlet point to the outlet point is determined by assuming a normal or allowable velocity of flow to be 0.3 to 1.5 m/sec. The time of concentration or the duration of flow of storm may be taken as the sum of inlet time and the time of flow through the drain.

4.6 HIGHWAY DRAINAGE DESIGN PROCEDURE

Before the actual design of drains the following information may be collected:

1. Total road length and the catchment area of the proposed road drain.

2. Run off coefficients for different surface i.e., different type of pavement surfaces.

3. Distance from the farthest point in the catchment or drainage area to the inlet of the side drain along the steepest gradient and the average value of slope.

4. Roughness coefficient, allowable velocity of flow through the drain.

5. Rainfall date including average intensity of rainfall.

6. Frequency of return period which varies from 10 years to 25 years depending upon the financial resources.

The design of the drain should consist of the following steps :

1. The run off coefficients and the drainage area of various surfaces and various cross sections are computed to get the weighted value of *C*.

2. Time of flow along the longitudinal drain is determined for the estimated length *L* upto the nearest cross drainage or water course.

3. Total area of drainage is calculated in units of 1000 m^2.

4. Calculate now the run off and then the cross sectional area of the drain.

5. The depth of water flow in the drain is calculated by assuming a suitable bottom width of the drain and the side slopes.

6. The longitudinal slope is calculated by using Manning's formula adopting suitable roughness coefficient.

7. The inlet time for the drainage surface is worked out from the graph.

Example 4.1. *The distance between the turf covered drainage area farthest point and point of entry to the side drain is 250 m. The average slope of the area is 2.0%. The average value of run off coefficient is 0.25. The length of the longitudinal drain is 540 m (from the inlet point to the cross drainage). The value of soil is sandy clay. The velocity in the drain is assumed to be 0.6 ml sec. Design a suitable section of the drain for a 10 year return period basis.*

Solution

Inlet time for turf with 2.0% slope for 250 m corresponding distance from the chart is 33 minutes.

Time taken by the storm water to flow through the drain upto the cross drainage @ 6 m/sec or 36 m/min will be

$$\frac{540}{36} = 15 \text{ minutes.}$$

$$\text{Total duration time} = 33 + 15 = 48 \text{ min}$$

$$\text{Drainage area} = 540 \times 250 = 1{,}35{,}000 \text{ m}^2$$

or 135 (1000 m^2 units)

Corresponding rainfall intensity for a 10 year period for 48 minutes comes out to be 70 mm/hour or 70/36000 mm/sec (from the corresponding chart)

$$C = 0.25 \text{ (given)}$$

$$Q = C.i.A$$

$$= 0.25 \times \frac{70}{3600} \times 135$$

$$= 0542 \text{ m}^3\text{/sec.}$$

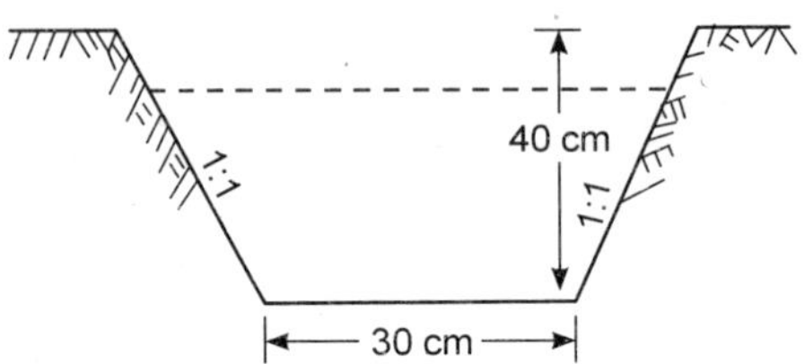

Figure 4.4

Assuming base width of the drain to be 30 cm with side slopes 1 : 1, the depth of drain comes out to be 35 cm. Assuming 5 cm free board the cross section of the drain will be 40 cm (Fig. 4.4)

4.7 SUB-SOIL OR SUB-SURFACE DRAINAGE

Water which penetrates into the ground will continue to flow underground until it meets with some impermeable material, then the flow ceases and the water accumulates. The top surface of this underground water is called *Water Table*. The ground above this level will remain unsaturated and below this level, saturated. The water table generally tends to be parallel to the ground level, but the depth of water table below the ground level may vary and it depends upon the geological conditions.

If this water table is very near to the sub-grade of the road, the consistency of the soil will change from dry to plastic state and the bearing capacity will consequently decrease. Another adverse effect which results from the increased moisture content is the volumetric increase which will ultimately cause cracks in the road surfacing. Hence it is essential to drain off this water to prevent the sub-grade, which is the foundation of the road from these adverse effects of sub-soil moisture. The method of removal of the sub-soil moisture is called *Sub-soil drainage.*

Following are the conditions under which sub-soil drainage should be provided :

(1) If the road is through flat country and water from adjacent lands stagnates and makes the road bed soft and unstable.

(2) When the road is in cutting and there is considerable seepage in the slopes.

(3) When the surface of the road has the normal underground water table sufficiently below the road crust even then due to capillary action moisture may rise to the surface of the road or the sub-grade.

(4) When the soil is subjected to the action of springs.

(5) When the road is at the foot of a hill and water therefrom damages the road.

4.8 METHODS OF PROVIDING SUB-SOIL OR SUB-SURFACE DRAINAGE

Pipe drains. This method of sub-soil drainage is suitable when a road runs in a flat country with low embankments and where the sub-soil water accumulates below the sub-grade. In this method pipe drains are placed

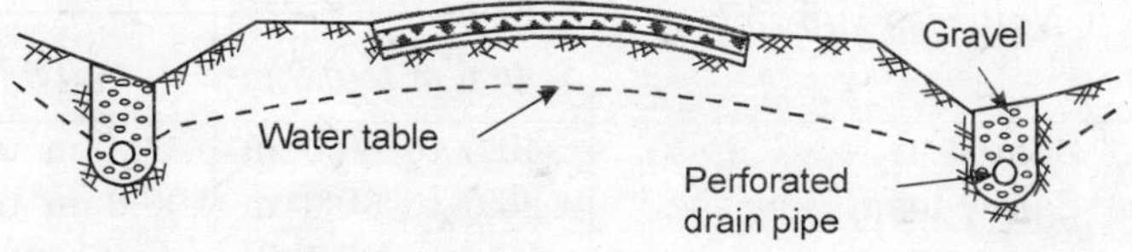

Figure 4.6 *Longitudinal pipe on the sides*

below the sub-grade. The pipes are usually made of vitrified clay and are placed on a bed of sand, crushed stone, or clay 15 cm thick,.with open joints butting against each other. To prevent the earth from above entering the pipes, through open joints, a tarred paper is sometimes provided to cover the joints. These pipes are placed in the trench with proper slopes both cross and longitudinal. The pipes are 15 to 20 cm in diameter. Cross or transverse pipes which are 6 to 10 cm in diameter are also laid at a distance of 6 m to 20 m apart. The trench is then filled with hard porous materials as shown in Figs. 4.6, 4.7 and 4.8.

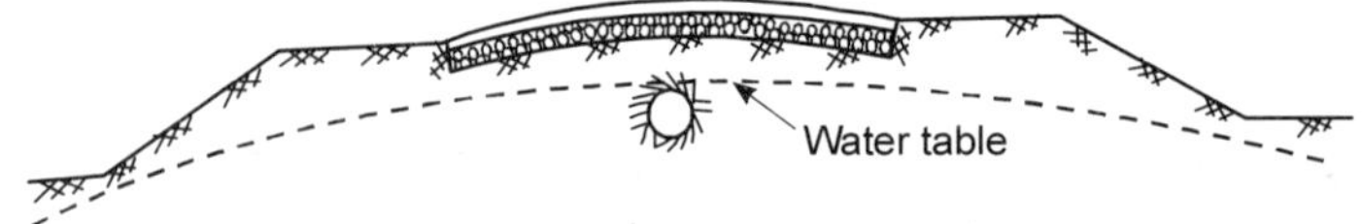

Figure 4.7 *Longitudinal pipes in the centre of road*

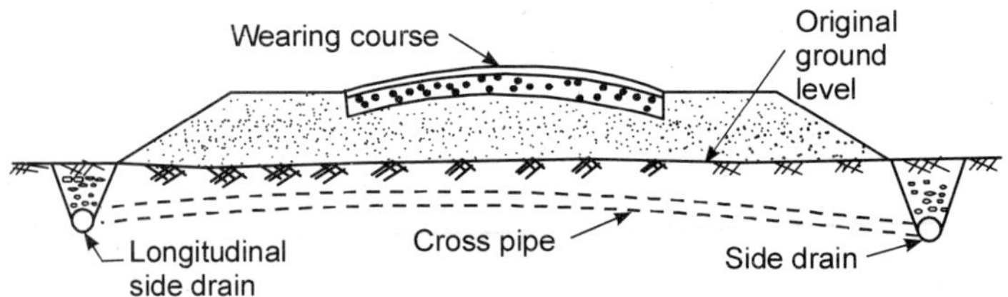

Figure 4.8 *Cross pipe*

The main longitudinal pipes may be laid in the centre or on both sides of the road depending upon the moisture conditions. The main pipes discharge their water into the surface drain.

Table 4.1 gives the recommended spacing of sub-soil drains, which are used for draining off the sub-soil moisture.

TABLE 4.1 *Centre to centre distance of sub-soil drain pipes*

S.No.	*Nature of sub-soil*	*Depth of the invert of the sub-soil drain from the ground level*	
		0 .6 m to 0.9 m	*0.9 m to 1.2 m*
1	Sand	49.5 to 60.0 m	60.0 m to 91.5 m
2	Sandy loam	26.0 to 30.0 m	30.0 m to 46.0 m
3	Loam	18.0 to 27.5 m	27.5 to 30.5 m
4	Clay loam	13.5 to 17.0 m	17.0 to 20.0 m
5	Sandy clay	10.0 to 12.0 m	12.0 to 14.0 m
6	Clay	7.5 to 9.0 m	9.0 to 11.0 m

As a thumb rule a drain line draws the moisture from the sub-soil of 120 metres from each side for each metre of depth.

C.G. Elliot gives the following (Table 4.2) spacing of the sub-soil drains for guidance:

Table 4.2 *Spacing of sub-soil drains as per C.G. Elliot*

S. No.	*Nature of sub-soil*	*c/c spacing of the sub-soil drains*
1	Close dense soil, having largely clay	9.0 to 12.0 m
2	Coastal plain lands composed of mixed clays with fine sand and of uniform structure	18.0 m
3	Alluvial grounds or heavy soils but having granular structure	21.0 to 24.0 m
4	Alluvial glacial drift and sandy loam silts with clayey sub-soil	38.0 m
5	Sandy lands and soils containing considerable quantity of vegetable matter	45.0–60.0 m

Drainage in Cutting. When the road is in cutting the sub-grade and the surfacing get closer to the sub-soil water level, and hence the stability of cutting will depend upon the efficient removal of water from the slopes.

To remove this water, burnt clay pipes having a diameter of 5 cm to 10 cm or ordinary drains having suitable cross-sections are provided on the slopes at an interval of 10 to 12 m, which discharge their water into the central drain.

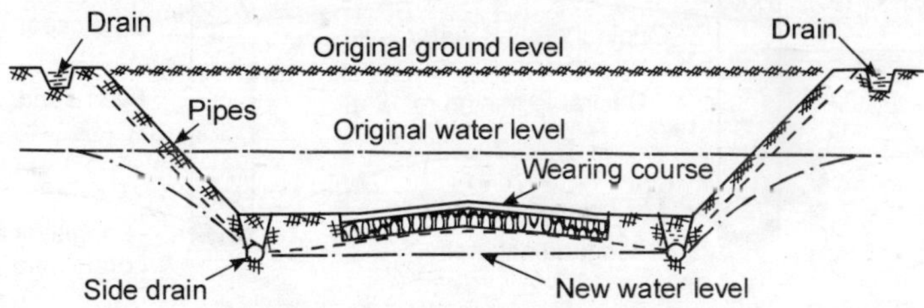

Figure 4.9 *Drainage in cutting*

Drainage in Side-long Ground. In embankment the stability of the side slopes is usually more important than the bearing power of the sub-grade or formation. If the embankment is built on ground with considerable

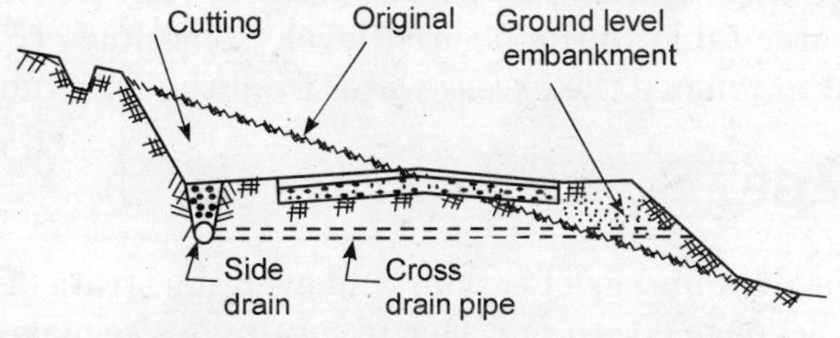

Figure 4.10 *Drainage in slopy ground*

crossfall or the road has a section partly in embankment and partly in cutting, the drainage problem should be carefully considered. If the water which comes down the slopes as surface water or sub-soil water, is not drained off immediately, it will reduce the bearing power of the soil and will almost invariably cause an embankment to slip. The arrangement of providing drains in a road section in partial cutting and partial embankment is shown in Fig. 4.6.

4.9 WATER TABLE

Water table is responsible for the stability and strength of a highway. Lower the water table more the stability. The level of water table should be 1 to 1.2 m below the sub-grade of die road and pavement layers are not subjected to moisture from below. In places where water table is high, sometimes as high as the ground level, the alignment should be diverted to avoid such a site. But sometimes it is un-avoidable and the alignment has to be taken through such places. In that case, the water table has to be lowered. *Lowering of water table* is comparatively very easy in permeable soils but in the case

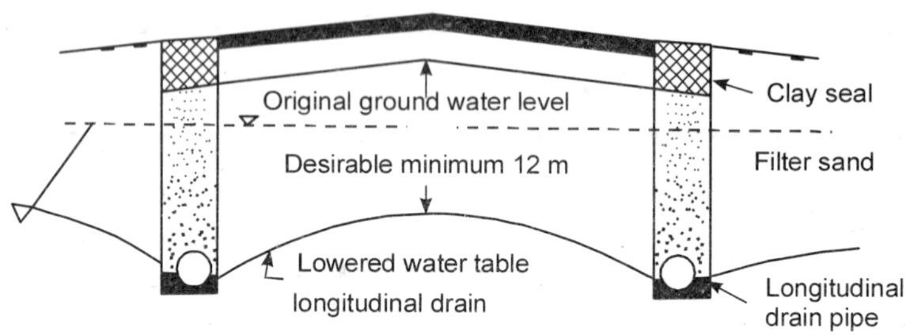

Figure 4.11 *Lowering of high water table*

of impermeable soil it is difficult. For permeable soils, porous pipes are laid along the edges of the road longitudinally and over the pipes the trenches are filled with filter sand. The pipes are placed with longitudinal grades parallel to the road gradient. The porous pipe suck the water from the sub-soil thereby water table to the desired level. Sometimes transverse pipes are also placed to remove the excess water from the longitudinal pipes.

4.10 SEEPAGE

When the general ground level is slopy and even the strata of soil is sloping, the problem of seepage does occur. Due to continuous seepage the sub-grade gets weaken and the road structure is exposed to failure. When the seepage zone is at a depth of less than 0.8 m or 0.9 m, seepage control measures are to be adopted. Seepage zone is ascertained by the general level of water table in the adjoining area. The usual method of seepage control is to provide

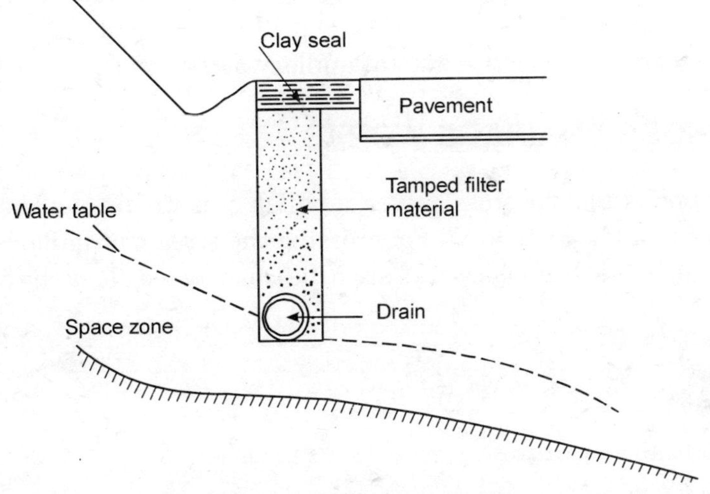

Figure 4.12 *Seepage control Method*

longitudinal porous pipe at the foot of the slope, in a trench later filled with sand as shown in Fig. 4.12.

4.11 CAPILLARY CONTROL

The water uses from the moist sub-soil due to capillary action which reduces the bearing capacity of the sub-grade. This problem of capillary use occurs when the road is in bank and the water table is close to the ground level.

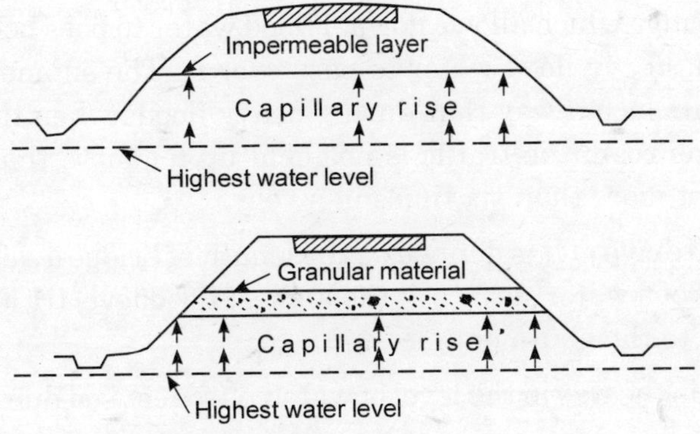

Figure 4.13 *Capillary control*

This can be prevented by providing a layer of granular material of a suitable thickness or by providing a layer of some impermeable material or a bitumen layer between the sub-grade and the ground level. This layer of granular material or impermeable material layer will act as a cut off layer beyond which the water will not rise due to capillary action.

4.12 CROSS DRAINAGE WORKS

A road alignment has to cross a river, a nallah or a stream or a sheet flow of rain water in a low lying area. For crossing the stream or nallah or a river the following cross drainage works are to be constructed along the alignment of road:

1. Bridge
2. Culvert
3. Irish bridge or causeway

The detailed study of those works is beyond the scope of this book. The readers are advised to refer some book on Bridges. Some important definitions will be useful for ready reference.

Bridge. It is defined as a structure that affords over low ground, water or other obstructions to cross a roadway, railway, footpath etc.

Culvert. It is defined as a small structure, having a maximum span of 6 m. Structurally there is very little difference between a bridge and a culvert.

Causeway. It is defined as a dip which allows flood water to pass over the road.

Submersible Bridge. It is slightly different from a causeway. It is defined as a normal bridge which allows normal flood water to pass below it through the vents and, heavy flood water to pass over it. The submersible bridges are designed in such a way that during heavy floods when the flood water passes over the road, the traffic is not held up for more than 3 days at a stretch and not more than six times in a year.

Linear Waterway. It is defined as the length of bridge available between extreme edges of a water surface at the highest flood level (H.F.L.) measured at right angle to the abutment faces.

Afflux. It is the rise in the level of watercourse caused due to obstruction of bridge piers.

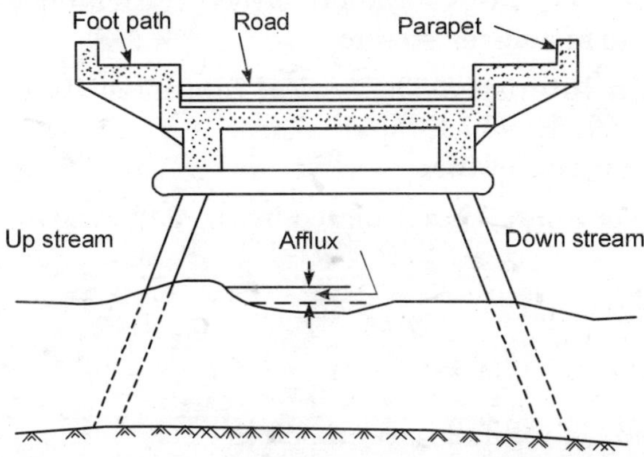

Figure 4.14 *Afflux*

Run-off. It is defined as the portion of rain water on the catchment area which flows to a water course. The run off is calculated by the following formula:

Dicken's formula

$$Q = CA^{3/4}$$

Rynes's formula

$$Q = CA^{2/3}$$

where

Q = discharge in cubic metres/sec

A = catchment area in square kilometres

C = an empirical constant depending upon the nature of catchment area

The values of C for Dicken's formula area

C = 11.3 for Northern India

= 22 for Western Ghats

= 13.8–19.3 for Central zone

Value of C for Ryne's formula

C = 6.7 for areas 24 km within the sea cost

= 8.42 for areas 160 km within the sea cost

= 10 for hill areas.

Catchment Area. It is defined as the area from which rainfall or storm water flow to a nearby stream, a drainage line or a low lying area. The boundary line of this basin or area is called watershed.

Viaduct. It is a very long continuous structure carrying a roadway over a dry valley instead of a water stream.

Critical Rain Intensity. The critical rain intensity is defined as the intensity of rainfall for which the discharge is maximum when it remains for concentrated period of time.

The intensity of rainfall is calculated by the following formula

$$i = \frac{T+1}{t+1} \times \frac{F}{T}$$

where

i = intensity of rainfall in centimetres

T = duration of rainfall in hours

t = the time in hours during which period the rainfall remains maximum

The critical intensity of rainfall is

$$I_c = \frac{T+1}{T_c+1} \times \frac{F}{T}$$

where T_c is the concentration time in hours.

The concentration time can be calculated by the following formula

$$T_c = \phi \frac{L^3}{H}^{1/3}$$

where

L = distance from the farthest point to the site of the bridge or culvert

$$\phi = \frac{9}{4} \frac{5280}{c^2 I.k}$$

where

k = coefficient of runoff

c = Bazin's coefficient

I = intensity of rainfall.

The value of ϕ for most of the cases calculated comes out to be 0.89.

Scour. When the depth of water below the bridge increases due to vertical cutting of the bed, it is called scour. In other words further deepening of river or cutting of bed due to the action of water is called *scouring*. The horizontal cutting of the banks of a river is called *Erosion*. Scouring is very harmful to the structures and hence it must be prevented or minimised. Depth of scour can be calculated by the following formula Lacey's formula:

$$D = 0.473 \left(\frac{Q}{f}\right)^{1/3} \times R$$

where,

D = scour depth in metres

Q = discharge in cubic metres/sec

f = Lacey's constant

R = 1.27 for straight reach
1.50 for moderate bend
1.75 for sharp bend
2.00 for right angle bend
2.00 for at the pier nose

The value of Lacey's silt factor f is approximately calculated from the relation

$$f = 1.76\sqrt{m}$$

where

m = weighted mean diameter of inner bed particles in mm. The value of m and f are given in the following table.

S.No.	*Nature or type of rivers bed material*	*Mean weighted dia. in mm*	*Value of f*
1	Very fine sand	0.052	0.40
2	Fine silt (western India)	0.081	0.50
3	Fine Silt	0.120	0.60
4	Fine Silt (Delta type)	0.158	0.70
5	Medium silt	0.233	0.85
6	Standard silt	0.323	1.00
7	Medium sand	0.505	1.25
8	Coarse sand	0.725	1.50
9	Fine bajri	0.988	1.75
10	Large size sand	1.290	2.00

In quasi alluvial soil stream having rigid banks but erodible beds, the normal depth can be calculated by the following formulae

$$D = \frac{Q}{W.V}$$

$$D = \frac{1.2\ Q^{0.63}}{f^{0.33} \times W^{0.80}}$$

where

W = surface width of stream

V = velocity of flow

S = slope

f = silt factor

Example 4.2. *Calculate the maximum scour depth of an alluvial stream carrying maximum discharge of 450 cumec and silt factor 1.1 with the following conditions :*

(i) when the road bridge to be constructed has three span of 45 m each.

(ii) when the road bridge has four spans of 25 m each.

Solution

Regime surface width of stream

$$W = 4.8\sqrt{Q}$$

$$= 4.8\ \sqrt{450}$$

$$= 101.81 \text{ m}$$

Case I :

Since the proposed bridge consists of 3 spans of 45 each, therefore

$$L = 3 \times 45 = 135 \text{ m which is} > W$$

therefore normal scour depth

$$D = 0.473\ \frac{Q}{f}^{1/3}$$

$$= 0.473\ \frac{450}{1.1}^{/13}$$

$$= 3.5 \text{ metres}$$

Maximum scour depth will be at the noses of the piers and is equal to 2D.

Maximum scour depth = 2 × 3.5 = 7.0 m.

Case II :

In this bridge consists of 4 spans of 25 m each.

$$L = 4 \times 25 = 100 \text{ m} < W$$

As the waterway is less than the natural stream width, hence the normal scour depth will be

$$D = \text{Regime depth} \times \left(\frac{W}{L}\right)^{0.61}$$

$$= 0.473 \left(\frac{Q}{f}\right)^{1/3} \times \left(\frac{W}{L}\right)^{0.61}$$

$$= 0.473 \left(\frac{450}{1.1}\right)^{1/3} \times \left(\frac{101.81}{100}\right)^{0.61}$$

$$= 3.5 \text{ m}$$

Maximum scour depth = 2 × 35 = 7.0 m.

Waterway. For deciding the waterway the first thing to be decided is the maximum expected discharge which will pass under the bridge. Waterway is the area of the opening under the bridge which should be sufficient to pass the maximum flood discharge that would everpass under the bridge widiout increasing its velocity to a dangerous level. The maximum velocity depends upon the nature of the bed of the river. Generally speaking the maximum permissible velocity under normal conditions should not exceed 3 m/sec.

After calculating the maximum flood discharge, the waterway can be calculated

$$Q = V \times A$$

where

Q = maximum flood discharge

V = permissible velocity

A = area of waterway

If h and ha be the head causing normal stream velocity and afflux respectively, then

$$Q = \text{Length} \times \text{Depth} \times \sqrt{2a(h + ha)}$$

The length is divided into number of spans. More the number of spans, more the cost of bridge and more the obstruction to the normal flow of the stream.

Length of a bridge. The length of a bridge can be calculated after determining the waterway.

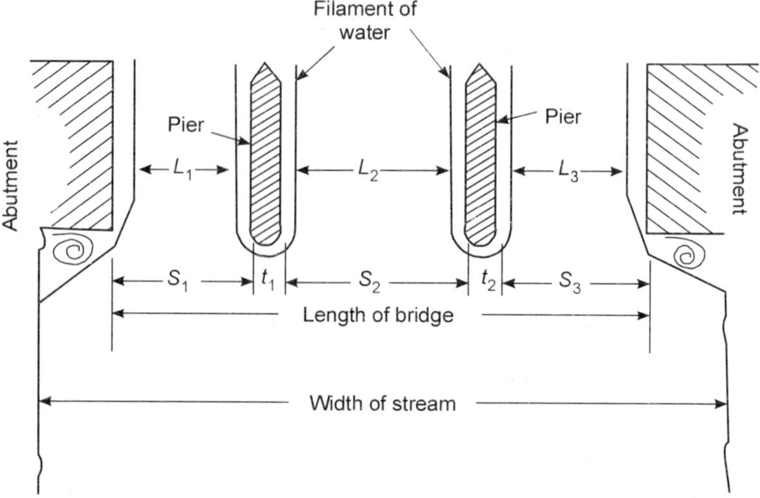

Figure 4.15 *Length of a bridge*

Length of a bridge = waterway + width of all piers.

Depth of foundations of a bridge. For small bridges and culverts the depth of foundations is calculated by the following rules as suggested by I.R.C.

(a) For erodible soils or river beds the foundation should be taken down to a depth one third greater than the maximum scour depth, subject to a minimum of 2 m below the scourline for arched bridges and 1.2 m for other bridges.

(b) For hard beds, when rock or other non-erodible materials at maximum material is available, the foundations should be firmly secured to the hard stratum about 60 cm in the hard material and 30 cm in the rock.

(c) The pressure on the foundation material should be well within the bearing capacity of the material and the bed. This rule is applied when no bed floor is provided and the stream is free to scour.

(d) In case of culverts, when the bed floor is usually provided, keep the top of the floor about 30 cm below the bed, carry the foundations of abutments 1.2 m below the top of the floor. Provide an upstream curtain wall 90 cm to 130 cm deep and down stream curtain wall 1.3 to 2.2 m deep from the top of the floor.

Width of roadway on a bridge or culvert. In the case of culverts across National Highway, State Highway and District Roads, the width of culvert between the faces of parapets, should be equal to the full designed formation width of the road. For other district roads and village roads, the width of

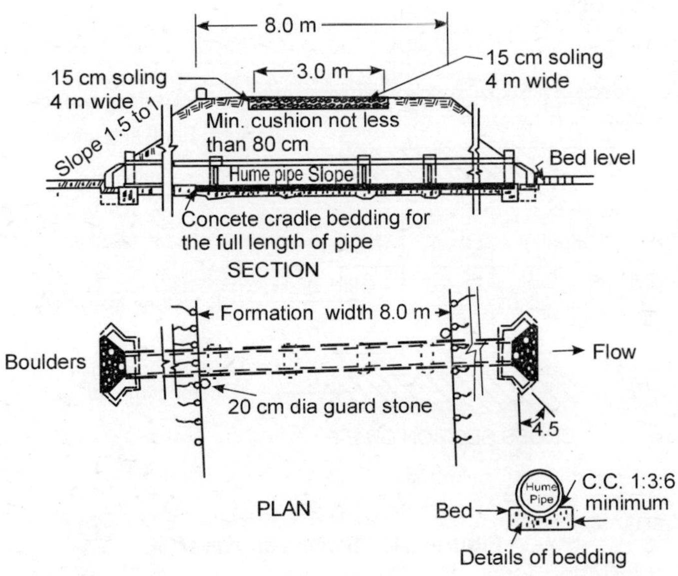

Figure 4.16 *Pipe culvert*

roadway at the culvert should be 4 m. Any future extension of the formation of the road, should also be taken into consideration while fixing the width of culvert.

In case of small bridges across National Highway, State Highway and major district roads, the minimum clear roadway between the inner faces of the kerbs or wheel guards should be 8 m.

Culverts. A culvert is a cross drainage structure to carry a roadway over a small streamlet or nallah. When the linear waterway required is 6 m or less, it is called a culvert. Culverts are in fact closed conduits used for the drainage of highway with the exception of storm drains. The two main functions of culverts are:

1. Collecting and leading the water across the road so as not to cause any damage to the road bank or stream bed by scouring.

2. Allowing sufficient waterway to prevent heading up of water above the road surface.

Following are the common types of culverts :

(i) Pipe culvert

(ii) Slab culvert

(iii) Box culvert

(iv) Arch culvert

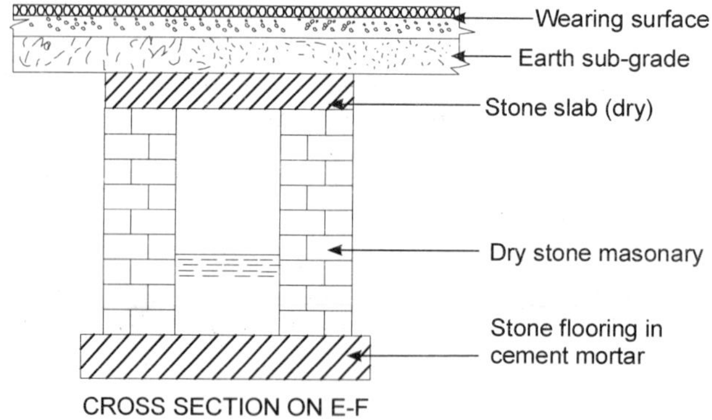

Figure 4.17 *Stone slab culvert*

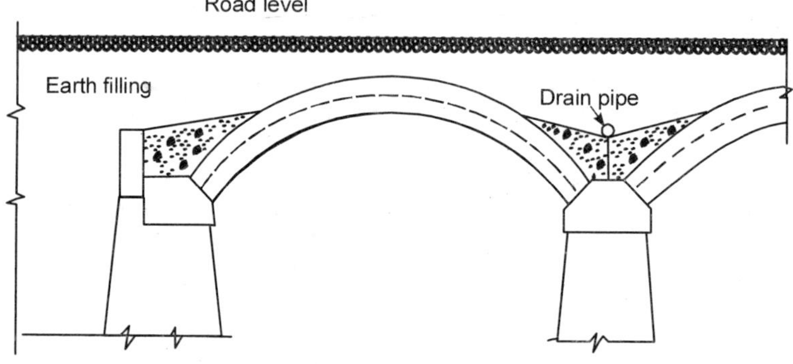

Figure 4.18 *Arch culvert*

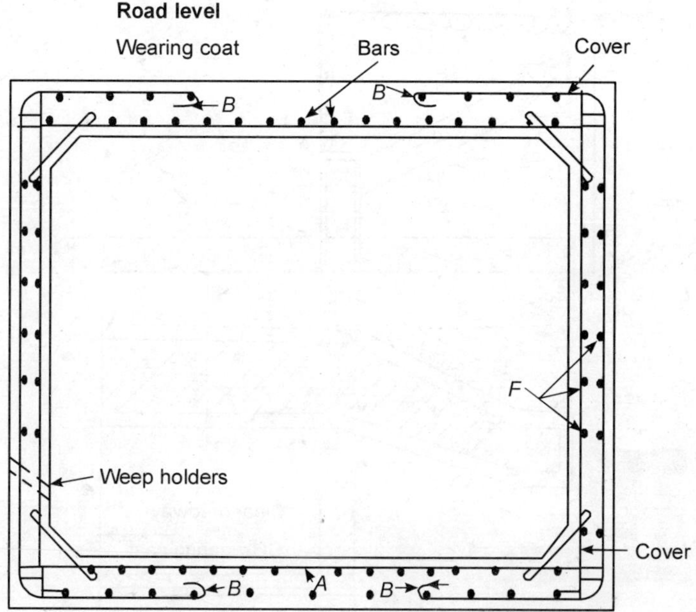

Figure 4.19 (a) *Cross-section of a box culvert*

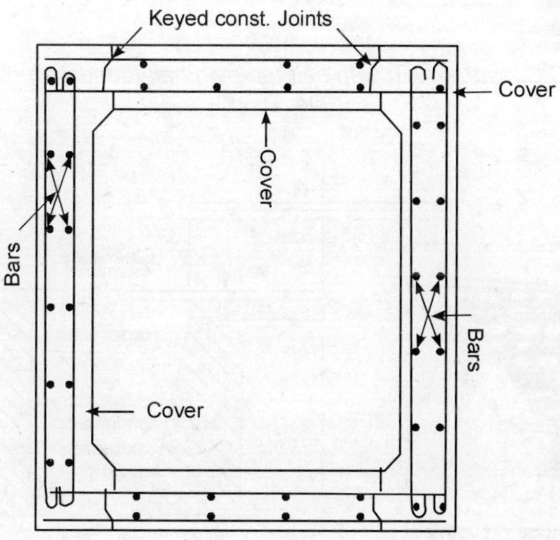

Figure 4.19 (b)

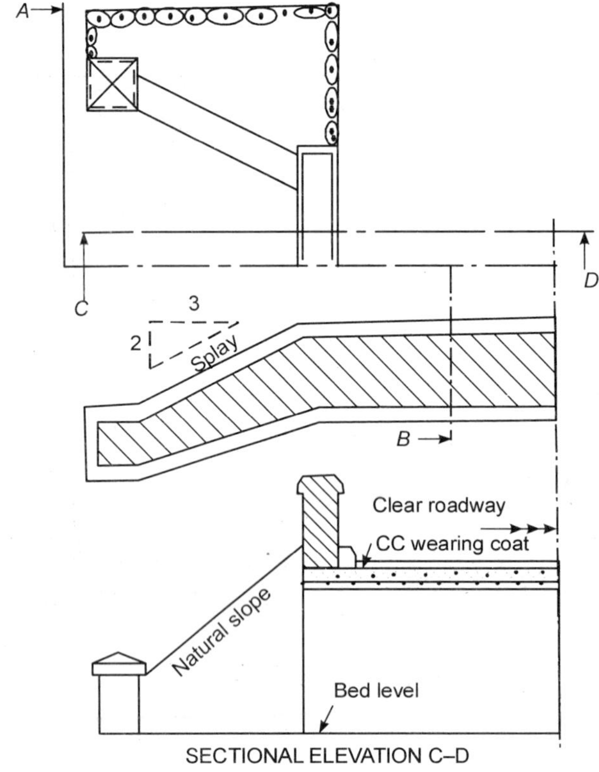

Figure 4.20 *R.C.C. slab culvert with half top plan half foundation plan and sectional elevation*

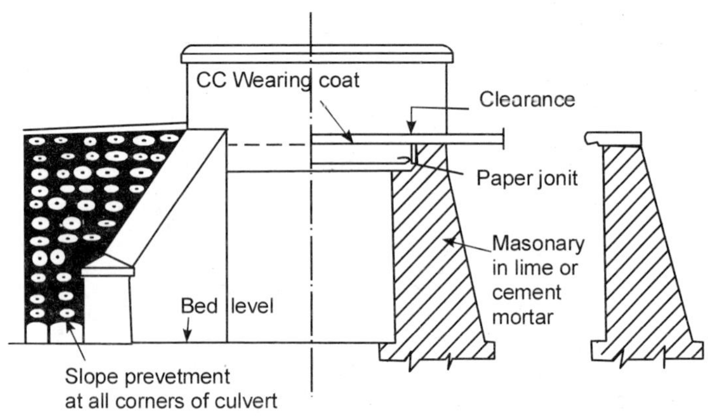

Figure 4.21 *R.C.C. culvert and wing wall*

4.13 DRAINAGE IN CITY OR URBAN ROADS

The drainage of road in urban or built up areas is different from that of other places. Open drains in built-up areas cannot be provided as they are unsightly, and occupy a lot of space ; and are a source of danger to the traffic. The drainage is, therefore, provided through vents and gratings. The surface water enters through those gratings and vents (through kerbs) into an underground pipe system.

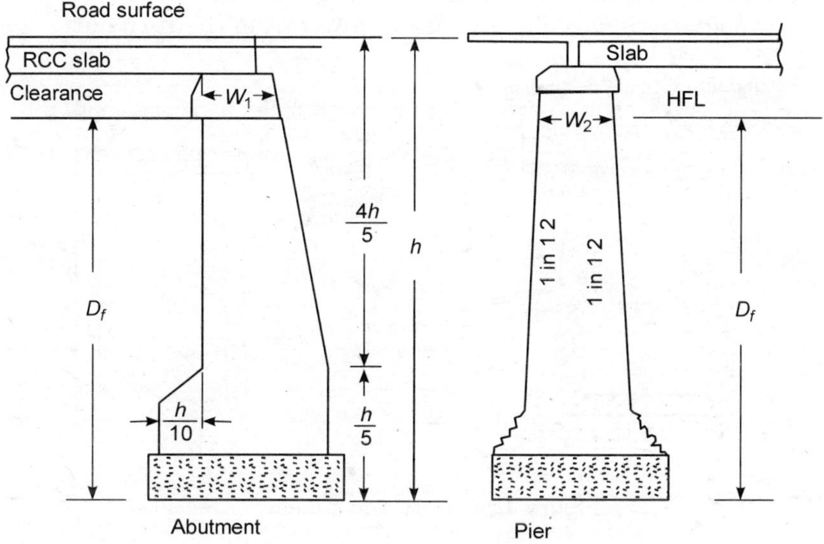

Figure 4.22 *Abutment and pier*

Pipes are laid underground and water is led to them through transverse pipes. Water is admitted through the gullies by means of hollow kerbs known

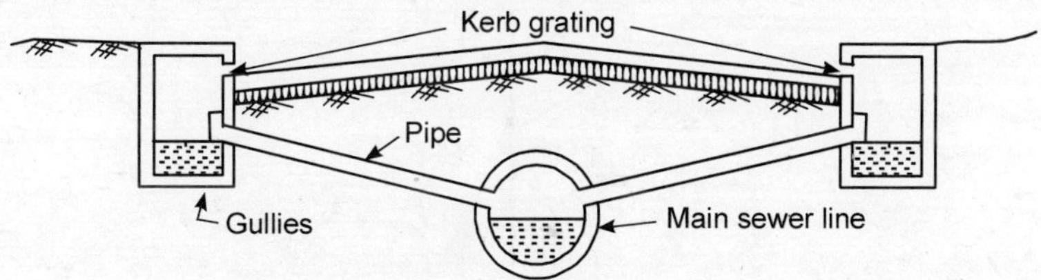

Figure 4.23 *Road Inlets*

as *Kerb inlet* or by means of gratings which are fixed at the edge of the road. Kerb inlets are generally preferred to the gratings as the gratings reduce the carriage width and sometimes the gratings are broken due to heavy impact of vehicles. Surface drains are generally provided on one side only as it is uneconomical to run drains on both sides of the road unless the house drains are to be pitched up.

4.14 DRAINAGE IN CROSS-ROADS

Surface drainage in cross-roads, that is, at the road junctions or intersections requires careful planning. When two roads are crossing each other, the grade of underground drains, or channels and the cross slope of roads is so adjusted that the low spot is displaced a few metres away from the turn out.

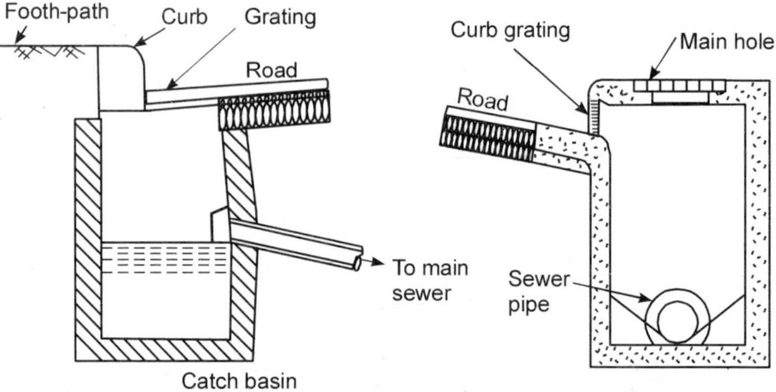

Figure 4.24 *Road side gullies*

Figure 4.24 shows the arrangement of gullies for road junction where both carriageway ways have the normal arched camber. It should be noted that

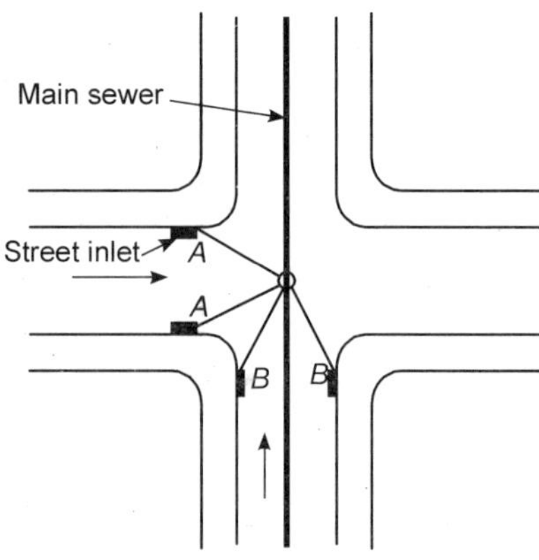

Figure 4.25 *Position of road-side gullies*

the water should not be allowed to collect at the crossing but should be intercepted at points *A, A′, B, B′* etc., before it reaches the crossing (Fig. 4.25).

PROBLEMS

4.1. Give neat sketches in plan and section, illustrating fully the surface drainage arrangements that you would propose for a hill road half in cutting and half in embankment.

4.2. Why is surface drainage of roads necessary. Outline in brief the circumstances under which the necessity of sub-soil drainage occurs while setting out a new road.

4.3. What methods do you suggest for making such a road which should remain dry throughout the year. Illustrate your answer with neat sketches where necessary and possible.

[**Hint.** *The road surface should remain dry, means that the water from, the surface of road should be drained off easily, and immediately*]

4.4. Describe in brief the methods of providing surface drains in city roads. Illustrate your answer with neat sketches.

4.5. Under what circumstances sub-surface drainage is provided ? What are the different methods of providing surface drainage ?

4.6. Draw a neat sketch showing the arrangement of drainage system in city roads crossing each other. Describe the method in brief.

4.7. Discuss in brief the damages caused by water to highway subgrades and pavements. What precautions would you suggest to safeguard the road against these damages ?

4.8. Distinguish between the rural and urban highway drainage systems. State the factors which control the design in each case.

4.9. What are the causes of changes in moisture contents in the sub-grade ? How can you improve the drainage of a road and control higher water level ?

4.10. Explain the need for efficient drainage of the road surface. Describe with sketches the methods used for draining a road on a steep hill slope.

4.11. Calculate flood discharge from a catchment area of 65 sq. kilometre when the rainfall intensity during the storm is 15 cm in two hours. The time of concentration is 20 hours and the run off coefficient in 0.35.

4.12. Explain briefly the different methods of estimating the flood discharge of stream. Which of the method is more accurate ?

4.13. The catchment area of a stream is of sandy soil with light vegetation cover and its area is 12,000 hectares. The length of the catchment is 25 km and the fall in level from the critical point to the bridge site is 480 m. Calculate the peak run off for designing the bridge, if the severest storm as recorded yielded 18 cm of rain in 4 hours. Assume area factor = 0.70 and co-efficient absorption as 0.20

4.14. Define and explain the following terms : Afflux, catchment area, scour, run off.

4.15. The flood discharge under a bridge is 4,764 m^3/sec. If the normal width and waterway are 924 m and 900 m respectively, calculate the scour depth and afflux. Lacey's silt factor is 1.5 and the bridge is on straight reach.

4.16. The maximum quantity of water expected in one of the longitudinal drains on a clayey soil is 0.9 m^3 sec. Design the cross section and longitudinal slope of a trapezoidal drain, assuming the bottom width to be 1.0 m and cross slope 1:1.5 (Vertical : Horizontal). The allowable velocity of flow in the drain is 1.2 m/sec and Manning's roughness coefficient is 0.02.

4.17. Discuss the importance of highway drainage.

4.18. Write a short note on the cross drainage works.

BIBLIOGRAPHY

1. The AASHO Road Test, Highway Research Board No. 33 – 1955.
2. Croney D and J.A. Loe Drainage in pavements. Institution of Civil Engineers 1965–30, 225–270.
3. Whiffin A.C. and N.W. Lister – Asphalt pavements – Ann Arbor, Michigan 1963,499–521.
4. Recommended practice for treatment of embankments to prevent erosion I.R.C–36 – 1974.
5. Recommendations for the construction of roads in water logged area I.R.C.: 34 –1930.
6. Guidelines for the design of small bridges and culverts I.R.C. : 13 –1973.
7. Armo Drainage and Metal Products – Handbook of Drainage Construction for Roads.
8. Pichworth H.F. – Concrete Pipe Culverts Handbook – Concrete Association U.S.A. – 1958.
9. Code of Practice No. 5 London Institute of Municipal Engineers.

5

Earth Roads and Soil Stabilization

In this Chapter you will study,

• Construction of Earth Roads • Compaction of Earth • Construction of Embankments • Settlement of Embankments • Preparing Sub-grade • Soil Stabilization • Methods of Soil Stabilization • Principles of Soil Stabilization • Investigations of Soil Stabilization • Mechanical Stabilization • Types of Mechanical Stabilization • Stabilization by Compaction • Impact Computer • Specification and Control of Compaction • Stabilization by Consolidation • Stabilization by Electrical and Thermal Methods • Soil Aggregate Stabilization • Chloride Stabilization • Lignin Stabilization • Molasses Stabilization • Cement Stabilization • Mechanism of Cement Stabilization • Mixing and Compacting Soil-Cement • Curing • Construction • Bitumen Stabilization • Stabilization of Black Cotton Soil • Desert Sand Stabilization • Stabilization by Grouting • Moorum Roads • Stabilized Moorum

GENERAL

An ordinary earth road is one whose foundations and wearing surface are composed of natural soil, available along the alignments of the road. Earth roads are the most used ones in this country. In India, earth roads are called kachcha roads. They are cheap, easy in construction and maintenance. They provide a link between villages and villages, a village with district roads or a railway station. These roads are fair weather roads as they become muddy in rainy season and dusty in dry weather and are only suitable for iron-wheeled traffic.

There are two types of earth roads, viz., (i) ordinary earth roads which are constructed from the local material available at the site, and (ii) stabilized earth roads. The stabilized earth roads are better in wearing conditions. They afford less resistance and give neat appearance.

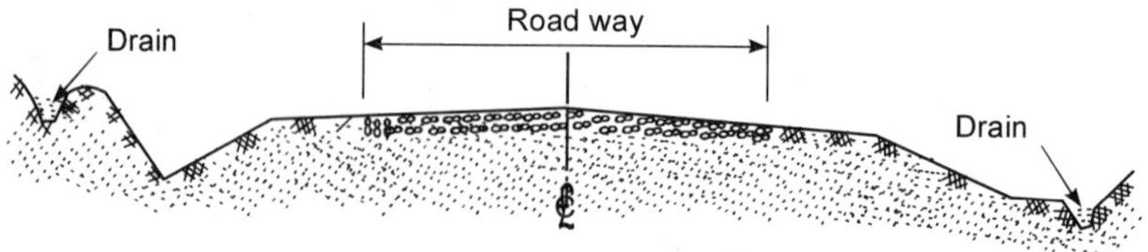

Figure 5.1 *Village road*

5.1 CONSTRUCTION OF EARTH ROADS

The first stage in the construction of earth roads in the preparation of formation. All grass and weeds should be removed from the site. As per the configuration of the country, these roads may be taken through a flat country, in cutting or in embankment as the case may be. If the road is to be in cutting the excavated earth should be deposited in the form of spoil banks. On the other hand if the road is going in embankment, earth is obtained from the barrow pits. In any case the formation level should be kept higher than that of the surrounding land for easy drainage. The width of formation varies from 6 m to 9 m according to the requirement.

The *sub grade* is given a camber of 1 in 24 and the surface is rolled and watered. A layer of 10 cm graded soil is spread evenly and rolled at optimum

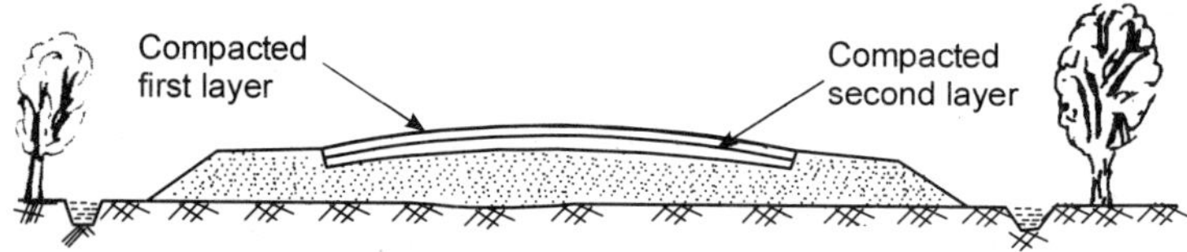

Figure 5.2 *Earth road*

moisture content (O.M.C.) with steep foot rollers and finally finished with a lighter roller. Sometimes another layer of nearly 10 cm thick is also spread and rolled properly to act as *wearing course.*

The surface is then watered for 4 to 5 days for curing and no traffic is allowed to pass on the road during this period. Then for a period of 10 to 15 days the surface is sprinkled with water until completely cured or set, but the traffic is allowed to pass over it during this period.

Maintenance. These roads require constant attention. If these roads are not maintained properly, they will wear out and will become unserviceable very soon. Hence periodic repairs of the pot-holes and ruts is very essential. The ruts and pot-holes are filled with earth and compacted by hand rammers. Side drains should also be repaired properly. Actually speaking, the life and efficiency of earth roads mainly depends upon the efficient drainage system.

5.2 COMPACTION OF EARTH

Compaction of soil or earth is the process of constraining the soil particles to pack more closely together through a reduction in air voids, generally by mechanical means. To object of compacting the soil is to improve its physical conditions and improve its bearing capacity, reduce its compressibility the percentage of voids are reduced due to which the soil will absorb water. Due to compaction resistance to frost action is increased and tendencies for volume change i.e., shrinkage and swelling.

Compaction is measured quantitatively in terms of dry density such as weight of soil solids per cu metre volume. The moisture content of the soil is the weight of moisture present expressed as the percentage of weight of dry soil. The important factors which influence the increase in the dry density of soil after compaction are moisture content and the method of compaction. With the same amount of compaction or compactive effects, which a maximum dry density is obtained, is known as optimum moisture contents.

In the laboratory, compaction is generally measured quantitatively in terms of the dry density produced after compaction. In 1933, Proctor first induced a laboratory compaction test. The test is believed to produce for most soils in the laboratory the results which can be obtained with compaction equipment in the field, although a precise reproduction of field conditions is not possible, the laboratory test is used as a basis for determining the water content suitable for field compaction. The test is called Proctor's mould.

Soil compaction in the field is achieved by rolling, ramming or by vibration.

5.3 CONSTRUCTION OF EMBANKMENTS

The embankments are constructed either by rolling thin layers of soil or by hydraulic fills. The method of compacting embankments by rolling is called Rolled Earth method. Each layer of earth is rolled to a satisfactory thickness and to a desired density before the next layer is placed. Compaction is carried out at optimum moisture content so as to take maximum advantage of dry density, using a specified compaction equipment or effort. The thickness of layers varies from 10 to 30 cm depending on the various factors such as soil type, equipment and specification etc.

The practice of dumping the earth without compaction and allowing the earth to get consolidated under weather conditions during few subsequent seasons. The setlement under such conditions should be avoided as it takes long time and if the pavement is constructed before the settlement is complete, the pavement will fail ultimately.

Each layer of soil is properly compacted before the next layer is applied over it. On the slopes of the embankment turf or grass is grown so as to prevent erosion. Granular soil is generally preferred for embankments.

5.4 SETTLEMENT OF EMBANKMENTS

After compaction or consolidation, the embankments may settle down. If the embankment foundation consists of compressible soil with high moisture content, the consolidation can occur due to increase in load. Excessive settlement occurs due to improper compaction during construction and hence can be eliminated by proper compaction. Sometimes though rarely, for accelerating the rate of consolidation and to reduce chances of excessive settlements, vertical sand piles are driven in marshy soils. The sand piles are 25 to 30 cm in diameter and are placed at 2.5 to 3.0 m spacing. The top of the piles or sand columns is covered with a thick layer of sand called sand blanket which is generally 40–60 cm thick.

5.5 PREPARING SUB-GRADE

The preparation of sub-grade includes clearance of jungles, preparation of grade etc. The sub-grade may be on bank, cutting or on the natural ground. After clearing the ground of shrubs, grass or rubbish and other organic matter,

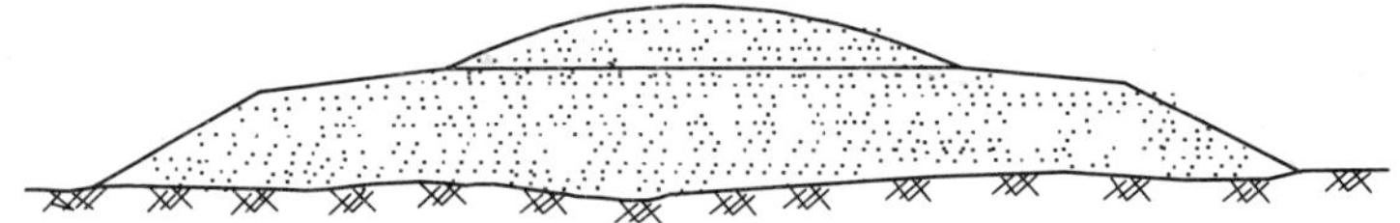

Figure 5.3 *Feather edge*

the top of the ground is brought to the desired profile either by cutting or embankment with the desired camber and gradient. Bulldozers, scrapers, graders etc. are used for the purpose. The top of the sub-grade should be adequately compacted. The camber and gradient of the sub-grade will depend on the intensity of rainfall, nature of and the general profile of the ground including drainage facilities. In case the sub-grade is to be constructed on a slopy ground stepped cutting should be provided so as to prevent slippage of the sub-grade.

For making the desired sub-grade either in cutting or a bank care should be taken to provide proper drainage as the life, stability and strength of the road mainly depends on proper and efficient drainage system of the area.

Soils of the following characteristics are considered satisfactory for the construction of earth roads:

Contents percentage	*for base course percentage*	*for wearing*
Clay	< 5	10 – 18
Silt	9–32	5 – 15
Sand	60–80	65 – 80

The earth should have less than 35% liquid limit and 4–10% plastic limit.

5.6 SOIL STABILIZATION

Soil stabilization is a technique aimed at increasing or maintaining the stability of a soil mass or otherwise improving its engineering properties. Soil stabilization is used in the varieties of engineering works e.g., for the construction of earth road or other cheap roads, for making an area easily trafficable within a short period of time for military use, for reducing the permeability and compressibility of soil and so on.

Earth roads, constructed as described above, will wear out very soon and will become unserviceable within no time. Constant repair of earth roads in villages is also not practicable. Hence to reduce the perpetual headache of maintenance and to keep the road in serviceable condition for the major part of the year, *stabilized* earth roads should be constructed. Soil stabilization is defined as the process of treating a soil in such a manner so to improve or alter its physical conditions or properties so that it may become more stable and durable. The soil so stabilized may form the sub-grade or wearing layer of a road.

5.6.1 Objects of Stabilization

1. To increase compressive strength irrespective of moisture content.

2. To increase resistance to softening action of water.

3. To reduce shrinkage due to withdrawal of moisture, and swelling due to wetting.

(4) To increase flexibility to take the wheel load without deformation and cracking.

(5) To increase shear-strength and resistance to punching.

5.7 METHODS OF SOIL STABILIZATION

Following are the different methods of soil stabilization:

1. Mechanical stabilization
2. Cement stabilization
3. Lime stabilization
4. Bituminous stabilization
5. Chemical stabilization
6. Grouting
7. Electrical stabilization
8. Complex stabilization

1. Mechanical Stabilization. This is the simplest method of stabilization. It has been observed that if a soil consists of coarse grains and fine grains in the ratio of two and one or if a soil consists of two parts of sand and one part of clay, and is compacted at optimum moisture content (O.M.C.), it will have better wearing qualities and will not deform easily under normal wheel load. So in this method the soil is tested. If the soil is coarse-grained, fine-grained soil is so added that the proportion if coarse and fine grains is 2 and 1. Similarly, if the soil is sandy, requisite quantity of clay is added to adjust the proportion.

The soil is ploughed to a depth of nearly 15 cm, and pulverized ; and then the required quantity of fine or coarse grains is added. After sprinkling the surface is compacted by light rollers and then left to be cured for about 4 to 5 days.

2. Cement Stabilization. Cement is a binding material. When mixed with soil, it forms a sort of low strength concrete in which the soil acts as aggregate and cement as matrix. So the soil is excavated to a depth of nearly 15 cm and 8 to 12% of cement is mixed. Sufficient quantity of water is then added and the soil-cement mixture is compacted properly by road stabilizers. After it has been compacted it is then cured for about 7 to 8 days by simply sprinkling water over it.

The primary object of soil stabilization is to improve the wearing conditions of the soil and reduce the deformation either in the road surfacing or its sub-grade due to wheel load irrespective of moisture content. Because, if the soil deforms, it will not distribute the load on sub-soil and due to repeated deformations, the road surface is likely to crack or break. However the objects of stabilization may be enumerated as follows.

The cement binds the particles of clay and forms a flexible dustless and durable road surface. The cement stabilized roads are water-tight, require less maintenance and are very suitable for light traffic.

3. Lime Stabilization. In this case the process of stabilization is similar to that of cement stabilization. The soil is loosened, pulverised, sieved and mixed with 5 to 10% by weight of hydrated lime. The two are thoroughly mixed. Sufficient quantity of water is added and the surface is compacted. The lime helps in reducing the shrinkage and swelling of soil.

4. Bitumen Stabilization. In this method the soil is treated with about 8 to 10% of road oils, cut backs or emulsions, according to the nature of the soil. Their object is to glue together the soil particles and fill up the voids.

The soil is graded and then compacted properly at O.M.C. About 5.5 litres of oil is applied on 1 sq. metre of the surface so that the oil may penetrate nearly 1.25 to 2.5 cm in the soil.

This method of soil stabilization is unsuitable for soils containing more than 50% of fine grained particles. Before applying this method it is essential that the soil should be well graded and compacted properly, otherwise more quantity of bitumen will be required.

5. Chemical Stabilization. Hygroscopic materials such as calcium chloride, sodium etc. are mixed with the soil at the rate of 1 kg per 5 sq. metres of the surface and the soil is thoroughly compacted. These hygroscopic materials help in retaining proper amount of moisture in the soil and add to its stability. However, dampness in the surface reduces shrinkage and cracking.

6. Grouting. Grouting or injecting is a process of introducing a stabilizer of fluid consistency into soil and rock formations. The stabilizer used is known as *a grout*. Grouting of soil formations is usually done (i) to reduce the permeability, (ii) to increase me strength for the purpose of increasing the bearing capacity, and (iii) to reduce the compressibility.

The usual grouting materials are cement, soil, bitumen and chemicals. Holes are driven at regular intervals and of desired depth and the grouting material of fluid consistency is injected under heavy pressure with the help of a grouting pump. The grout having the cementing properties will bind the soil particles.

7. Electrical Stabilization. Electrical stabilization is a method of drawing out the fine-grained soil by passing *direct current* through them. It is also sometimes called *Electro-osmosis*. With the damage of the fine particles the volume of the soil decreases i.e., the soil is consolidated and the shear strength is increased. During the *electro-osmosis,* electro-chemical decomposition of the electrodes takes place and the metal salts are deposited in the soil pores. There may also be an *ion-exchange,* and an alteration of soil particle arrangement This will ultimately lead to hardening of soil and process is sometimes known as *electro-chemical hardening.*

8. Complex Stabilisation. Complex stabilisation is defined as the method of stabilization with more than one stabilizer. Difficult soils such as organic soils, highly plastic for clays and soils with *easy-soluble* salts require more than one stabilizer for their effective treatment. Complex stabilization involves the use of binding material and surface acting additives or electrolytes. At present the following combinations are considered to be the best:

(i) Cement + calcium chloride + lime

(ii) Cement + bituminous emulsions

(iii) Cut-back + lime

(iv) Cement + naphtha soap

5.8 PRINCIPLES OF SOIL STABILIZATION

As pointed out earlier, soil stabilization means improving the bearing capacity and stability of soil by the use of controlled compaction, proportioning or addition of suitable stabilizers. The soil stabilization should precede the following:

(i) Evaluating the properties of the soil.

(ii) Deciding the method of supplementing the desired characteristics.

(iii) Method of designing the stabilized soil mix to achieve the desired characteristics or qualities of the soil.

(iv) Deciding upon the adaptation of the appropriate method of construction.

The following changes are expected after the stabilization :

(i) Stability will increase and change in the properties like density, shrinkage and other physical characteristics.

(ii) Changes will occur in the chemical properties.

(iii Bearing capacity will increase and the soil will become more resistant to water.

5.9 INVESTIGATIONS OF SOIL STABILIZATION

Undisturbed and disturbed soil samples are collected from shallow pits after removing the top layer and organic matter along the alignment and from the sides of the alignment at regular intervals. These samples are then tested in the laboratory for physical chemical and engineering properties. Identification and classification tests such as plastic limit, liquid limit etc. are carried out. CBR tests are also conducted, the investigations are carried out in the following manner:

(i) Investigation of route and alignment.

(ii) Survey for the availability of materials to be used for stabilization.

(iii) Soil surveys and field investigation of soil,

(iv) Soil classification.

5.10 MECHANICAL STABILIZATION

When a granular structure such as a road base or surfacing has the property of resistance to lateral displacement under load, it is said to be mechanically stabilized or stable. In mechanically stabilized soil materials, the resistance is provided by natural forces of cohesion and internal friction that exist in the soil. Cohesion is mainly associated with the silt and clay content of the material, while internal friction is a characteristic of the coarser particles. For a soil to meet the criterion of being stable, it must fulfill the requirements with respect to shear strength, resistance to abrasion, rigidity, incompressibility, freedom from swelling shrinkage and frost action. Each one of these requirements must be complied with in a mechanically stabilized road. However the requirements will vary depending upon the functions of the soil material within the permanent and the loads to which the pavement will be subjected.

Mechanical stabilization is still the method that is most widely used in road construction throughout the world. Its popularity for this purpose is based on the fact that it makes possible the maximum use of locally available materials in highway construction of embankments, sub-bases and surface course. Mechanically stabilized roads may range from simple earth roads which have been cleared of vegetation and compacted by traffic but which still meet the stability criteria to the highways which utilize highly sophisticated blends of different materials.

In developing countries, roadways in which mechanically stabilized locally available materials are used for the entire pavement are primarily used for rural areas where the critical need is to provide access to remote areas but the traffic demands are not sufficient to necessitate construction of more expensive roads. The mechanically stabilized road provide a stage construction for other roads. Stage construction means step by step improvement of the road.

5.11 TYPES OF MECHANICAL STABILIZATION

Mechanical stabilization can be divided into two categories viz. Mechanical stabilization by treatment and Mechanical stabilization with the help of additives. The first category is of three types:

(a) Compaction
(b) Consolidation and
(c) Electrical and thermal methods

The second category is of four types:

(a) Soil and aggregate
(b) Chlorides
(c) Lignin
(d) Molasses

5.12 STABILIZATION BY COMPACTION

The stabilization by compaction is effected by the following factors:

1. Moisture content
2. Soil type
3. Method of compaction

The selection of the proper type of compaction equipment and method is vital to the production of long term economy expected and desired from a road. The soil can be compacted by rollers kneading, vibrating and pounding.

Rolling. Smooth steel wheel rollers have long played an important role in road construction and are acknowledged to be the oldest form of mechanical compactors. The rollers generally have three wheels. The tender rollers have two or three rolls behind each other so that each roll follows the track of the front roller.

Sheep foot rollers consists of steel drums with steel projections or feet extended in a radial direction outward from the surface of the cylinder. The drums are manufactured from 100 to 180 cm in diameter, weighing 2500 kg to 25000 kg when loaded. Sheep foot rollers can be self propelled or towed by a tractor. The sheep foot roller compacts from the bottom. During the first pass over a lift of loose soil, the feet penetrate to near the base of the layer so that the bottom material is compacted first. As additional passes are made, the feet penetrate to lesser depths. The sheep foot roller is designed to break up the surface crust on a soil by compacting small areas at high load concentration. Since compaction is obtained by the feet penetration and applying a high vertical pressure while at the same time prounding lateral pressure the sheep foot roller is most efficient in compacting fine grained soils. Sheep foot rollers produce larger amount of air voids which in turn will absorb larger amount of moisture.

For many years, it was a common practice for earth works to be compacted by directing the regular construction units to travel over freshly placed materials until it was quite dense. This led to the development of *pneumatic rollers*. The pneumatic rollers are arranged so as to distribute equally and uniformally the compaction load, regardless of die ground profile. Full ground coverage is obtained by having the wide face tyres arranged so that there is always an overlap between the front wheel and the rear wheel. In addition many pneumatic tyred roller have a *wobble wheel* motion which provides a type of kneading action to the soil.

When the compacting soil contains silt and clay rollers of any kind are used but when the soil is of a coarse grain type purely, vibratory compactors are used. Vibrators consists of a vibratory unit of either the out of balance weight type or a pulsating type mounted on a screed plate or roller in such a way that not effect is an up and down vibratory movement of the compactor at a frequency of about 20 to 30 cycles per second. The frequency of the natural frequency of soils particularly coarse grained soil and so the particles are literally shaken about when the compactors move over the ground. When the coarse grain soils are being compacted, the result is that the particles are continually rotated until they slip over each other and fall into whatever gap is available between the other two particles so a compact mass is obtained. With pure sand and gravel, the particles are usually most easily vibrated into position when the soils are saturated. Not only the excess water act in providing the lubrication necessary to allow particles to slide easily but it breaks the apparent cohesive bonds between the particles when the materials are damp.

Even with non-cumec grained soils, vibratory rollers are very effective.

5.13 IMPACT COMPACTOR

The compacting equipments so far described involve fairly substantial amount of maneuvering space for satisfactory results when the trenches are to be back filled such as foundation trenches, it will not be possible to use compactors and vibrators. In such situations hand operated compactors are generally used. Nowadays mechanically operated compactors are also being used in our country. These rammers are called *impact compactors* because they compact the soil by impact.

5.14 SPECIFICATION AND CONTROL OF COMPACTION

Stabilization by means of compaction process constitute one of the major items of cost in the construction of highways and it is important therefore, if initial costs are to be kept low the state of compaction specified and obtained should not be greater than, is really necessary. From a long term cost project it is equally important for a desirable state of compaction to be obtained, as failure to do so can lead to settlement and consequent damage to the highway structures.

It is a common practice to specify the state of compaction required as a percentage of maximum dry density obtained with one of the standard laboratory tests. This method of compaction specification has been criticized by many agencies including I.R.C. on the following grounds.

1. If the moisture content selected for compaction in the field is much lower than the optimum moisture content obtained with the laboratory compaction test, a poor state of compaction measured in terms of air voids contents of the soil, will result even if a high relative compaction is actually obtained.

2. If the natural moisture content of the soil in the field is much higher than the laboratory optimum moisture content, then the specified relative compaction may not be attainable.

To overcome these disadvantages the I.R.C. suggests that the moisture content of the soil for compaction purpose should be that which is likely to be naturally attained beneath the road pavement and that the state of compaction be controlled by specifying a maximum allowable percentage of air voids at that moisture content. These recommendations have been adopted as they have the particular advantage that they virtually eliminate the difficulty which often arises in distinguishing between variations in the dry unit weight resulting from changes in the state of compaction and variations resulting from changes in the state of compaction and variations resulting from changes in the soil type and moisture content.

5.15 STABILIZATION BY CONSOLIDATION

Stabilization by consolidation means preloading the soil before the construction of embankments with time allowed to the soil for partially or completely consolidated under the increased load. It helps in increasing the beaming capacity of soil, decreases the amount of subsequent settlements. Methods of preloading are by means of direct construction of embankments with or without sand piles, lowering of the ground water level, stage construction and consolidation by atmospheric pressure.

Preloading by direct construction of an embankment is by far the most common way of *prestressing* the soil. The soil is being compacted layer by layer and then left to the atmosphere. Lowering of the water table is also a method of consolidation whereas stage construction is simply another method of preloading of soil by direct construction of embankment in stages. It is a long term process which results in the consolidation of a weak underlying soil without displaying it. The embankment is first built to a height which is limited by the shearing strength and further construction is delayed until the excess pore water stress induced in the weak layers by the weight of the embankment has dissipated over the course of consolidation proces. When this pressure has been reduced to a safe value, its additional height of fill is placed and the process is repeated until the embankment is finally completed.

A most interesting low cost method of soil stabilization by consolidation has been developed by Royal Swedish Geotechnical Institution. It used the atmospheric pressure as a temporary surcharge and vertical sand drains to accelerate consolidation. In the field the soil area which is being stabilized is covered with a sand filter 45 to 60 cm thick and on top of this a light impermeable membrane of plastic sheet or a layer of clay or a bituminous covering is placed. Outside the width of the filter the covering membrane is connected tightly to the underlying soft clay which is free from cracks and road channels and is beneath the normally dry soil at the very surface of the ground. A suction pipe is then carried through the membrane and the pump starts sucking air and of the filter and gradually out of the clay. In any horizontal plane the total normal pressure i.e., the atmospheric pressure and the weight of the overlying pressure and the weight of the overlying soil, remains constant and therefore the grain pressure increases on the pore pressure decreases. Thus the pump sucks the water out of the pores which forced to consolidate accordingly.

5.16 STABILIZATION BY ELECTRICAL AND THERMAL METHODS

This method of stabilization has a very limited scope and is not prevalent in our country. This method has the practability in areas where there is acute shortage of natural aggregates. The method relies on the fact that when a clay soil is heated it looses weight. This reduction in weight is due to the evaporation of moisture absorbed by the clay. This method is being used for producing natural aggregates in Great Britain, U.S.S.R., Sweden, Australia and some other countries.

The stabilization process depends primarily on the clay minerals and the extent to which they are crystallized.

5.17 SOIL AGGREGATE STABILIZATION

Stabilization by altering the gradation of a soil constitutes soil aggregate stabilization. The purpose of changing the gradation is to obtain the proper proportions of particle size fractions that will give a dense homogenous mass. The main use of this type of stabilization is for the sub-base construction and for road surfaces of earth roads to attain the following:

1. Stability to support the weight of the traffic.

2. Resistance to abrasive action of traffic.

3. Ability to shed a large portion of the rain which falls on the surface since a large amount of water penetrating the surface might cause loss of the stability in the surface course or softening of the sub-grade.

4. Capillary properties to replace the moisture lost by surface evaporation and thus maintain the desirable damp conditions in the soil aggregate particles are firmly bound together.

Considering the above facts of stability and abrasion the material must be hard tough and durable pebbles of stone or gravel and a filler of sand or other finally divided material together with sufficient binding material to permit the consolidation and formation of a tightly bonded layer. A certain amount of clay binder must be present to help prevent smaller particles from being whipped off under the action of traffic. The clay content should be just sufficient to perform this function since any additional clay will smell beyond this limit and may cause dislocation of the granular material.

5.18 CHLORIDE STABILIZATION

It has been observed that dust can be prevented more effectively by using sea water than by fresh water and for a much longer period of time. This resulted to the investigation of using chlorides as additives for stabilization. Calcium and sodium chlorides are generally used as they are hygroscopic in nature and are widely available in oceans as rocksalt, in lakes etc. Due to their hygroscopic nature, they absorb moisture from the air and being a deliquiscent matter get dissolve in water. These two properties of chlorides are their most important ones as they retain the desired amount of moisture during dry climate and hence prevent dust and instability of the soil mass.

Chlorides present in the soil-water mixture cause an increase in surface tension of water. This has two important aspects viz. by increasing the water surface tension the rate of evaporation is reduced and by evaporating some more pore water this in turn will increase more surface tension and the water film so produced will tighten the soil particles thereby increasing the unit weight and stability of the soil. On account of their moisture retention property, compaction in dry weather is also possible. The presence of chlorides in the soil may change its plasticity characteristics. However chlorides are good cementing materials and therefore the material being stabilized must already have high stability. They are therefore most effective as additives to soil aggregate surfaces. Best results are obtained when well graded materials are compacted to dense impervious mass as these reduce considerably the ease with which chlorides may be leached downwards from the upper portion of the pavement.

5.19 LIGNIN STABILIZATION

Lignin is a spent sulphite liquor obtained from paper industry and plywood industry as a waste product. In the sulphite paper making industry, the paper mill retains the fibres and cellulose while the cementing materials are wasted from the process in a water solution called *spent sulphite liquor* which is used for stabilizing the mechanically stabilized roads and surfaces. Though it is widely used as a stabilizing agent in some western countries very little is known about the mechanism except that it has binding properties.

5.20 MOLASSES STABILIZATION

Molasses used for soil stabilization purpose is a waste residue known as blackstrap which is obtained as a by-product of the sugar industry. It is very thick syrupy liquid which contains resinous and some inorganic constituents which render it unfit for human consumption. Black strap molasses is a hygroscopic materials and helps the soil to absorb moisture from the atmosphere and to control the evaporation of water from the soil-aggregate pavement at the time of compaction. Molasses is also a cementing material but the cement so formed is dissolved in water itself, but if water can be kept away its binding action is very strong.

5.21 CEMENT STABILIZATION

Of all the methods of stabilization now in use, cement stabilization is only second to mechanical stabilization in importance and usage. Dr. J.H. Amies of U.S.A in 1917 was the first to come out with soil cement products and have since been popular throughout the world. Some of the factors which have helped to make the use of Portland cement as a soil stabilizer are

1. Cement is readily available everywhere.
2. Use of cement generally involves less care and control as compared to other methods.
3. Almost any soil can be stabilized with cement.

Cement treated soil is a simple intimate mixture of pulverized soil, cement and water. There are three type of cement treated soil mixtures viz. soil-cement, cement modified soil and Plastic soil-cement.

Soil Cement. This is a hardened material joined by curing a mechanically compacted intimate mixture of pulverized soil, Portland cement and water. Sufficient cement is added to the soil to harden it and sufficient water is added to hydrate the cement. Soil cement has a number of uses of which for the most important is as road surfacing for rural areas. Another important use is foundations for heavy structures, earthen dams, levees, high embankments and bedding for pipes in culverts.

Cement Modified Soil. Some soils cannot be stabilized economically with cement. For instance the amount of cement that may be required to stabilize clay soil so that it can be used for road surfacing may be sometimes 20–30%, which is comparatively very high. In such cases some other material is used in which only 2–3% cement will be required to modify the physical properties of the soil. Alternatively, a normally sub-standard granular material having relatively high plasticity and low bearing capacity may be made acceptable as a road material by adding sufficient quantity of cement to lower its plasticity and improve its bearing capacity. In both the cases the amount of cement added is not sufficient to harden or even semi-harden the soil. The cement added is just sufficient to modify the physical properties of the soil. Usually the soil chosen for modifying have high water retention capacity and volume change characteristics which may cause excessive distortion of pavements hence these soils have low supporting values as their moisture contents will vary with weather conditions.

Plastic Soil Cement. This is a hardened natural form by curing an intimate mixture of pulverized soil, cement and water to produce a mortar like consistency at the time of mixing and placing. By comparison soil-cement is mixed and placed with only enough moisture to permit adequate compaction and cement hydration. Soil cement and plastic soil cement are similar in that they are both capable of meeting specified criteria of durability and stability.

Plastic soil-cement is used primarily as an erosion control material by paving steep, irregular and confined areas and for lining of canals. It is also used for the face of dam. In general plastic soil-cement mixture require about 4 percentage points more cement than soil-cement mixture. It is extensively used as a low cost lining material for canals.

5.22 MECHANISM OF CEMENT STABILIZATION

When water is added to neat cement, the major hydration products are basic calcium silicate hydrates, calcium aluminate hydrates and hydrated lime. The first of two these products constitute the major cementatious components while the lime is deposited as a separate crystalline solid phase. When the cement is present in granular soil, the cementation which takes place is probably very similar to that which occurs in concrete except that the cement does not fill the voids between the soil particles. In other words the process of cementation is purely by mechanical bonding of the calcium silicate and aluminate hydrates to the rough mineral surfaces. When cement is used to stabilize fine gradient soil, the cementation becomes a combination of mechanical bonding and chemical bonding which involves a reaction between the surfaces of soil particles and cement.

5.23 MIXING AND COMPACTING SOIL-CEMENT

In general, the more efficiently cement, water and soil are mixed, better the results and more stability and durability of soil-cement product. Efficiency of mixing cement water and soil will reduce the amount of cement for attaining the desired results. For a given soil water cement mixture it can be said that the greater the unit weight, the more stable and durable is the hardened structural material. Studies indicate that a change in dry unit weight of only 1% is liable to produce a change in unconfined compressive strength of between 8 to 12%. It is useful to define the minimum state of compaction required in terms of a maximum state of compaction required in terms of a maximum permissible air contents of the compacted material. The maximum air voids content which is usually specified are 5% i.e., the mixture must be compacted to a dry soil unit weight of 95%. Regular laboratory tests should be conducted to arrive at a final dry unit weight of the compacted soil-cement mixture.

5.24 CURING

The environmental conditions under which curing is done has a considerable offect on the extent to which a soil may be stabilized with cement. Like concrete, the soil cement must be moist cured during the initial stage so that sufficient moisture is available to meet the demand of hydration of cement. The temperature of curing has also a considerable effect on the strength. Higher strength results can be obtained at higher curing temperature. This factor is of considerable economic importance as lower cement content will be required when the curing temperature is between 35 40%.

5.25 CONSTRUCTION

The objectives in soil-cement construction are simple, they are to obtain an intimate mixture of the specified amounts of soil, cement and water and to compact this mixture so as to achieve a high dry unit weight. Best results are obtained by the following construction procedure.

Before construction the sub-grade use properly compacted and brought to the desired camber and gradient. The soil, if it is not loose, is pulverized and spread over the sub-grade. Now spread the designed quantity of cement over the pulverized soil surface. If during the process of spreading cement, rain falls, the process should be stopped. The dry cement and soil are mixed intimately sandy soils can be easily dry mixed. The mixing is taken to a depth of 8 to 15 cm. Add water to just wet the mix. If the soil already contains the desired amount of moisture as per design addition of water is not required. After adding water and mixing the soil-cement is compacted and cured.

5.26 BITUMEN STABILIZATION

This type of the stabilization is believed to be added one used for the first time for the dust *palliatime* way back in the year 1898 but because popular only during 1920–30 but still its use is restricted to places where coarse mineral aggregates are scarce due to high cost of bituminous materials. The incorporation of bitumen in a cohesive soil results in a water resistant layer of soil. Such treated soils will have good bearing capacity at low moisture contents. The following soil gradation of soils are generally prescribed for bitumen stabilization.

4.75 mm	50%
425 micron	35 –100%
75 micron	10 – 50%

Emulsions, cut-backs and coaltar is generally used for bitumen-soil stabilization. Cut-backs are generally preferred. Slow curing cut-backs are used for highly plastic soils whereas rapid curing cut backs are used for sandy soils. Road tars at different grades are also used. For relatively less plastic soils, emulsion can be used. The amount of binder depends upon the moisture content type of soil and the type of binder. The combined volume of binder and moisture should not exceed the pore space at the desired density. Normally 4–8% binder is used (Percentage by dry weight). For higher clay mineral content higher percentage of binder is required. The binder in the form of liquid is used for spraying. Soils with liquid less than 30 and plastic limit less than 18 can be satisfactorily stabilized. Soils to be stabilized should contain no acid or very little acids. Organic matters are also detrimental to the stability of soil. It is generally considered that the best results are obtained with more fluid medium cured cutbacks.

It is appropriate here to discuss the role which the moisture plays as an ingredient. Water plays an important role as it helps in mixing and dispensing bitumen and compaction of soil-bitumen mixtures. During mixing the moisture facilitates distribution of bitumen uniformally over the surface. The amount of bitumen required to stabilize the soil varies between 4–8%.

The clay minerals present in the soil have quite an effect on the stability of soil-bituminous mixtures. The more the silica content of the soil more the amount of stabilizer required. The rapid curing cut backs number RC-1, RC-2 and RC-3 or Road tar RT-4, RT-5 and RT-6 are generally used. For earth grand roads slow curing cutback MC-0, MC-l, SC-0 or SC-1 are generally suitable. Bitumen stabilization has also been preferred for black cotton soils. The addition of bitumenous binder to a soil improves the basic properties of soil and hence makes it more dependable material for highway construction either as a base course or surface course, but the process is costly.

5.27 STABILIZATION OF BLACK COTTON SOIL

Black cotton soil is a trelchorous soil and is available in nearly one-third of our country. They are highly clayey in nature, greyish to blackish in colour. In places where is black cotton soil, road aggregates are generally not available and are to be transported from distant places which considerably increases cost of conventional type of roads. The black cotton soil has a typical behaviour, it swells enormously when come in contact with water and shrinks when dries up resulting in wide cracks so the construction of roads in particular and other structures in general pore difficult problems. The colloidal content of black cotton soil varies upto 50% and the fraction passes 0.075 mm sieve varies from 70% to 100%. The liquid limit is between 45 to 95 and plastic limit 20–60. The optimum moisture content is nearly 30 percent Following are some of the typical problems of Black Cotton soils :

1. Its volume changes tremendously with the change in moisture contents.
2. It is difficult to pulverize the soil due to its high dry strength.
3. The wet soil is very sticky.
4. The black cotton soil also shrinks even when compacted at O.M.C. after it is dried up.
5. Due to excessive swelling it exerts high pressure from below the sub-grade.
6. Inspite of its normal bearing capacity it is highly un-dependable.

Stabilization of black cotton soil is a difficult task. Several methods have been tried but did not prove very satisfactory of the various methods tried so far lime stabilization seems to be better. By the addition of suitable quantity of lime and allowing it to hydrate properly for a day. On account of treatment with lime, plastic index decreases to almost zero, affinity with water decreases, shrinkage limit increases volume changes decreases and the soil behaves as a non-plastic soil. The complete process of stabilization of black cotton soil is to excavate the soil to a designed depth, pulverize the soil and mix lime and sufficient quantity of water required for complete hydration of lime, leave for a day or two and then compaction is done to the desired degree. Sometimes 3–5% by dry weight of cement is also added for better results. Generally the stabilization of black cotton soil is done in two stages. In the just stage only 2–3% lime is added and allowed to hydrate. This is the pre-conditioning of soil. This preconditioned soil is now again pulverized and then the remaining quantity of lime is added and allowed to dry and then compacted.

5.28 DESERT SAND STABILIZATION

Large deposits of sand in the form of deserts spread over large stretches in many parts of Rajasthan in our country. Desert sand particles are round and finely graded due to shortage of water in deserts. Hence lime and cement stabilization are not possible because in both the methods more water is required for hydration and curing. Bitumen stabilization has given satisfactory and encouraging results. Use of hot sand bitumen would result in satisfactory mix provided some material to be used as a filler is used. The hot sand bitumen stabilization is not very economical.

The most promising method of stabilization of desert sand is by treating the sand with emulsion as the additional quantity of water for curing required is quite low. The emulsion contains nearly 50% water and hence very little water is required for mixing also.

5.29 STABILIZATION BY GROUTING

Grouting is the process of injecting a stabilizer of fluid consistency into soil. The stabilizer so injected is called grout. The object of soil stabilization by grouting is to improve permeability, increase bearing capacity, stability, strength and reduce compressibility. The usual grouting materials are cement, bitumen, lime, soil or some other chemicals or combination of two or more of the above materials. The type of grout to be used will depend on the formation, purpose of grouting and the cost. The ability of grout to penetrate a formation is called *groutability*.

Clay-cement grouts are the cheapest type of grout. Clay cement grouts consists of mixture of clay and cement in varying proportions. Clay-chemical grouts are also used for arresting water movement in medium and fine sands. Chemical grouts are very expensive as compared to other grouts. The chemical grouts generally used are sodium silicate, calcium chlorides etc. There are two methods of chemical grouting viz. Jooster method and the K.L.M. method. In the Jooster method two fluid consist of injecting grouts separately by inserting injection pipe into the ground. The first solution consists of sodium silicate and the second solution consists of sodium chloride.

The second solution is injected after the pipe is withdrawn. The K.L.M. method consists of injecting single mixture of two chemicals through a pipe which is gradually withdrawn. Single fluid injection or grouting is comparatively cheaper but it does not increase the shear strength of the soil.

Use of simple tools such as disc harrows, disc ploughs, grade blades etc. should be encouraged to be used for soil stabilization. The Central Road Research Institute has developed a versatile simple equipment known as Rotiroller which can be conveniently towed by an agricultural impliment. This is a multi-purpose machine suitable for agriculture purposes as well as for soil stabilization. For stabilization the rotiroller dig the top soil upto the required depth, pulverizes the soil and mixes it with the stabilizer. The use of simple impliments for road making and soil stabilization is very much benefited in many ways as the impliments are labour intensive methods and provide reasonable control over the quality of work.

5.30 MOORUM ROADS

Moorum is a disintegrated rock, gritty and silicious in nature, having grains not exceeding 18 mm in size. It is generally mixed with natural clay of calcarious origin.

Moorum roads are also cheap type of roads and are mostly used in villages. In this type of roads, the traffic-way has the surface of moorum which is consolidated in layers to get the required finished surface. The surfacing material is laid directly on the sub-grade. No foundation of soling is required. The surface is properly compacted by rolling. Proper camber and gradient is maintained during the process of rolling. Generally 1 in 24 camber is provided.

5.31 STABILIZED MOORUM

Stabilized moorum may be used for sub-base of a road. This will fulfill the requirement of a material to be usefully used as a sub-base material. Moorum is abundantly available in many parts of our country. Various tests are conducted on different samples of moorum and were classified as per I.S. 1948–1970 classification and was found that most samples belonged to 'SC' group of I.S. classification. Procter's compaction test and soaked CBR tests are conducted and was found that the test results are quite encouraging, Moorum samples passing through 20 mm sieve and retained on 0.425 mm sieve, it is found there is appreciable increase in the CBR value. The CBR values so obtained varies from 16.5% to 40.4%. Although sufficient improvement in the CBR values had resulted by excluding fractions passing

0.425 mm sieve, but is not a practical solution of improving CBR value. The next alternative of improving CBR value is treating the moorum with lime because it is found that the fines of moorum are reasonably plastic. It is found through extensive, experimentation that moorum treated with 3–5% lime and cured for seven days give 27.6% – 75% CBR values which are really encouraging and appreciable. So lime stabilized moorum can be used as a sub-base material or as a surfacing for moorum roads. The studies are conducted by U.P. P.W.D. Research Institute.

PROBLEMS

5.1. Outline the importance of earth and other cheap roads in India. Where are these roads used in India ?

5.2. What is meant by soil stabilization ? What are the different methods of soil stabilization ? Describe in brief.

5.3. Describe in detail the construction of an earth road. Describe the procedure of construction step by step.

5.4. What is the difference between stabilized and unstabilized road surface ? What is the object of stabilization ?

5.5. On what factors the stabilization of soil depends. What is the role of moisture in stabilization ?

5.6. Discuss the method of bitumen stabilization of soils and state also the limitation of the method.

5.7. Elucidate in brief the method of Moorum Stabilization for Road construction.

5.8. Elucidate the principles of soil stabilization.

5.9. State the procedure and compaction of earthen embankments.

5.10. Differentiate between compaction and consolidation. How soil can be stabilized by compaction.

5.11. Slate the principles and procedure of soil-cement stabilization.

5.12. Differentiate between soil-cement, cement modified cement and soil plastic cement.

5.13. Compare the results of compaction with sheep foot rollers, pneumatic rollers and flatfoot rollers.

5.14. State the mechanism of soil bitumen stabilization. Name the various bituminous materials used for soil bitumen stabilization.

5.15. State the procedure of stabilizing BC soils.

5.16. State and discuss how desert sand is stabilised ?

5.17. State the process of grouting for soil stabilization. Name the various methods generally used.

5.18. State the advantages of soil stabilization and discuss bitumen stabilization in brief.

5.19. Describe in brief the construction of stabilized moorum roads.

BIBLIOGRAPHY

1. Mehra S.R. 'A Method of Soil Stabilization for Road Construction, Indian National Society on Soil Mechanics and Foundation Engg., 1960.
2. Report of the Seminar on Low Cost Roads and Soil Stabilization. Economic Commission for Asia and the Far East (Bangkok) Inland Transport Committee, Highway Sub-committee, New Delhi, 1958.
3. Lambe T.W., 'Soil Stabilization—Foundation Engg., McGraw Hill 1962.
4. O' Flaherty—Soil Stabilization—Highway Engineering – Edward Arnold, London.
5. Alam Singh, Soil Engineering in Theory and Practice – Asia Publishers, 1962.
6. Justo and Khana, Highway Engg., 1984.
7. Kahyali L.R., Highway Engg., 1984.
8. Sharp B.R., 'Permanent Design', Transport Communication No. 144, 145 – 1939.
9. Cooper G.S. 'The influence of multiple wheel under carriages' Technical Paper No. 16, London, 1952.
10. Alam Singh and G. Ranjan 'Some Investigation in California Bearing Ratio and North Dakata Cone Test' Road and Road Construction, London, 1960.
11. Mehra S.R., 'A Method of Stabilized Soil Construction for Roads', I.R.C., 1968.
12. Soil Mechanics for Road Engineers—Road Research Laboratory, London, 1952.
13. Grainger G.D., 'A study of burnt clay as road making material', Technical Paper, 1962.
14. A.A.S.H.O., Manual of Highway Construction, Practices and Methods 1958.
15. Robinson P.J.M. 'British Studies on the Incorporation of admixtures with Soil' a technical Paper of Seminar on 'Soil Stabilization', 1982.
16. Glossary of Highway Engineering Terms Journal of Indian Road Congress Volume XVIII–3, New Delhi.

17. Kadiyati L.R., 'A critical study of the Theory and Practice of Lime—Soil Stabilization' a seminar paper.
18. Tentative specifications for the construction of stabilized soil roads with soft aggregates in areas of moderate and heavy rainfall I.R.C. : 28–1957.
19. Recommended Design Criteria for the use of soil lime mixes in Road Construction I.R.C. : 51–1953.

6

Gravel and Waterbound Macadam Roads

In this Chapter you will study,

• Gravel Roads • Waterbound Macadam Road • Sizes of Aggregate for W.B.M. Road • Granding of Coarse Aggregate • Thickness of Layers or Courses • Construction of W.B.M. Road • Causes of Disintegration of Waterbound Macadam Roadc • Corrugationc on W.B.M. Roadc • Maintenance • Improvements on W.B.M. Road Servicing • I.S.I Tests for Aggregates • Code of Practice for W.B.M. Base Course and Wearing Courses

GENERAL

Gravel and waterbound macadam (W.B.M.) roads are cheap quality roads and are mostly used in Indian villages. The construction of gravel roads is very much similar to that of moorum roads. The waterbound macadam road is a superior road which is used in many cases as a foundation layer for other superior roads. The W.B.M. road, as it is known, is the mother road for all types of road construction.

6.1 GRAVEL ROADS

It is a cheap type of road and is mostly used in Indian villages. Gravel can be obtained from the river beds or by crushing the stone. The surface layer of gravel roads consists of 6 mm to 36 mm size gravel mixed with sand and clay. It works as a binder. The function of binder is to fill up the voids and bind the particles of gravel.

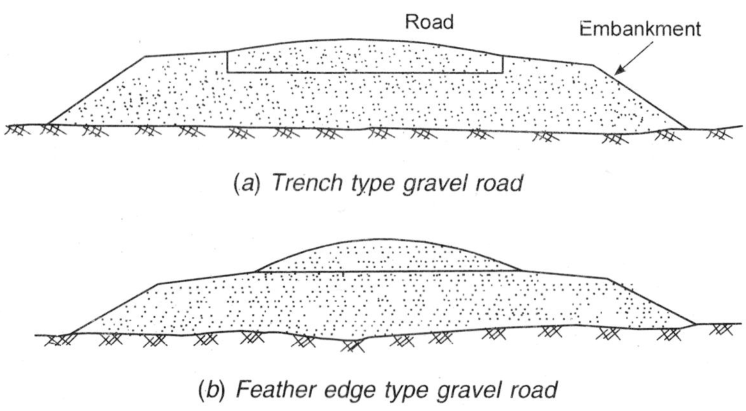

(*a*) *Trench type gravel road*

(*b*) *Feather edge type gravel road*

Figure 6.2

The sub-grade is prepared and properly compacted. Proper camber and gradient is given to the sub-grade surface. The mixture of gravel, earth and sand is spread on the sub-grade, in a thickness which varies from 15 cm to 30 cm according to the requirement of the traffic. The layer of gravel is rolled by light rollers. During the process of rolling proper camber and gradient is maintained. The gravel is generally spread in two layers. During rolling some water is sprinkled on the surface. Gravel roads may be constructed either *Trench* type and *Feather* type as shown in Fig. 6.1 (a) and (b). Trench type gravel roads are more common in this country. In the feather type gravel roads, the gravel is spread on the formation directly without excavating a trench.

The surface drainage in the gravel roads should be very efficient, so that the surface must not be worn out by water. Periodic repairs are necessary to maintain proper camber and gradient.

6.2 WATERBOUND MACADAM ROAD

The waterbound macadam road is one of the oldest methods of road construction known to mankind. The road metals used for the construction of this road are known as *macadam.* The W.B.M. road is used as village road, or as a base for the important modern roads such as surface painted roads, premixed carpet, bituminous macadam and cement concrete roads.

This type of road consists of road metals bound together with water only. The smaller pieces of the metals are used to fill up the voids.

6.3 SIZES OF AGGREGATE FOR W.B.M. ROAD

Crushed or broken stones, crushed slag, overburnt brick aggregate or other types of naturally occurring aggregates such as kankar, laterite etc. are used for the construction of the water bound macadam roads. The size of aggregate mainly depends on the intensity of traffic, weight of the vehicles to pass over the road, bearing capacity of the soil and the hardness of the aggregate to be used. As a general rule the maximum size of the aggregate may not exceed three-quarters of the total consolidated thickness of any one layer of construction. Some engineers recommend that the maximum size of the coarse aggregate may be equal to the consolidated thickness of each layer, the reason being that larger stones engage the base course and the surface has no tendency to 'roll' or wave.

Following nominal sizes of coarse aggregate are recommended for W.B.M. roads:

(a) For limestone, flint, kankar, laterite and vitrified brick ballast aggregates 40–63 mm.

(b) For granite, trap, basalt, diorite etc. hard stone aggregates 20–50 mm.

The size of the aggregate should be reduced with its hardness. Coarse-grained rocks are not suitable for use in roads in small sizes.

6.4 GRADING OF COARSE AGGREGATE

Table 6.1 gives the various recommendations for the grading of the road metal for W.B.M. roads.

TABLE **6.1** *Recommended metal grading*

Nominal size of Aggregate	*% of Aggregates*					
	1st Recommendation	*2nd Recommendation*	*3rd Recommendation*	*4th Recommendation*	*5th Recommendation*	*6th Recommendation*
63 mm	–	–				
50 mm	60%	–		50%	50%	60%
40 mm	30%	60%	50%		25%	
32 mm	–	–			–	
25 mm	–	30%			–	
20 mm	10%	–	30%		15%	30%
12 mm	–	10%		50%		
6 mm	–	–	20%		10%	
Stone Dust	–	–				10%

Table 6.2 gives the size and grading of coarse aggregate for waterbound-macadam roads as per I.R.C. 19–1972.

TABLE 6.2 *Size and grading of coarse aggregate for W.B.M. road (as per I.R.C. 19–1972)*

Grading No.	*Nominal size of Aggregates and material*	*Sieve Designation in mm*	*Percentage weight of Aggregate passing the Sieve*
1.	90 mm to 40 mm	100	100
	Jhama Brick	90	65 – 85
	Aggregate	63	25 – 60
		40	0 – 15
		20	0 – 5
2.	63 mm to 40 mm	80	100
	Soft-Stone:	63	90 – 100
	Lime stone, flint, kankar,	50	35 – 70
	quartzite,laterite and	40	0 – 15
	vitrified brick ballast etc.	20	0 – 5
3.	50 mm to 20 mm	63	100
	Hard Stone:	50	95 – 100
	Gramite, trap,	40	35 – 70
	basalt,diorite etc.	20	0 – 10
		10	0 – 5

Grading No. 1 is more suitable for sub-base course, but it should not be used for courses having compacted thickness less than 90 mm.

6.5 THICKNESS OF LAYERS OR COURSES

The thickness of the consolidated metal layer should not normally be less than 11.5 cm, except where the road in future is to be treated with tar or bitumen. If hard stone metal is used the thickness of the consolidated layer may be 7.5 cm for light traffic. I.R.C. has recommended that the thickness of a single compacted layer in case of hard metal should not exceed 10 cm.

As the pressure intensity of a loaded iron-tyred cart is more than due to a 12 tonne road roller, therefore for such traffic the thickness of the course can be increased.

For all important roads on medium sub-grades two courses of stone metals (7.5 cm thick each) over a sub-base of 7.5 cm of gravel or clinker should be constructed. In case of light traffic one layer of 11.5 cm consolidated thickness should be provided instead of two layer of 7.5 cm thickness.

The thickness of one consolidated layer for hard metal should not be more than 7.5 cm and 11.5 cm for soft metal. If the designed pavement thickness is more it should be laid in two or more layers. Care should be taken that lower course should not be fully consolidated but nearly consolidated, so that while compacting the next layer above it, it can be fully consolidated. No binding material should be used on the first layer.

6.6 CONSTRUCTION OF W.B.M. ROAD

The stages of construction of a W.B.M. road are as follows:

(i) *Subgrade and Kerbs.* After the construction of embankments or cutting is completed, the subgrade is prepared in the shape of required camber and gradient. It is properly compacted by light rollers or hand rammers. Earthen kerbs of 15 cm × 15 cm section are prepared along the edges of the proposed road to hold the materials.

(ii) *Base or foundation layer.* The base or foundation course consists of 12 to 18 cm size boulders or broken pieces of stone, large size kankar, overburnt brick-bats or flat bricks. Stone boulders or bricks are each placed by hand, firmly bedded in position with as little voids as possible. In the case of bricks the larger edge of bricks is placed transverse to the direction of the roadway. In the case of stone base the voids if any should be properly filled with stone pieces and the surface by lightly rolled with a 10 tonne roller. The width of the foundation should be 60 cm wider than the pavement, projecting 30 cm on each side.

(iii) *Wearing course.* This is also known as metalling course and can be laid in one or two layers depending upon the total thickness required. The thickness of the loose metal in each layer should not exceed 15 cm. It is because of the reason that ordinary rollers will not be able to roll more than 15 cm thick layer. If very heavy rollers are used, the metal will be crushed. The wearing course is prepared in the following order:

(a) First of all stone metal is evenly spread over the proposed surface in the required thickness. After spreading, the metal is hand packed properly to the required camber and gradient. Generally 1 in 30 to 1 in 40 camber is provided.

(b) The layer of stone metal is then rolled by 8 ton rollers. Rolling should first be started from the edges and gradually shifted towards the centre. This dry rolling will provide interlocking of aggregates.

(c) After dry rolling, stone screening or 12 mm grit is evenly sprinkled on the surface at the rate of 1 cu metre per 100 sq m. It is known as bindage layer. This surface is now rolled dry, and then sufficient quantity of water, to wet the aggregates, is sprinkled, and the surface is again rolled. During the process of rolling proper camber and gradient should be maintained.

(d) The next day final coating of sandy soil containing clay and moorum and 75% sand is spread in a thickness of 5 mm and the surface is wetted by copious quantity of water. This layer of sandy soil is known as Bindage layer. Nearly 40 to 60 litres of water is used per sq. metre of the surface. The surface is then rolled. During rolling water is sprinkled on the roller wheels so that the metal should not stick to the roller surface. The object of spreading so much of water is that the material such as sand, clay etc., should reach interstices of stone.

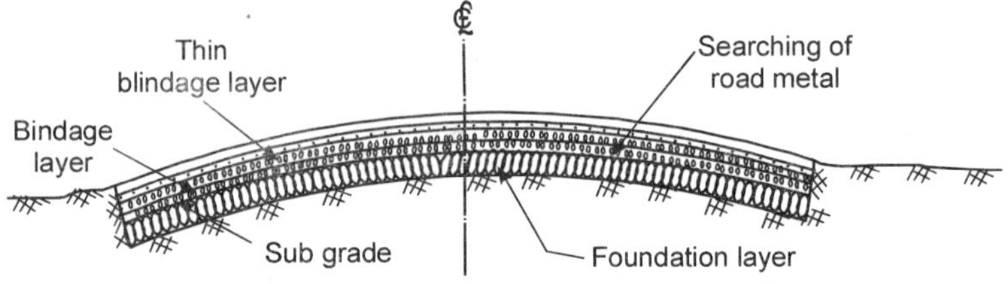

Figure 6.3 *Waterbound macadam road*

The rolling is continued on the next day also. Some quantity of sand at the rate of 1 cu metre per 160 to 180 sq m of road surface is spread. The surface is then kept wet for about seven days and then it is opened for traffic.

6.7 CAUSES OF DISINTEGRATION OF WATERBOUND MACADAM ROADS

Following are the main causes of disintegration of W.B.M. roads.

(a) In the W.B.M. road the stone aggregates are binded together by means of grit, sand and clay. The rain water wash away the soil binder and the stone aggregates protrude out or get loose on the surface layer, thus it forms the pot-holes and ruts.

(b) The high temperature of the sun during day and cool temperature during night causes disintegration.

(c) The heavy mixed traffic and adverse weather conditions damage the water-bound macadam road in no time.

(d) Steel-tyred bullock carts crush the road metal and the fast moving pneumatic-tyred vehicles create a partial vacuum and loosen the surface metal. The pulverized metal is blown by the wind and the fast moving vehicles.

6.8 CORRUGATIONS ON W.B.M. ROADS

Corrugations are the wave-like deformations formed in the surface of the roads. The corrugations make the travel uncomfortable and cause great damage to the vehicles. Vehicles cannot move with high speed on the corrugated road surfaces, due to heavy vibrations.

Following are the main causes of formation of corrugations on W.B.M. roads:

(a) The wheels of the vehicles due to spun action throw up the loose surface blindage and cause pots and depressions in the surface.

(b) During the construction of the road, the defective rolling of thick layers of the metal, causes corrugations. While rolling by road roller, due to the effect of the metal creeping in front of the roller, a hard ridge is formed followed by a depression.

To prevent the formation of the initial waves while roiling, the road roller gearing should be in good condition to prevent jerky rolling. Similarly long runs should be taken to minimize stopping and reversing the roller, which also cause corrugations. While rolling the longitudinal evenness should be checked with string.

(c) Imperfect grading of coarse aggregates also cause uneven displacement of stones resulting into corrugations under the traffic load.

(d) Improper consolidation of the embankments also causes corrugation.

(e) Heavy loaded vehicles than the designed load also cause corrugations.

(f) The braking and starting action of the vehicles also causes corruption.

(g) During rains when the bearing capacity of the road reduces, the movement of traffic causes sinking of road at various places.

6.9 MAINTENANCE

To keep the road surface in a serviceable condition, periodical maintenance of W.B.M. road is very essential. The blindage is sucked up by fast moving traffic and the road surface without bindage disintegrates in no time. Also the surfacing gives way in isolated patches known as *pot holes*. These *ruts* and *pot holes* should be repaired as soon as they appear on the surface. The *pot holes* and *ruts ate* cleaned with a wire bruch and filled with fresh metal, properly hand packed and compacted by hand rammers. Bindage is now spread to fill up the voids and the surface is again rammed by sprinkling water. If, however, the *pot holes* occur on more than one-third area of a road surfacing in a particular reach, this area should be repaired by scarifying the surface, rolling the old road metal, spreading the required fresh layer of loose metal, rolling and preparing the surface in conformity with the old road. This sort of repairing is called *partial resurfacing*.

6.10 IMPROVEMENTS ON W.B.M. ROAD SERVICING

The waterbound macadam roads (and other cheap type of roads such as earth, moorum and gravel roads), produce dust when in service and produce more dust when they are badly worn out. Therefore to prevent this dust nuisance the following improvements are suggested:

1. Water should be sprinkled on the surface periodically. (Too much of water will make the road muddy and wet road surface will get worn out quickly).

2. A coat of road oil may be given to the road surface. This coat will help in keeping the dust down and will increase the load carrying capacity of the road.

3. Some hygroscopic material such as calcium chloride is sprinkled on the surface of W.B.M. road. This powder will absorb moisture from the atmosphere and thus the bindage will also remain moist and intact

6.11 I.S.I TESTS FOR AGGREGATES

The properties of the mineral aggregates which form the pavement structure are of great importance in the highway construction. The following laboratory tests are conducted on the mineral aggregates to determine their essential properties important to a highway engineer.

1. Sieve Analysis Test. This test determines the grading of aggregate particles. A weighed quantity of mineral aggregates is placed over a set of standard sieves of square opening and being shaked by a standard sieve shaker for a specified time. The aggregates oft each sieve is weighed and the percentage retained on each sieve is calculated.

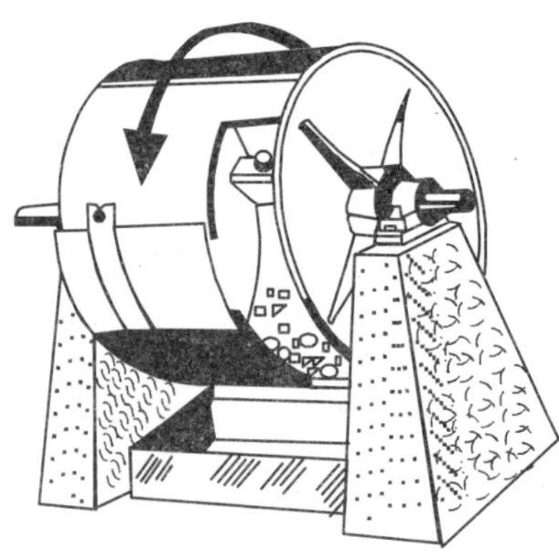

Figure 6.4 *Loss Angel's abrasion testing machine*

2. Particle Shape Test. With the help of this test, the shape of particles is determined as *flaky* or elongate. Those particles whose least dimension is less than 0.6 times of their mean size are termed *as flaky* and those particles whose greatest dimension is more than 1.8 times their mean size, are termed as elongated.

3. Abrasion Test. This test is conducted in a Loss Angel's abrasion testing machine. With the help of this test, it is found as to how much percentage of the aggregate will wear out in a specified time. The Loss Angel's abrasion testing machine consists of a hollow steel cylinder having an internal diameter of 700 mm and inside length 500 mm. It is fixed on strong steel axils. It is fitted with a dust proof opening with a cover. Stone aggregates are heated in an oven for 4 hours at a temperature of 100 to 110°C. The weight of aggregates which are fed in the cylinder is 5 kg or 10 kg. Alongwith the stone aggregates, steel balls of 40 mm diameter are also fed and the drum is rotated at the rate of 30 to 33 revolutions per minute for 500 to 1000 revolutions. After the test the aggregates are sieved on Indian Standard Sieve No. 170 and the percentage loss on wear is calculated.

4. Weathering Test. This test is performed on cylinders of diameter 6 cm and length 7.5 cm cubes of 6 cm side to find out the effect of atmospheric action. The test specimens are heated for 24 hours at a temperature of 105 ± 2°C. The specimen is then weighed. Let this weight be W_1. After this the specimen is kept in a desicator for 24 hours at a temperature 20–30° C. After cooling the specimen is again weighed. Let this weight be W_2. After that the specimens of stone are kept in a saturated solution of sodium or magnesium sulphate at a temperature of 105 ± 2° C and heated till all water is evaporated. The process is repeated 29 times and each time the specimen is weighed. Let the average weight of the specimen is W_3.

Now the specimens are kept in pure water for 24 hours. Let the weight of the specimens before and after immersion in water is W_4 and W_5 respectively.

∴ Percentage water absorption

$$J_1 = \frac{(W_2 - W_1)100}{W_1}$$

Volume of specimen

$$V_1 = \frac{(W_3 - W_3)}{D}$$

where

D = sp. gravity of the stone.

Percentage water absorption after 30 times immersion in saturated solution of sodium or magnesium sulphate $J_2 = \dfrac{(W_4 - W_1)100}{W_1}$

Now volume $V_2 = \dfrac{W_4 - W_5}{D}$

Increase in the percentage of water absorption $= \dfrac{100(J_2 - J_1)}{J}$

Percentage increase in volume

$$= \frac{V_2 - V_1}{V_1} \times 100$$

From the above experiment various varieties of stones are compared and the best suited variety is selected.

6.12 CODE OF PRACTICE FOR W.B.M. BASE COURSE AND WEARING COURSES

Water bound macadam base course shall consist of clean, crushed aggregates mechanically interlocked by rolling and bonded together with binder such as stone screening and water on a prepared sub-grade or base course and shall be constructed in accordance with specifications.

Materials for Base Course. The coarse aggregates and screenings shall be either stone crushing or crushed slag of suitable quality unless adhesive provided screenings shall be of the same material as the coarse aggregate. The crushed stone shall consist of hard, durable particles or fragments of stone free from excess of flat, elongated, soft or integrated particles dust or other objectionable matter whatsoever. The crushed aggregate shall meet the following requirements:

Flat or elongated pieces not more than 15% wear at 500 revolutions of Los Angle Test not more than 50%.

Size and grading requirements of the aggregates are as follows:

Sieve designation	*% by weight passing*
650	100
450	90–100
350	25–70
250	0–15
200	0

The screenings used to fill the voids in the coarse aggregates should be 20 mm maximum size graded.

Wearing Course. The crushed aggregates and the screenings should have the same specifications as for the base course.

In order to prevent ravelling of the surface course, sufficient clayey material is necessary on the binder material which should plastic index of 4 to 9. If limestone formations are available nearby, limestone dirt or kankar modules may be usefully employed for the purpose.

The coarse aggregate shall be spread uniformally and evenly upon the prepared sub-grade. But in no case shall the aggregates be dumped in piles directly on the sub-grade within the area over which it is to be spread nor shall hauling over the partly completed base course be permitted. Whenever possible approved mechanical devices shall be used to spread the aggregates uniformally so as to minimise hand operation.

The surface of the aggregate shall be carefully tuned up and all high and low spots remedied by removing or adding aggregates as may be required. The surface shall be checked continuously during the spreading and rolling of the coarse aggregates variation greater than 10 mm from a 3 m straight edge laid parallel to the centre line of the road.

The base course shall normally be constructed in layers of not more than 8 cm of compacted thickness. Each layer shall be tested by depth blocks. No segregation of large or fine particles be allowed and the course aggregates as spread shall be of uniform thickness/gradation with no pockets of fine particles or material. The course aggregate shall be spread no further in advance of rolling, screening and bonding of the preceding section, than can be completed in one day. Curbs and gutters and other structures shall be completed and railway tracks shall be adjusted to grade before any aggregate is spread adjacent to it.

Rolling the Aggregate. Immediately following the spreading of the course aggregate, it shall be compacted to full width by rolling with either 3 wheel roller of 5–10 tonne capacity or tender roller. The roller shall begin from sides with the rear wheel of the roller covering equal parts of shoulder and coarse aggregate. The roller shall run forward and backward until the shoulder and the metal are firmly bound together.

Application of Screening. After the coarse aggregate has been thoroughly keyed and set by rolling, screenings in an amount that will completely fill the interstices shall be applied gradually over the surface. Dry rolling shall be continued while the screenings are being spread so that the jamming effect of the roll will cause them to settle into the voids of the coarse aggregate. The screenings shall not be dumped in piles on the coarse aggregates but shall be spread uniformally in successive thin layers.

The screening shall be applied at a uniform and slow rate so as to ensure filling of all voids. Rolling and brooming shall continue with the spreading of the screening. If the screenings are in a damp and wet condition, then shall not be rolled and broomed after spreading until the surface of the particles has dried off the spreading, brooming and rolling of the screenings shall be performed on sections which can be completed in a day and shall continue until no screenings can be forced into the voids of the coarse aggregate when the screenings are dry.

Sprinkling and Grouting. When all the voids of a section of the course have been filled with screenings, it shall be sprinkled with water, swept and rolled. Hand brooms shall be used to sweep the wet screenings into the unfilled voids and to distribute them evenly. The sprinkling, rolling and sweeping shall be continued and additional screenings applied where necessary until the coarse aggregate is well bonded and firmly set for its entire depth and until a grout has been formed of screenings and water that will fill all the voids and will form a vane of grout before the wheels of the roller.

In case of water-bound macadam wearing course construction after the application and dry rolling of screening, for the sprinkling and grouting operations, instead of application of additional screenings, bending material shall be spread. When a section has been thoroughly filled and grouted, it shall be left to dry before it is spread to traffic.

Quantity of Materials. (i) The approximate quantity of loose materials for 10m^2 sub-base are

(a) Coarse aggregate of 90 to 140 mm size
= 1.21–1.43 m^3

(b) Stone screenings 12.5 mm size
= 0.40–0.44 m^3

(c) Binding material = 0.08–0.10 m^3

(ii) *7.5 cm thick compacted layer*

(a) Coarse aggregate 63-40 mm size
= 0.91–10.7 m^3

(b) Stone screening for base coarse 12.5 mm size
= 0.18–0.21 m^3

(c) Stone screening for surfacing
= 0.15–0.17 m^3

(d) Binding material for base course
= 0.05–0.09 m^3

for surfacing = 0.10–0.15 m^3

PROBLEMS

6.1. Describe in brief the construction of gravel road. Where this type of road is used ? What improvements can you suggest to improve the working conditions of this road ?

[**Hint.** See article 6.6. The importance which have been suggested for W.B.M. road can also be applied to gravel roads.]

6.2. Describe in detail step by step the construction of a waterbound macadam road. What improvements do you suggest to reduce the dust nuisance of W.B.M. road ?

6.3. Write down detailed specifications for road metalling, bihdage and blindage layers to be used in W.B.M. road.

[**Hint.** See artcle 6.3. Detailed specifications means constructional details to be followed, including thickness of each layer or course, size and quantity of metals to be used etc.]

6.4. What do you understand by the waterbound macadam road ? Name the different metals which are generally used for the construction of such a road. Draw a neat sketch of W.B.M. road having a carriage way of 4 m.

6.5. Give in brief the circumstances under which a waterbound macadam road is used.

6.6. Write short note on the grading of the coarse aggregate for W.B.M. roads.

6.7. Write short notes on :

(i) Size of aggregate for W.B.M. roads.

(ii) Thickness of W.B.M. road layers.

(iii) Causes of disintegration of W.B.M. roads.

(iv) Construction of W.B.M. roads.

BIBLIOGRAPHY

1. Traffic Census on Non-urban Roads I.R.C : 9–1972.
2. Tentative Specifications for construction of stabilized earth roads with soft aggregates I.R.C.: 28–1967.
3. Tentative Guidelines for use of low grade aggregates and soil aggregate mixtures I.R.C.: 63–1973.
4. Standard specifications and code of practice for W.B.M. Roads I.R.C.: 19–1977.

7

Bituminous Roads

In this Chapter you will study,

• Type of Bituminous Roads • Bitumen, Coal-tar and Asphalt • Cut Backs and Emulsions • I.S.I. Tests for Bitumen • Characteristic of Bitumen • Choice and Recommendations for Bitumen Binders • Temperature Susceptibility of Bitumen • Selection Criteria for Bitumen • Surface Painting or Surface Dressing • Two Coat Surface Dressing • Renewal Coat of Surface Dressing • Quantity of Bitumen, Aggregate for Surface Dressing Roads • Bituminous Macadam • Bituminous Concrete • Sheet Asphalt • Mix Defects • Aggregate Gradation for Bituminous Roads • Causes of Disintegration of Bituminous Roads • Maintenance and Repair of Bituminous Road • Grades of Road Tar IS : 215–1961 • Bituminous Mix Design • Tests for Asphaltic Mix • Types of Bitumen • Control of Bitumen Application Temperature • Bituminous Road Equipments • Safety Measures while Working with Bitumen Mix Equipments • Methods of Reduction of Consumption of Bitumen

GENERAL

Bituminous roads are those roads in which some binding material such as bitumen, coal-tar or asphalt is used in surfacing. Such roads are also known as *black top* roads. These roads are common and popular in cities and towns. National highways and State highways are also generally bituminous roads.

7.1 TYPE OF BITUMINOUS ROADS

There are various types of bituminous roads which are useful for different intensities of pressures of wheel loads and different types of traffic. The following are the various types of bituminous surfaces :

(1) Surface painting or surface dressing.

(2) Bituminous macadam.

(3) Bituminous concrete.

(4) Sheet asphalt or asphaltic mat.

7.2 BITUMEN, COAL-TAR AND ASPHALT

It will not be out of place to give some idea of bitumen, coal-tar and asphalt, before proceeding further.

Bitumen. There are various definitions of bitumen. The *international definition* of bitumen is that the bitumen is a mixture of natural pyrogenous hydrocarbons and their non-metallic derivatives which may be gaseous, liquid, viscous or solids but must be completely soluble in carbon disulphide. Bitumen, in other words, is a hydro-carbon compound in a solid or semi-solid state. It contains 85 parts of carbon, 12 parts of hydrogen and 3 parts of oxygen. The properties of bitumen depend upon its source and the method of preparation. However, it is blackish-brown in colour, sticky in nature, semi-solid and melts when heated.

Bitumen is obtained by the partial distillation of crude petroleum. The distillation may take place over centuries as Nature Work, or the distillation may be carried out in refineries, where the lighter oil fractions, which hold the bitumen in solution, are distilled off. Thus bitumen is an important by-product of the fractional distillation of crude petroleum.

Asphalt. Asphalt is a substance containing a high percentage of *bitumen* and some mineral substances. Bitumen is the basic constituent of asphalt. When crude petroleum is fractionally distilled in nature, the residual product is called asphalt. (If distillation of crude petroleum is done in refineries the product is called Bitumen).

It is blackish brown in colour, non-inflammable but burns with a smoky flame at a temperature of 250° C.

Asphalt can be classified as (1) Rock asphalt, and (ii) Lake-asphalt.

(i) *Rock asphalt.* It is a bituminous lime stone. It is a compound containing 30 to 90 parts of lime stone and 20 to 10 parts of bitumen. It is used for pavements.

(ii) *Lake asphalt.* It is a natural asphalt found in lakes and contains 40 to 70 per cent of bitumen and about 30 per cent of water, lime, clay and sand. Lake asphalt is refined by boiling it in a tank when water evaporates and the impurities float on the surface and are removed. Refine lake asphalt is the best material for road making.

Coaltar. Coaltar is a jet black, viscous liquid. It is the residual product obtained by the destructive distillation of coal. Coaltar is usually obtained as a by-product in the manufacture of *coal gas.* It is used as a preservative for timber and for making roads.

Table 7.1 gives the application temperatures for different grades of bitumen and tars.

TABLE 7.1 *Application temperatures for bitumen and tars*

Sl. No.	*Grade of the bitumen tar*	*Application temperature for sprating purpose*	*Application temperature for mixing with aggregate*
	(A) Straight run or paving Bitumen		
1	30/40 penetration	170 –185°C	170 – 185°C
2	60/70	165 – 180°C	155 – 165°C
3	80/100	160 – 175°C	150 – 165°C
4	180/200	150 – 165°C	135 – 150°C
5	A – 90 and S – 90 grades	175 – 161°C	150 – 175°C
6	(B) Cut back Bitumen Note : Some cut backs are for cold application and do not require heating	161 – 161°C	150 – 165°C
	(C) Road tars		
7	Grade RT –2	93– 104°C	93 –104°C
8	RT –3	104 – 116°C	116°C
9	RT –4	132°C	132°C
10	RT –5	138°C	138°C

7.3 CUT BACKS AND EMULSIONS

Cut Backs. They can be defined as the solution of bitumen (asphalt or coaltar) in solvents. The solvent is the distillate of the petroleum or coaltar.

into a liquid form so that they can be used without heating or require only a light heating. When the cut backs are mixed with aggregates or spread on road surface, the volatile oil evaporates leaving behind the thick oil. Cut backs contain nearly 80 per cent asphalt and 20 per cent solvent. The type of solvent to be used, depends upon the rate of evaporation required of the cut backs, to suit different types of road construction.

Emulsions. Emulsions are defined as a mixture comprising two unmixable liquids, the one dispersed in the other in the form of fine globules or droplets. Oil and water when mixed and vigorously shaken, the oil is broken up into small globules floating in water. This is a water oil emulsion. Bitumen emulsion consists of asphalt or tar particles of size about 3 microns $\frac{1}{1000}$mm dispersed in water in the presence of an emulsifying agent which is added nearly 1 per cent. The emulsifying agent is absorbed on the surface of each globule forming a film on its surface. Upon the permanency of the film and its degree of homogeneity, it depends on the stability of the emulsion. Generally soap and resinous substances are used as emulsifying agents. Emulsions contains nearly 55 to 65 per cent of bitumen by weight.

Bitumen emulsions can be applied at normal temperature without heating. When emulsion is spread on the road, it changes from brown to black colour while the water soaks in or evaporates, allowing the bitumen particles to reunite. Emulsion is useful for damp surface. Before applying emulsions, the road surface shall be thoroughly cleaned and slightly damped with water.

7.4 I.S.I. TESTS FOR BITUMEN

Bitumen is available in a variety of types and grades. Different grades of bitumen are used for various types of road construction. Following tests have been prescribed by I.S.I, to test the workability of various grades of bitumen :

1. **Penetration Test.** This test indicates the hardness of bitumen used in road construction by increasing the distance in *tenths of a millimetre* to which a standard needle will penetrate vertically into a sample of bitumen under stipulated conditions of temperature, loading and time. (I.S. : 1203 – 1958) unless otherwise stated, the bitumen sample is penetrated for 5 seconds at a temperature of 25 °C and the load on the needle is 100 gm. The softer the bitumen the greater will be its number of penetration units.

The test is being performed by an instrument known as *penetrometer*. The sample of bitumen is placed in the bottom pan and the needle is lowered

* In refinery, the volatile which are distilled off, are known as cuts. To manufacture *cut backs*, the *cut* is put *back* into the prepared residue and hence the name.

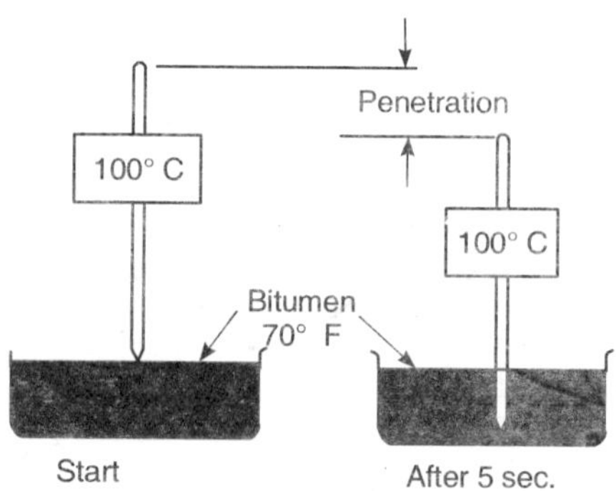

Figure 7.1 *Penetration test*

down and left for 5 seconds. The temperature of the sample is kept at 25°C. The weight of the needle is 100 gm.

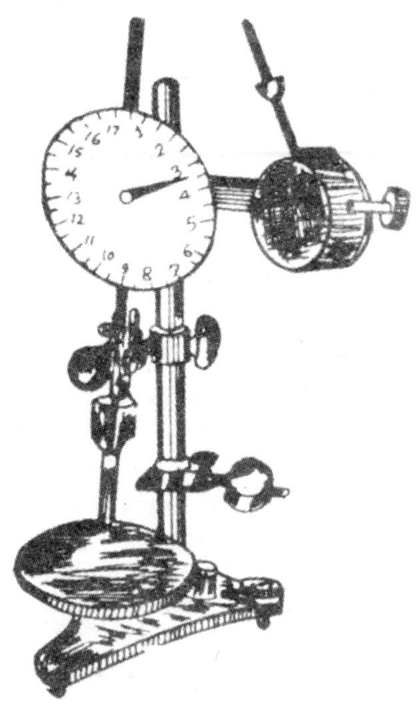

Figure 7.2 *Penetrometer*

2. Softening Point Test. Softening point is also sometimes termed as melting point. As the bitumen is heated it becomes softer and softer until it flows readily with the rise of temperature. This test is usually performed by the ring and ball apparatus (I.S. : 1205–1958). A brass ring containing the test sample of bitumen is suspended in water at a given temperature. The

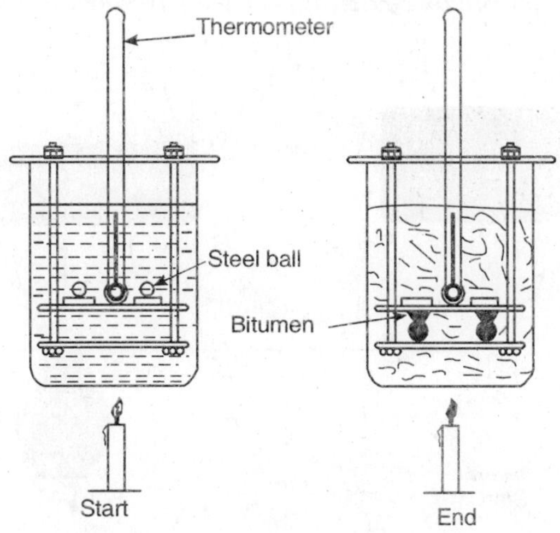

Figure 7.3 *Softening point test*

brass ring has an internal diameter of 15 mm and thickness 8 mm. The sample is poured in the ring and 10 mm dia. ball is placed over it. The water is then heated at the rate of 5 degrees per minute, till the bitumen is soft enough to allow the ball to pass through it and touch the bottom plate placod at a specified distance of 2.45 cm. If the melting point of bitumen is higher than that of water, glycerine should be used in place of water.

3. Ductility Test. The ductility of bitumen is defined as the length or distance in centimetres to which a standard briquette of bitumen can be

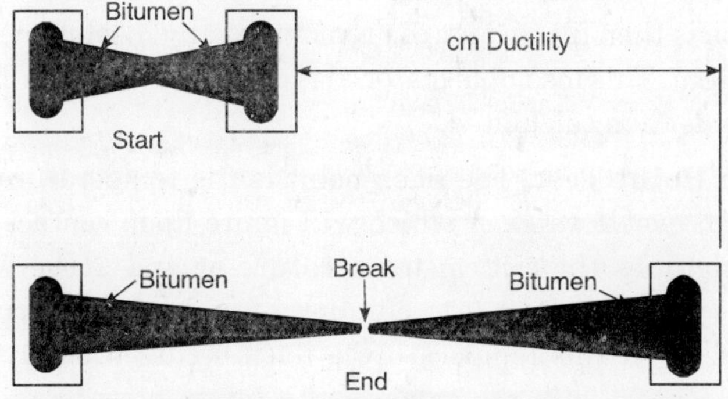

Figure 7.4 *Ductility test*

stretched before the thread breaks, at a temperature of 25°C, when the ends of the briquette are stretched at the rate 5 cm per minute (IS : 1208–1958).

4. Viscosity Test. Viscosity is defined as the resistance to the flow of liquid bitumen i.e., cut-back or emulsion. Viscosity at any specified temperature is measured by recording the time in seconds for a given quantity

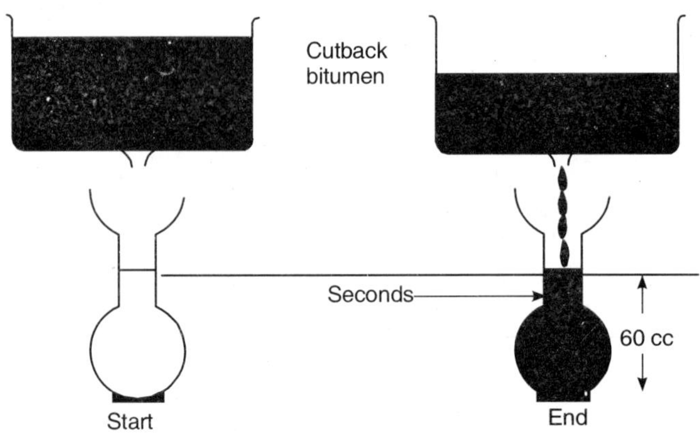

Figure 7.5 *Viscosity test*

of the product at the same temperature to flow through an orifice of standard dimensions into a receiver as shown in Fig. 7.5. The longer the time required, the higher is the viscosity of the material. Test values are expressed in seconds (I.S.: 1206–1958).

5. Solubility Test. This test measures the extent to which bitumen goes into solution either in carbon disulphite (CS_2) or carbon tetrachloride ($CC1_4$). A weighed sample of the material is treated with an excess of the solvent. The solution is then filtered for the removal of any insoluble material which may be present. This insoluble material is dried and weighed. The percentage of solubility is then calculated.

6. Flash Point Test. The flash point is the temperature at which the bitumen will evolve vapours which will ignite upon contact with a flame. This test signifies the critical temperature at and above which suitable precautions should be taken to eliminate free hazards during its heating and handling. At this temperature a flash occurs when a small flame is brought in contact with the vapours of a bituminous product. The lest is performed by heating a sample material in an open cut at a specified rate

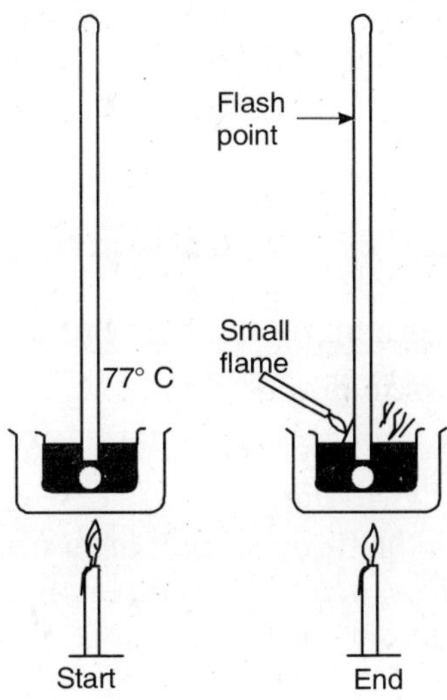

Figure 7.6 *Flash point test*

and determining the temperature at which a small flame passed over the surface will cause the vapours to ignite or flash. (IS : 1209–1958).

7. Distillation Test. It indicates the rate which a cut-back will cure after application. This test determines the relative proportions of bitumen and

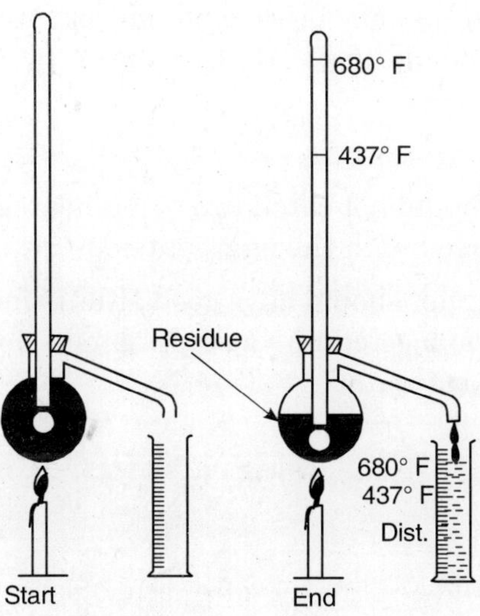

Figure 7.7 *Distillation test*

the volatile flux present in cut-back. It also measures the amount of volatiles driven off at certain specified temperatures. The residual bitumen is tested for penetration, ductility and solubility. I.S. : 123–1958 describes the standard test method of this test.

7.5 CHARACTERISTIC OF BITUMEN

The bitumen to be used as a binder for the construction of roads should have the following characteristic properties :

1. The bitumen in the boiler or the distributor should be sufficiently fluid for easy and efficiently spraying.

2. The bitumen should be fluid enough after application on the pavement to *wet* thoroughly both the surface and the cover aggregate when the later is spread. Too hard a grade of bitumen will not maintain the necessary or desired fluidity.

3. Once the cover aggregate has been spread and lightly rolled. The bitumen should revert back rapidly to its original viscosity.

4. It should have very little affinity for temperature or in other words it should not be susceptible to temperature so that it can maintain same viscosity within a reasonable range of temperature to which a road surface is subjected. Bitumen binders with low temperature susceptibility will not become soft during summer and hard and brittle during winter.

5. To ensure durability of a bitumen binder it should have a good amount of volatiles and it should not loose this property even when subjected to higher temperatures.

6. It should be ductile.

7. The bitumen should not catch fire when heated at a high temperature to enable it to be mixed with the aggregates.

8. The bitumen binder should have good affinity for aggregates and should not get loose when subjected to water or should not be stripped off in the continuous presence of water.

7.6 CHOICE AND RECOMMENDATIONS FOR BITUMEN BINDERS

1. The bitumen should be slightly softer and of less viscous grade to be used in colder climates and at high altitudes.

2. For surface treatment of old bituminous surfaces which show evidence of drying out, cut back have an advantage for the light volatile assist

penetration of the old surface and replace some of the material that has evaporated.

The following grades of bitumen are considered suitable for surface dressing:

Straight run bitumen

(a) Mexphalte 80/100

(b) Spramex 180/200

Cutback bitumen

(a) Shelspra B-5

(b) Shelmac RC-3

7.7 TEMPERATURE SUSCEPTIBILITY OF BITUMEN

The rate of change of viscosity of a bitumen decides its temperature susceptibility. To design a bituminous mix under different conditions of temperature, temperature susceptibility plays an important role. As stated above the bitumen should show very little or no change in its viscosity or consistency in a given reasonable range of temperature. Highly susceptible bitumen will lead to bleeding in summer and become hard and fragile in cold weather.

7.8 SELECTION CRITERIA FOR BITUMEN

The selection of type of bitumen will depend on the following factors :

1. Climatic conditions of the region in which the bitumen is to be used particularly the temperature range.

2. Nature of job i.e., whether the work to be executed is of a smaller magnitude, huge construction or a maintenance work.

3. Nature of application of binder by spraying or by pre-mixing.

4. Availability of plant and equipment for the work as hot mix plant, paver, spreader etc.

5. Nature and volume of traffic and its type i.e., pneumatic and iron wheeled traffic.

Following grade of bituminous binders are generally recommended :

Name and grade binder	*Use*
1. 30/40 penetration grade	Bituminous macadam, Bituminous or asphaltic concrete etc. in normal summer temperature in plains.

Contd.

2. 60/70 grade	Bituminous macadam in winter and high altitude, hot mix asphaltic concrete in summer, hot mix bituminous macadam in summer and in plains.
3. 80/100 grade	Surface painting, seal coat, premixed carpet summer temperature (32°C to 42°C) hot mix asphaltic concrete and bituminous macadam in winter and at high altitude.
4. 180/200 grade	Surface painting or surface dressing, Seal coat, premixed carpet, penetration or grouted macadam in winter and at high altitude.
5. Cutback RC-3	Surface dressing and seal coat in summer as well as winter in plains and even at high altitude.
6. Emulsions work, sealing of fine	Wet and cold climate, maintenance and patch cracks.
7. Medium and slow curing cutbacks MC and SC	Priming and tack coat.

7.9 SURFACE PAINTING OR SURFACE DRESSING

It is a kind of bituminous road in which a film of tar or asphalt is applied on the prepared top of road foundation and then on this film is spread a thin layer of stone chippings or other fine mineral aggregate and the surfacing is then rolled. The thickness of this surface varies from 2 cm to 3 cm and the purpose of this coat is to seal the foundation layer, which may be a W.B.M. surface, gravel layer or stabilized earth. Usually the surface treatment is given to a new or an existing W.B.M. road surface. If the surface treatment is to be given to a new W.B.M. road surface then in the construction of W.B.M. road itself, bindage and blindage layers are not provided.

In the case of new W.B.M. road, the road metals are properly rolled by sprinkling water and when road metals are properly interlocked, the surface dressing is applied. In the case of an existing W.B.M. road, the surface is brought to the required camber and gradient i.e., the road is reconditioned. It is then cleaned by wire brushes and surface dressing is applied.

Methods of Applying Surface Dressing. The surface dressing can be applied either by cold process or hot process. In the hot process the tar or asphalt is heated. Tar, if used, is heated to a temperature of 120°C and asphalt, if used, is heated to a temperature of 180°C.

A uniform thin layer of hot binder is laid over the *clean and dry* surface of W.B.M. road. The binder is spread in longitudinal strips, starting from the edges of road. Nearly 2 kg of binder is used per square metre of the surface. The layer of binder should be thin but uniform. 12 to 20 mm size stone chippings should be spread evenly on the binder when it has become tacky. About 1.8 to 2 m^3 of chippings will be required per 100 m^2. After *broadcasting* the stone chippings, the surface is lightly rolled with light roller. Usual precautions of rolling (i.e., the rolling should be started first from the edges and the roller should move to and fro), should be taken.

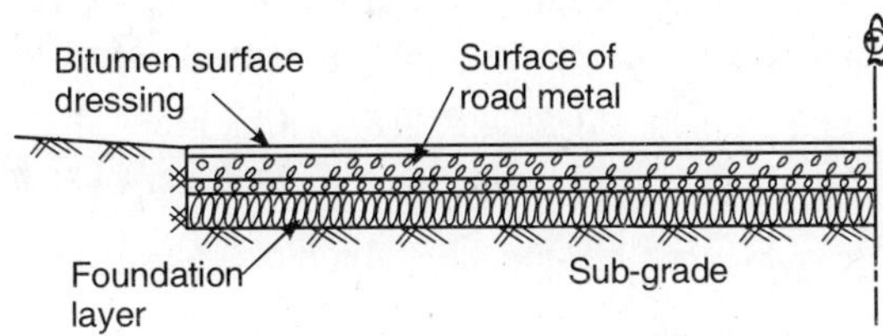

Figure 7.8 *Bitumen surfacing over macadam road*

In the case of cold process, the rapid setting emulsions are spread on a clean and dry surface at the rate of 2 kg/m^2 of the surface and then gritted with 6 mm size stone chippings before the emulsion breaks. (When the emulsions changes its colour from brown to black the emulsion is said to have been broken). The water of the emulsion will evaporate as time passes and the binder will become thicker and thicker. Cut-backs are applied at the rate of 1.5 kg/m^2. Cut-backs are used when surface is damp.

7.10 TWO COAT SURFACE DRESSING

The method of surface dressing or surface treatment explained above is called *single coat surface dressing*. For heavy traffic of mixed type i.e., consisting of fast moving traffic and bullock carts, two coat surface dressing will be required. The thickness of two coat dressing is nearly 3 cm. There are two methods of laying two coats. In one method the two coats are applied simultaneously as explained above. In the other method, the first coat is applied and the road is opened to traffic for one to two weeks. The surface is then cleaned thoroughly, and the second coat is applied. In the hot process the second coat will require bitumen at the rate of 1.5 m^2/100 m^2 of road surface. Similarly surface dressing may be given with emulsions or cut-backs in a similar manner.

In some cases bitumen and stone chippings are first mixed in a hot state and the mixture is spread and rolled to the desired thickness. This is known as *premixed surface dressing*. In this case 64 kg of bitumen is taken for 1 cu. m of stone chippings.

7.11 RENEWAL COAT OF SURFACE DRESSING

Due to constant wear and tear, the surface is worn out under the mixed traffic. If pot holes and ruts appear on more than two-third of road surface in a particular portion of the road, a renewal coat is necessary. The surface is cleaned and worn out portion scrapped if necessary. After cleaning the surface bitumen is applied in a thin uniform layer. When the bitumen has become tacky, stone chippings are broadcast at the rate of 1.5 m^3/100 m^2. The size of chippings should not be more than 9 mm. The bitumen to be applied should be 1.5 kg/m^2 of the road surface. The surface is then properly rolled. During rolling proper camber and gradient should be maintained.

7.12 QUANTITY OF BITUMEN, AGGREGATE FOR SURFACE DRESSING ROADS

Methed of construction	*Bitumen*		*Aggregate*	
	Grade	*Qty*	*Size*	*Qty.*
Single Cost	80/100	2 kg/m^2	1.25 cm	1.5 m^3/100 m^2
	SSBS	2 kg/m^2		
	RC 3	2.25 kg/m^2		
Two coat				
1st coat	80/100	2 kg/m^2	1.25 cm	1.8 m^3/100 m^2
	SSBS	2 kg/m^2		
	RC 3	2.25 kg/m^2		
2nd coat	80/100	1.25 kg/m^2	6 mm	1.8 m^3/100 m^2
	SSBS	1.25 kg/m^2		
	RC 3	1.25 kg/m^2		
Seat coat	80/100	1.25 kg/m^2	6 mm	1.8 m^3/100 m^2
	SSBS	1.25 kg/m^2		
	RC 3	1.25 kg/m^2		

7.13 BITUMINOUS MACADAM

The surface painted roads are only suitable for light traffic. Under heavy and mixed traffic, such type of road surfaces are worn out very easily. For heavy traffic Bituminous Macadam surfaces are very suitable. There are two types of bituminous macadam surfaces viz., (i) grouted macadam, and (ii) premixed macadam.

(i) Grouted Macadam. The grouted macadam surfaces may be semi-grouted or full-grouted.

When the binder is allowed to penetrate to a small depth of 2 to 2.5 cm below the surface of macadam, it is called *semi-grouted macadam.* In this method cold process can also be used. When the penetration is substantial i.e., upto 5 to 7.5 cm, it is called full grouted macadam. Cold process is generally not used as the bituminous macadam roads constructed by hot process are superior to that constructed by cold process.

Construction of full-grouted macadam. The base consists of a W.B.M. of sufficient thickness to withstand the traffic load. Generally 15 to 20 cm thickness is recommended.

Earthen kerbs are constructed along the edges of the base course. The base surface is cleaned and the aggregate is spread evenly to the required profile. The thickness of loose metal varies from 9 to 6 cm. The surface is now lightly rolled dry. After the rolling is complete, hot bitumen or asphalt is sprayed with a spraying machine or spout can. The rate of bitumen to be sprayed varies from 8 to 13.5 kg/m^2 of the surface, depending upon the gradation of aggregate used and the desired depth of penetration. After spraying the binder, grits of nearly 10 mm size are immediately broadcast over the surface at the rate of 1.5 to 2 m^3 per 100 sq. metres of the surface. The surface is now rolled by 10–12 tonne rollers. The rolling should be first started from the edges and should be shifted towards the centre. After the surface has been rolled to compaction, it is opened to the traffic for a month or so. After this, the seal coat is provided.

The top of such surface is rough and open textured and pervious also. To make the surface smooth and impervious, a *seal coat* is provided. The surface of the grouted macadam is cleaned and bitumen at the rate of 15 kg/m^2 and stone chippings of nearly 9 mm size are broadcast at the rate of 1.2 m^3 to 1 5 m^3 per 100 sq. metres. The surface is now rolled by 6–8 tonne rollers. After 24 hours the surface is opened to the traffic.

Construction of semi-grouted macadam. After the base has been cleaned and earthen curbs provided a 2.5 to 5 cm thick layer of coarse sand or moorum containing a very high content of clay is spread on the reconditioned surface of W.B.M. road. Over this, stone metal of 24 to 36 mm size is laid in a thickness of nearly 6 cm. The surface is freely sprinkled with water and the surface is rolled to a thickness of nearly 5 cm. The sand or moorum will form a very thin paste with water and will *work up* during the process of rolling. After rolling, the surface is allowed to dry. The slurry having come upto a distance of nearly 2 to 2.5 cm, the remaining depth of aggregates is grouted with heated tar or asphalt at the rate of 2.5 kg/m^2. Immediately after this grits of 12 mm size are broadcast at the rate of 1.2 to 1.5 m^3 per 100 sq. metres of the road surface. The surface is rolled to compaction and opened to traffic for a few days. Some coat is then given as explained earlier.

(ii) Premixed Macadam. In this method, the surface of the W.B.M. is first reconditioned (by doing necessary repairs of pot holes and ruts and is brought to the required camber and gradient). Kerbs are formed along the edges. Refined tar and road metal are heated to a temperature of 120°C and are mixed in bitumen mixers at the rate of 56 kg of tar or bitumen per cubic metre of road metal. The hot mixture is then laid between the kerbs in a thickness of nearly 5 cm. The surface is rolled by 8–10 tonne rollers. Water is sprinkled on the roller wheels to prevent adhering of road metals to the wheel surfaces. The surface is then checked. Depressions, if any, should be rectified.

Over this, a thin premixed sealing coat nearly 1.5 thick, consisting of 30 mm size chippings or coarse sand, is given and the surface is roller properly.

7.14 BITUMINOUS CONCRETE

This is a type of construction, in which the surfacing consists of fine and coarse aggregates mixed together with tar or bitumen in a mixer. This is a superior type of surfacing and can be used for heavier and mixed traffic.

The old surface of the waterbound macadam road is reconditioned. The surface is dried and cleaned. Side kerbs of earth or bricks are formed to support the bituminous surfacing. The thickness of the W.B.M. base should not be less than 15 cm. Well graded road metal of 6 to 30 mm size is heated to a temperature of 180°C in the drier of the mixing plant. Sand of 1 mm to 3 mm size well graded is also heated and mixed with the road metal in the ratio of one and two. Tar or bitumen is also heated to a temperature of 180°C and mixed with the hot aggregate. The binder is taken at the rate of 48 kg/m^3 of road metal and 128 kg/m^3 of sand. For proper mixing $\frac{2}{3}$ rd of the binder is mixed with the road metal and the remaining $\frac{1}{3}$ rd is mixed when the sand is added. The whole thing is thoroughly mixed. The bituminous concrete is spread on the prepared base of W.B.M. road, between the kerbs to a thickness which varies from 5 to 10 cm. The surface is then rolled by 6–8 ton rollers maintaining proper profile. The rolling should be continued for a few days till the surface is thoroughly compacted. After this premixed seal, coat consisting of coarse sand and bitumen is applied as usual. The surface is again rolled and is finally opened to the traffic

This type of surfacing is better than grouted macadam and can carry heavier traffic. It is also suitable for iron wheeled traffic. (*Sheicrete mix is a patented type bituminous concrete*).

7.14.1 Sheet Asphalt

Sheet asphalt surfacing is used on the W.B.M. base or cement concrete base. Sheet asphalt is elastic having less friction. It gives a better type of wearing surface. It consists of a 6 cm thick layer of asphalt concrete over which 2 to 4 cm thick layer of asphalt mortar consisting of coarse dry sand mixed with asphalt in a hot state. The method of laying and mixing is the same as explained for Bituminous Concrete.

7.15 MIX DEFECTS

Following are the common possible bitumen concrete mix defects :

(a) Lack of Bitumen. If the quantity of the bitumen is less, the mixes will have dry lean appearance. They will lack in the black lustre shining of the well coated mix. Such mixes will be difficult to spread and will have transverse cracks while rolling under road roller, and will not readily heal up.

(b) Mix too Cold. When the bitumen is not properly heated, such mixes shall be stiff and difficult to handle. The large size particles cannot be coated properly. The spreading will be difficult and there will be a tendency for the mix to tear under the screed.

(c) Excess Bitumen. The mix containing excess bitumen will have slopy mix, which will slump too readily even in the lorry or the container.

(d) Poor Mixing. When the mixing is not thoroughly done, after laying some areas will appear lean and brown and others as shiny and black having excessive bitumen.

(e) Poor Grading. If the grading of the aggregate is not properly done, the workability will be poor. The appearance after laying will be rough open textured if the quantity of coarse aggregate is more. If fine aggregate is more, the surface will show lean and too finely graded texture.

(f) Over-heated Mix. During heating in the boiler, if excess of blue smoke is seen coming, it indicates the over heating of the bitumen. Over heating of the bitumen should not be allowed, because after laying, such mix will have dull burnt appearance. The binding strength of such mix is also poor.

(g) Excess Moisture. If the aggregate used for mixing has excess of moisture, the appearance of the surface will be similar to those with excess of bitumen. The excess moisture will evaporate while heating and the stearn rising from the mix indicate its presence.

7.16 AGGREGATE GRADATION FOR BITUMINOUS ROADS

Table 7.2 gives the aggregate gradation for premixed bituminous carpets of various thicknesses.

TABLE 7.2 *Aggregate grdation for premixed bituminous carpets*

Sieve designation	*Semi-dense carpets of asphaltic concrete* — *Aggregate percent by weight passing the sieve*			
			25 to 40 mm compacted thickness	
	20 mm compacted thickness	*25 mm compacted thickness*	*Grading 1*	*Grading 2*
20 mm	–	100	–	100
12.5 mm	100	75–100	100	80–100
10.0 mm	75–100	60–85	80–100	70–90
4.75 mm	35–55	35–55	55–75	50–70
2.36 mm	20–35	20–35	35–50	35–50
600 micron	6–16	6–16	13–23	13–23
150 micron	4–12	4–12	8–16	8–16
75 micron	2–8	2–8	4–10	4–10
Binder content	The binder bitumen should be 5 .5% by weight of the total mix.		The binder bitumen should be 5% to 7.5% of the total mix.	

7.17 CAUSES OF DISINTEGRATION OF BITUMINOUS ROADS

Practically it has been seen that there are many failures with surface painting even with light traffic. Most of such road failures are not due to wrong use of the surface treatment but due to the improper preparation of the sub-grade and the foundations. The base of the road is usually the weak point, which causes failure. The water or dampness is the main enemy of the surface treatment. If the dampness reaches the road structure due to rain, due to capillary action, from side drains etc., it will weaken the sub-grade and will cause disintegration of the road.

The sub-grade is the main element of the road structure, which carries the load of the traffic through wearing course and distribute it to the sub-soil over wider area.

If due to any reason the water reaches the underneath into the sub-grade and the sub-soil, it will loose its bearing capacity, this will be the cause of the road failure.

Following are some of the main causes of failure of the treated surfaces :

(*a*) *Use of incorrect quantity of binder*. If the quantity of the binding material is in excess, it will cause smoothing and softness of road. On the other hand use of less quantity of binder material will make the surface brittle having the tendency to fretting or crumbling.

(*b*) *Incorrect Aggregate Proportion.* It will result in too dense or too open mix, which may fail.

(*c*) *Overheating.* During mixing if the bitumen is over-heated, the volatile oils will be lost and the binder material will loose cohesion and strength of binding and will become brittle.

(*d*) *Waving.* The waving in the road surface is caused due to excess of binder which act as lubricant or it may be due to the sub-grade being smooth and allowing the road crust to slide on it. Sometimes unsuitable metal, incapable of interlocking also becomes the cause of waving.

(*e*) *Sub-soil movement.* If the sub-soil under the road moves or slides due to any reason the road cracks and fails.

(*f*) *Bleeding or Flushing.* During hot weather, due to excessive heat the binding material comes out at the surface of the road from the surface course. It is generally caused due to (i) use of excess of binder, (ii) insufficient quantity of binding, materials, (iii) loss of cover aggregate. To stop bleeding, extra quantity of coarse sand and fine aggregate is spread on the surface.

(*g*) *Streaking or Striations.* This defect is caused due to (i) inexperience of the sprayer man, (ii) improper adjustment or careless operation of the binder distributor, and (iii) mechanical defects in the mixer.

(*h*) *Scabbing.* It is also called dislodgement of cover aggregate and is caused due to (i) use of too hard grade of binder, (ii) lack of adhesion of the aggregate with binder due to moisture, dust etc. on the surface of the aggregate, (iii) insufficient binder to hold the aggregate properly and firmly in position, (iv) loosing and whip-off of aggregate by fast-traffic, (v) raining immediately during or after the construction, (vi) undue delay in spraying the aggregate after spraying the binder material.

7.18 MAINTENANCE AND REPAIR OF BITUMINOUS ROAD

A bitumen-grouted macadam road, laid on sound sub-grade can carry upto 1000 or more medium light weight vehicles per lane per day. The life of the bituminous treated roads is about 12–15 years traffic upto 1200 tonnes per day. These roads should have routine maintenance of surface dressing etc. at every 2–3 years.

While doing the repairs and maintenance of the roads, the preventive maintenance should be carried out at the first indication of the failure, so that major defects may not develop.

The source of the trouble or reasons of the road failure should first of all be determined before starting the repair works. There is no use of doing surface repairs on the defective sub-grade or base. Before starting any repair it is better to first check the condition of the base. The patch repairing should be done properly. Undue strengthening of the weak spots should not be done, because it will create difference in traffic wear and impact, which will further damage the adjoining surface areas.

While doing repairing works every effort should be made to reduce the traffic interference. Warning signs and barriers should be placed at proper places. It is better to stack the maintenance materials along the road-side at suitable places, from where it can be taken.

All the pot holes and ruts should be repaired as early as possible, otherwise they will quickly cause further damage to the surface. The main cause of large disintegration area is untimely repair of the pot holes. The rain water is collected and get way through the pot holes into the sub-grade and causes failure of the road.

The old surface is loosened by pick axes to a depth of about 4 cm lor hard stone old aggregate and upto 6.3 to 7.5 cm for soft stones. The loosened metal is raked over to bring the metal from 20 mm gauge upwards to the surface. New metal of 40 mm gauge is added at the top of the old surface, to make a total thickness of atleast 11.5 cm, and is given the required shape. The aggregate so spread is consolidated and surface dressed.

On the roads which have thin crust of hard metal and it is proposed to lay the new metalling over it, without digging the old one, the old surface should be scored for the full width of the road with diagonal (criss-cross) lines about 40 mm deep and 30 cm apart. In case of soft metal top, the V-shaped trenches can be made at 60 cm intervals instead of scored lines for the purpose of keying between the old surface and the new surface.

As the main purpose of the scarifying is to get the proper bond between the old and the new layer, therefore the old surface should be disturbed only, what is necessary. In case the scarifying is done for the full depth of the metal, all the scarified metal should be removed and screened. Aggregate from 3 mm to 20 mm gauge are stacked separately. The mixture of the dust and fine grit are stacked properly. Now the extra required materials such as aggregate, earth etc. are added, and are spread along with the chipping materials on the sub-grade. The foundation so prepared is thoroughly watered. The old big size aggregate is now spread over it and combed through so that big stuff come on the top. Now new metal of 50 mm to 63 mm size are added according to the requirements and the whole surface is consolidated by road roller.

As the metal is properly graded and large percentage of old metal is used, the stabilization of the surface is done in less time. The wet soil underneath also help in the stabilization. While doing wet rolling, the water should be gradually added so that first the complete mechanical lacking of the metal is done, before the underneath earth cushion begin to be forced to the surface through the interstices. After doing the full consolidation, the fine material obtained from the screening is spread over the surface, thoroughly washed with copious quantity of water and allowed to stand for 24 hours. After its partial drying the surface is rolled with light roller. The road should be cured with water for 7 days.

The first coat of surface dressing should not be done before 14 days after the proper drying of the surface. Before applying top bitumenous coat, the surface should be cleaned out for atleast 10 mm depth.

7.19 GRADES OF ROAD TAR IS : 215–1961

Grade	*Type of work*
RT-1	Surface dressing at high altitude under very cold conditions
RT-2	Surface dressing or surface painting under normal conditions of temperature and weather.
RT-3	Surface painting and renewal coats, premixed carpeting.
RT-4	Premixed carpet and macadam concrete.

IS : Specifications for Road Tars
IS : 215–1961

Properties	*RT-1*	*RT-2*	*RT-3*	*RT-4*	*RT-5*
Viscosity					
Temperature	35°C	40°C	45°C	55°C	—
Range/sec	33–55	30–55	35–60	40–60	—
Equiscous					
Temperature EVT	32–36°C	37–41°C	43–41°C	53–57°C	63–67°C
Specific gravity at 27°C	1.16 to 1.26	1.16 to 1.26	1.18°to 1.28	1.18 to 1.28	1.18 to 1.28
Softening point	—	—	—	—	45–50°C
Water content	05%	0.5%	0.5%	0.5%	0.5%
Distillation					
Below 200	0.5%	0.5%	0.5%	0.5%	0.5%
200–270	4–12%	2–9%	1–6%	0–5%	0–4%
270–300	4–10%	4–8%	3–6%	2–7%	1–5%

7.20 BITUMINOUS MIX DESIGN

The bitumen mix design of a flexible pavement should aim at an economical blend with proper gradation of aggregates. The amount of bituminous binder should be adequate enough to cover the road metal or aggregates for proper binding and achieving the desired results. In any blend of all bituminous mix the following characteristics should be developed :

1. Durability. It is the resistance to the weathering action. Due to weathering the bitumen gets hardened and looses volatile and oxidation properties. Due to hardness it will not be able to sustain tensile stresses developed due to the wheel load and will develop cracks. Hence the bituminous mix should be such that it does not harden and crack.

2. Stability. It is the resistance to deformation under the wheel load. The unstable bitumen mix road surfaces will develop corrugations and soon the surface will become unserviceable. Stability is a function of friction and cohesion. Cohesion is caused on account of the mass viscosity of bitumen binder. Stability depends on the density of the mix also and density in turn depends on the voids in the compacted mass. If the voids are restricted the density will improve and consequently the strength and stability. The aggregates should be so graded that there are minimum number of voids but sufficient space for the binder expansion. If the aggregates are very densely packed up, the pavement under the traffic load will not have sufficient space for densification and the result will be bleeding of the bitumen, which will cause skidding.

3. Flexibility. The pavement under the wheel load will try to bend. If the pavement mix is such that it does not allow the pavement to bend, the result will be cracking of the surface and its ultimate failure. Flexibility is the property of the mix which allows the pavement to bend or it is the measure of the bending strength of the bituminous mix. So this is an important characteristic of the bituminous mix to be achieved to improve the serviceability of highways.

For proper and efficient designing a bituminous mix design the following points should be considered as follows :

1. Aggregates and its Grading. Aggregates must be strong, durable, hard and tough. Crushed stone aggregates are preferred over the round shape natural pebbles. But the selection of the type of aggregates will depend on the availability of the aggregates as the aggregates cannot be transported over a long distance. Grading to be selected should be such so as to form a compact mass. Large size aggregates are stronger and offer more resistance to crushing than the smaller aggregates but larger size aggregates will leave large voids. It is normally preferred to use large size aggregates of size 2.5 cm to 5.0 cm for the base course and for surface course 1.25 cm to 1.87 cm size aggregates are used. The grading of aggregates should be such so as to provide maximum compacted density. For a 40 mm thick bituminous concrete or premixed carpet road surfaces the IRC recommended grades are given in Table 7.2. IRC has recommended that the gradation should be such that minimum amount of binder is used.

It is usually found that the percentage of voids in an ungraded mass is almost 50% and in graded aggregates the voids arc reduced to 20% after rolling. So two or more available aggregates are proportioned so as to meet the specification limits. When two or three type of aggregates are proportioned, calculation of percentage of each one required to be blended is easy.

2. Specific Gravity. The specific gravity of the aggregates is generally determined and the specific gravity of the blended mass is determined by the following formula

$$\text{S.G.M.A.} = \frac{\text{weight of compacted aggregates}}{\text{volume of aggregates}}$$

where

S.G.M.A = Specific Gravity of Mixed Aggregates

W_1 = % weight of aggregate A

W_2 = % weight of aggregate B

W_3 = % weight of aggregate C

W_4 = % weight of aggregate D

W_4 = % weight of aggregate E

G_1 = Specific gravity of aggregate A

G_2 = Specific gravity of aggregate B

G_3 = Specific gravity of aggregate C

G_4 = Specific gravity of aggregate D

G_5 = Specific gravity of aggregate E

Compacted density of mixed aggregate (CDMA). It is she ratio of the weight of compacted mass of aggregates and the volume.

$$\text{CDMA} = \frac{\text{weight of compacted aggregates}}{\text{volume of aggregates}}$$

3. Voids in the Aggregates. Voids in the mixed aggregates (VMA) is calculated by the following formula

$$\text{VMA} = \frac{\text{SGMA–CDMA}}{\text{SGMA}} \; 100$$

Following methods are generally used for a mix design :

1. Marshal's method
2. Hubbard Field method

3. Heem method
4. Smith Tricaxial method
5. Taylor method

Marshal Method. The Marshal Method of mix design is generally used in our country. Before designing a bituminous mix a stability test is conducted. In the test known as Marshal Stability test, the resistance to plastic deformation of a cylinder mould of bituminous mix is measured by loading the specimen at the periphery at the rate 5 cm per minute. The stability of the mix is defined as a maximum load carried by a compacted specimen at a standard test temperature of 60°C. The flow is measured as the deformation units of 0.25 mm between no load and maximum load carried by the specimen. The test is aimed at obtaining optimum binder content for the aggregate with for a specific traffic intensity.

The Test. Marshal Stability Test is conducted in a cylindrical mould of 10.16 cm (4") diameter and 6.35 cm (21/2") height with a base plate and collar. The specimen is compressed radially at a constant rate of 5 cm per minute. The maximum load which the specimen can withstand is called the Marshal Stability value and the deformation in units of 0.25 mm is recorded as the Marshal flow value.

A pedestal compactor and hammer is used to compact a specimen by a weight 4.54 kg with a fall height of 45.7 cm. The specimen is extracted from

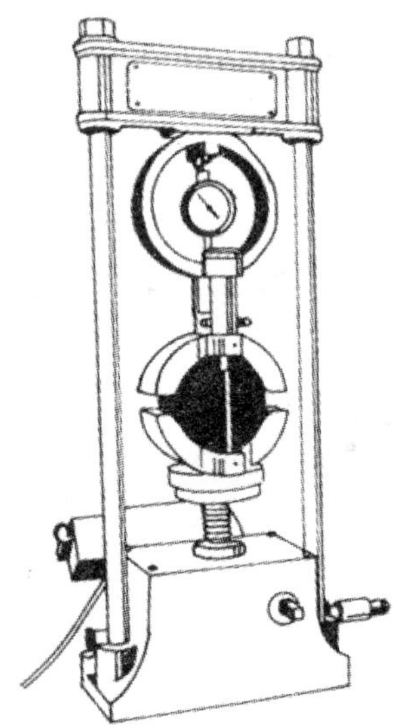

Figure 7.9 *Marshal stability test*

the mould by a sample extractor. The load is applied on the periphery of the mould in a loading machine of 5 tonne capacity at the rate of 5 cm per minute. The deformation is measured with the help of a dial gauge.

The aggregates and the filler material or binder is so proportioned and mixed that final mix has the gradation within the specified range. For

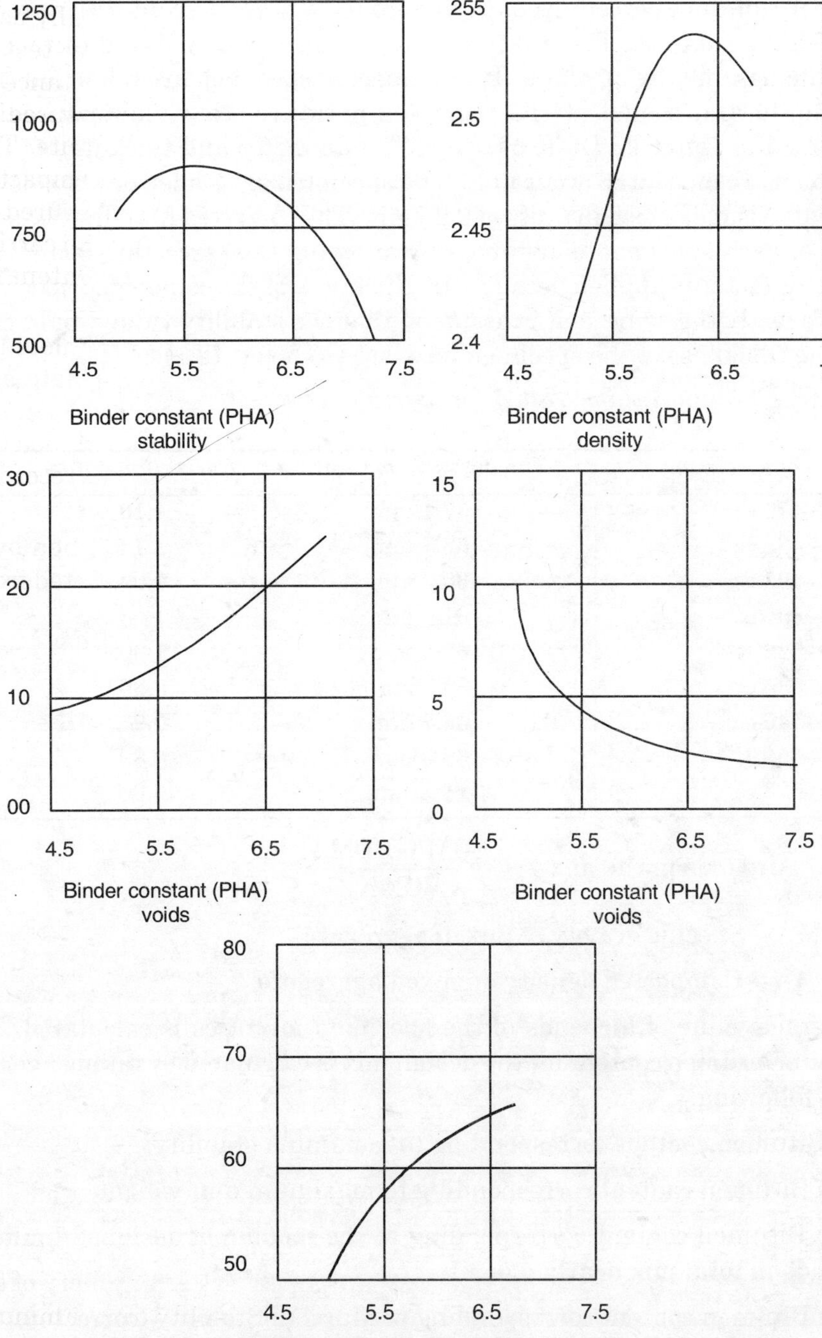

Figure 7.10 *Marshal test results*

conducting the Marshal test aggregates are heated at a temperature of 175–190°C and the bitumen is heated at a temperature of 121 to 140°C and then mixed in different percentages varying from 3.5 to 7.0 percent with different aggregate gradation. The mixture is placed in the mould and compacted with the hammer by 50 blows. The compacted specimen are taken out from the mould and cooled at the room temperature. Now the specimen is immersed in water ke Dt at 60 ± 1°C. The diameter and thickness of each specimen are measured accurately. The specimen are loaded one by one and flow value is measured in units of 0.25 mm. The corrected Marshal Stability value of each specimen is obtained by applying the correction factor. The corrector factor is applied because the thickness of the compacted specimen is not exactly the same so it is assumed that the stability value would have had the thickness of the specimen be exactly 6.35 cm (2½")

TABLE 7.3 *Correction Values for Marshal Test*

Value of specimen	*Thickness of specimen*	*Correction value*
457–470 cc	57.1 mm	1.19
471–482 cc	58.7 mm	1.14
483–495 cc	60.3 mm	1.09
496–508 cc	61.9 mm	1.04
509–522 cc	63.5 mm	1.00
523–535 cc	65.1 mm	0.96
536–546 cc	66.7 mm	0.93
547–559°C	68.3 mm	0.89
560–573 cc	69.9 mm	0.86

$$\text{Air voids in the mix} = \frac{\text{SGMA–CDMA}}{\text{SGMA}}$$

SGMA = Specific gravity of mixed aggregate

CDMA = Compacted density of mixed aggregate.

Average value of air voids of the specimen mixtures is calculated. The optimum bitumen content for the design mix is calculated by taking average of the following.

(i) Bitumen content corresponding to maximum stability.

(ii) Bitumen content corresponding to maximum unit weight.

(iii) Bitumen content corresponding to the median of designed limits of air voids in total mix nearly 4%.

(vi) Bitumen content corresponding to Marshal Stability (corrected),

(vi) Binder content corresponding to Marshal flow in units of 0.25 mm.

Inferences of the Marshal *Test.* Like many other tests the Marshal test has many pros and cons. A measure of the adjustment is made to suit a particular environment as the minimum stability requirement for a highway can be as low as 120 kg and as high as 950 kg. At the present Marshal design method is widely used in many countries including India. The following points are worth considering and are of interest:

1. The unit weight of the mixture increases with increasing bitumen content until a maximum value is obtained after which the unit weight decreases even with more bitumen content, this is because of the fact that at the first stage the binder acts as a lubricant and helps the aggregate particles to slide over each other. Once an optimum amount of binder is reached it acts only to disperse the particles from their space and so the wet unit weight decreases. If a dense mixture is to be obtained, the amount of binder to be used not be more than the optimum for the comparative effort applied.

2. The stability value of the mixture also increases with increasing binder content until a maximum value is obtained after which the stability decreases. Generally the optimum binder content for stability is close to the optimum binder content for unit weight. The optimum content for stability can be explained by noting that the Marshal test is actually a type of unconfincd compressive strength test in which some degree of lateral support is given to the mould/specimen, thus the maximum stability value occurs at the binder content at which the combination of the internal friction, a component of stability, providing the interlocking of aggregates and the cohesive component provided by the thin viscous layer or film coating the particles, is maximum under the test conditions. The test is a time measure of the cohesive component of stability. It is a well-known fact that each bituminous aggregate mixture placed in a pavement has an optimum binder content stability under the prevailing condition of environment.

3. The flow value increases as the binder content increases. In addition the rate of deformation change is slow at low binder content but increases rapidly as high binder contents are reached. For example, surfaces with low flow and high stability will not deform easily but are likely to be brittle but to those with low stability and high flow, deform easily under the traffic load.

4. The percentage voids in the total mix decreases with the increasing quantity of binder content until a value is reached at which it begins to bend off. The percentage voids in total mix is critical as regards durability, since greater the air voids content the more easily air and moisture can attack the binder and the binder aggregate bond. When the air voids contents are

too low the surface is likely to bleed or flush. Since the error in the air void determination may be as much as one percent the stability of the mix may be poor due to poor quality of aggregates, air void contents may be taken as a reason.

5. The percentage of aggregate voids filled wilh binders increases with the increasing binder content until a maximum value is reached. Again the rate of increase is fastest at low binder content and levels off at high contents. It is important to note that unless the percentage voids in the mineral aggregate (V.M.A.) is large, the paving mixture will be deficient in binder, they become brittle and crack. Thus the voids in the numerical aggregates may be large enough to ensure that there is sufficient space to accommodate air and binder. Hence the voids in the mineral aggregates framework should be filled as nearly as possible.

7.21 TESTS FOR ASPHALTIC MIX

The following tests are conducted on an asphaltic mix.

1. **Density or Unit Weight.** Density or unit weight of the mix is determined to compute percentage air voids and voids in the mineral aggregates optimum bitumen content and compaction.

2. **Percentage Voids in the Mix.** The determination of voids in the compacted bituminous mix form a part of the design method. Voids are essential in the compacted mix also to provide room for the expansion of binder during weather changes.

3. **Stability.** Stability may be defined as resistance to deformation due to shearing stresses and implies resistance to *showing* and rutting by traffic.

4. **Extraction.** It is defined as a procedure used for scrapping or separating the bitumen from the mineral aggregates in a bitumen mix. This will determine the percentage of bitumen present in the mix. The bitumen free aggregates are tested for gradation.

5. **Swell.** Bitumen mixes containing fines are sometimes measured for swell as a basis for judging the possible detrimental effects of water on the pavement. A sample of the mix is compacted in a metal cylinder 10 cm in diameter and cooled to room temperature. The specimen and the mould are placed in a water pan. The specimen is then slowly loaded and the compression is measured by a dial gauge.

6. **Stripping.** The failure of the bond already formed between the bitumen and the aggregate surfaces brought about but the displacement of the bitumen by water is commonly known as Stripping. There are many tests to determine or evaluate the stripping action of water.

In most of these tests, a sample of aggregate is coated with a known quantity of bitumen and immersing it in water under controlled conditions. The amount of stripping of the binder from the aggregate after a known period of time is measured. Methods differ in the type of specimen used in the test, the condition under which the sample is immersed in water and the method by which the degree of stripping is measured or assessed. One of the most commonly adopted method is the Static Immersion Test. In this method single size chippings are coated with a constant quantity of bitumen under a controlled temperature for upto 48 hours. The amount of stripped area of the stone surface as a percentage is estimated visually.

For designing a bitumen mix, as stated earlier various tests are performed

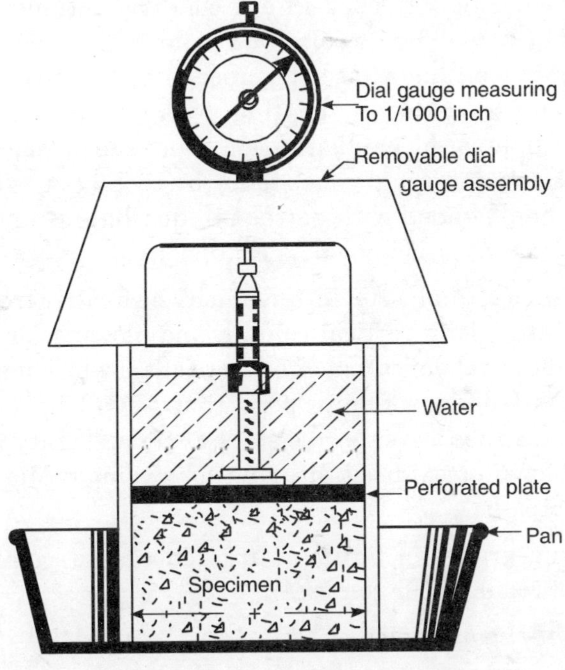

Figure 7.10 *Swell test*

to relate stability, void percentage, stripping etc. for a optimum bitumen content and stability of the mix.

7.22 TYPES OF BITUMEN

As stated earlier, bitumen is obtained by the distillation of crude petroleum. A selected grade of bitumen is heated in a distillation plant, operating at the atmospheric pressure and the volatile components such as gasoline, kerosene and light gas oils are removed by distillation and fractional condensation. The residue termed as *topped* crude, is then passed to a further distillation unit operating in order to remove the less volatile products such as heavy gas oil diesel and lubricating oil distillates. The highly viscous material obtained after processing is called Bitumen and is of two types :

1. Straight-run bitumen
2. Blown bitumen

Bitumen produced by distillation process is termed as straight-run bitumen and is of variety of grades and ranges from a very soft to vary hard consistency depending on the temperature and the rate of flow during distillation. The softer grades are designated by the penetration limits such as Mexphalte 80/100 which is commonly used in our country. Bitumen of a very hard consistency is identified by its softening point limits such as Mexphalte H 80/90. Straight-run bitumen has to be processed again to reduce its viscosity before it can be used either by heating, by addition of a flux (which is a petroleum distillate) or by emulsification. Once the bitumen has been used it reverts back to its original consistency by cooling or evaporation of the diluent. (Bitumen blended with petroleum distillate is known as cut-back and emulsified by water is known as Emulsion).

A soft grade of a straight-run bitumen may be further treated by running it, while hot, into a long vertical column and blowing air through it. The bitumen undergoes a chemical change as a result of which it attains a rubbing consistency, has a higher softening point than a straight run bitumen of the same hardness and has a greater resistance to flow. This type of the bitumen is known as blown or oxidised bitumen for example Mexphalte R 85/25, Mexphalte R 115/15.

The following grades of bitumen are generally recommended for the construction of roads in our country :

Straight-run bitumen

Mexphalte 30/40

Mexphalte 60/70

Mexphalte 80/100

Spramcx 180/200

Cutbacks

Shelmac RC3 – for cold application

Shelmac BS – for hot application

Shelpra BS – for hot application

7.23 CONTROL OF BITUMEN APPLICATION TEMPERATURE

The effective use of bitumen in any operation depends on its application at an appropriate viscosity, which differs for different operations such as spraying, mixing, pumping, compaction etc. Rapid and direct evaluation of binder viscosity in the field is however difficult. As such field control of bitumen application is related to its temperature. This is easily measurable directly. Following are the recommended temperatures for different grades of bitumen :

Grade of bitumen	*For spraying*	*For mixing*
Mexphalte 30/40	—	170–185°C
Mexphalte 60/70	165 – 180°C	155–165°C
Mexphalte 80/100	160 – 175 °C	150–165°C
Spranex 180/200	150 – 165°C	135–150°C
Shelmac RC 3	40 – 65°C	40–65°C
Shelmac BS	155 – 170°C	150–165°C
Shelspro BS	155 – 170°C	150–165°C

7.24 BITUMINOUS ROAD EQUIPMENTS

The following equipments are generally used for the construction of bituminous rocks :

1. Bitumen distributor
2. Aggregate spreader
3. Bitumen finisher
4. Hot mix plant

1. Bitumen Distributor. The bitumen distributor is a primary or key equipment for the mechanised bituminous road construction. It can be self propelled like a pneumatic lorry or can be trailor towed with a tractor. It consists of an insulated tank with a heating system, usually oil fired burners, with the direct heat passing from the flues to the tank. It is also equipped with a compressor and a rotary discharge pump. At the rear end of the tank, a spray bar unit is attached on which nozzle jets are fitted at specified intervals, through which the bitumen is sprayed under pressure. An arrangement is generally made for the full circulation of bitumen through the spray bar.

It is used for spraying bitumen on the surface evenly and under pressure for surface dressing, grouting, etc. Through this machine it is possible to apply the bitumen accurately in a measured quantity and specific rate of application for the entire road length regardless of any change in gradient or direction, can be maintained. The speed of the distributor and the speed of the bitumen pump can be controlled and determined. This controls and decides the quantity of bitumen to be spread or used. Normally the capacity

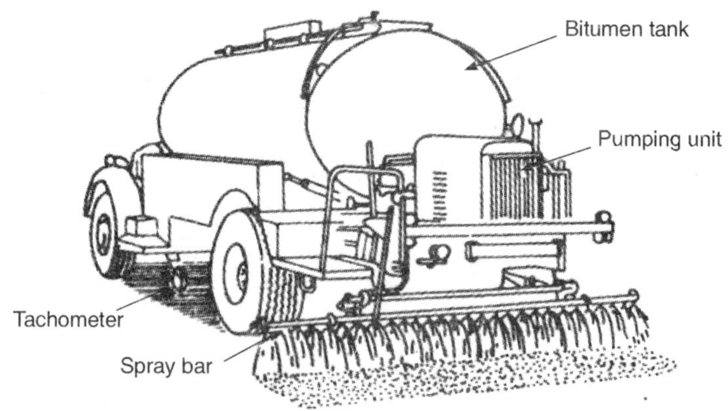

Figure 7.11 *Bitumen distributor*

of a bitumen distributor is 4.5 tonnes. For uniform application of bitumen the following precautions should be taken.

(i) The bitumen should be of the prespecified or correct viscosity.

(ii) The nozzles should be set at the correct angle with the spray bar and should be checked regularly for uniform and correct delivery.

(iii) The height of the spray bar should be so adjusted so as to give double or triple overlap of the spray as shown inFig. 7.12.

(iv) Correct pressure should be maintained throughout the length of the bar.

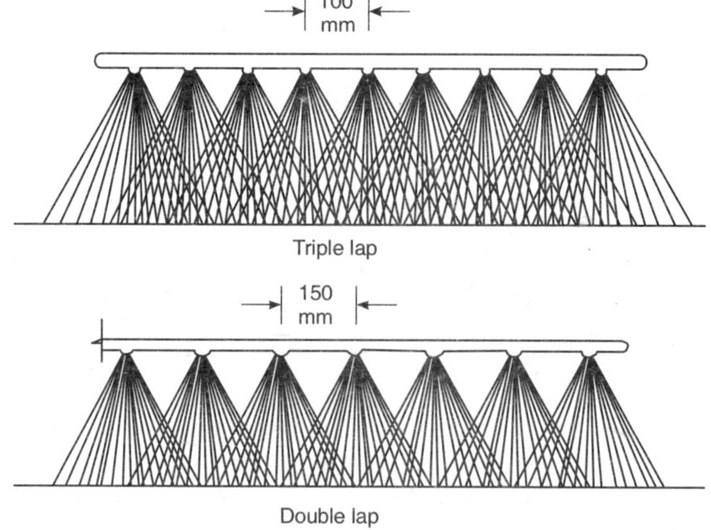

Figure 7.12 *Spraying of bitumen*

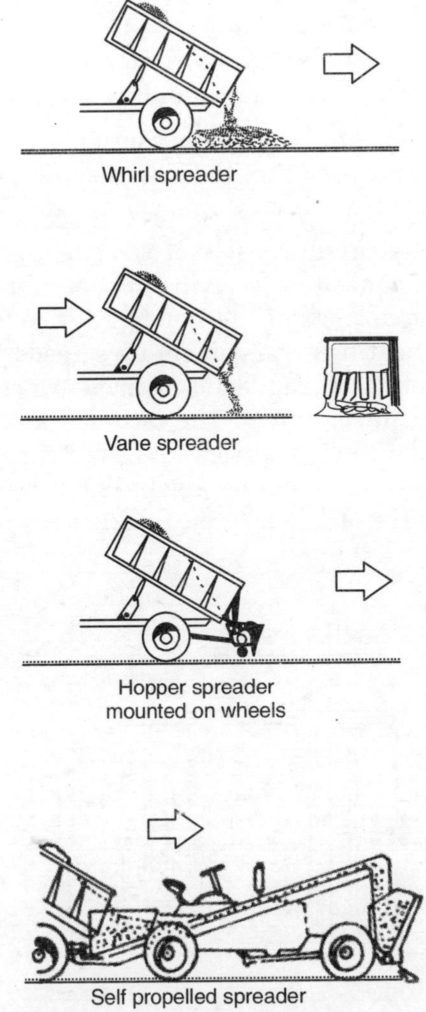

Figure 7.13 *Spreader*

(v) Before spraying, the nozzles and the spray bar should be heated.

(vi) The distributor should be towed or driven at a constant speed.

2. Aggregate Spreader. The aggregates are generally spread over the prepared base of the road pavement by hand in our country. If the aggregates are not evenly spread, it will pose quite a difficulty for the bitumen distributor. For heavy construction work it is advisable to use mechanised aggregate spreader. The aggregate spreaders are of many types. The main component of the aggregate spreader is the dumper self propelled or towed. As the spreader moves forward the dumper spreads the aggregate on the surface and is being evenly spread by a screeder attached at the rear end. The different type of spreaders are shown in Fig. 7.13.

3. **Bitumen Finisher.** The functions of a bitumen finisher is to produce a uniform level riding surface, correct the irregularities, if any, of the base course and to lay the bitumen mix in a uniform thickness throughout.

The bitumen finisher essentially consists of a tractor unit and the screeding unit. The tractor unit includes the receiving hopper and spreading screws while the screeding unit consists of the tamper, device for controlling the thickness of carpet and heating arrangement for screeds. The unit is towed by a tractor with the help of a long levelling arms which are pivoted to a point near the centre of the crawlers. The bitumen mix from the mixing plant is received from the tipping lorries into the hopper from which the material is carried to the screed unit by means of bar conveyor. A finished premixed carpctted surface of the desired thickness is obtained with the help of this machine.

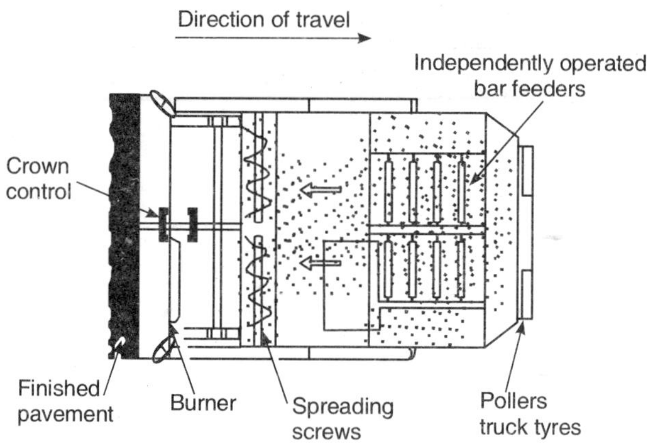

Figure 7.14 *Bitumen finisher (Plan)*

4. **Bitumen Mixing Plant (Hot Mix).** For pressurised carpetted roads and seat coats, the aggregates and the bitumen are pre-heated and mixed in the desired ratio. The hot mixplant also helps in drying the aggregates, screening the aggregates into the required size and measuring the required quantity of aggregates, the filler material and the bitumen too. The process consists of feeding the cold aggregates into the stock piles, separated into sizes by bulk heads and conveyed to the driers. The driers are rotating drums

where the aggregates are heated. The flow of aggregates can be maintained by inclining the drum away from the feeding hopper. A series of lifting flights of channels pick up the aggregates and allow them to slide through the burner flames. A pyrometre and a dust collector is attached at the discharge end. Rotating type screens are fitted directly over the hot aggregate bins. There are 3–4 aggregate bins each fitted with an over flow chute and a thermometer.

Bitumen storage tanks are insulated and fitted with heating arrangement. Aggregates are proportioned either by volume or by weight as may be convenient and desirable. The heated aggregates and the bitumen are mixed in mixing boxes with two horizontal shafts rotating in opposite direction. To each shaft are attached removable mixing blades or peddles. The bitumen aggregate mix is discharged from the mixing drum through an opening provided at the bottom, into the dumpers or lorries. The mixing plants can be either of continuous type or batch type. In the batch mixing plant the cold aggregate is fed accurately and conveyed to the driers through conveyor belts. Those mixing plants of the either variety is installed at central place and the mix is carried in lorries or dumpers to the site. The plant can serve a distance of 25 km both side.

7.25 SAFETY MEASURES WHILE WORKING WITH BITUMEN MIX EQUIPMENTS

The following safety measures should be adopted while working on a hot mix plant or allied equipments :

1. Faulty equipments and equipment with faulty brakes or brake system should never be used.

2. Belts, chains, driving shafts and other revolving parts should be enclosed in housing, gaurd or screen fencing.

3. No repair or replacement of parts, inspection or changing oil should be done when the engine is in motion.

4. The operator of the machine should not leave the machine in running condition.

5. No body should be allowed to stand under the frame of the loader, conveyors, cranes and such other machinery.

6. Before firing burner, it should be ensured that there is no leakage of fuel in the furnace as this may sometime cause explosion.

7. Do not fill bitumen tank completely. At the same time do not allow the bitumen level to fall so low to expose the flue tubes. If this precaution is not ensured flue tubes are likely to burst or bulge.

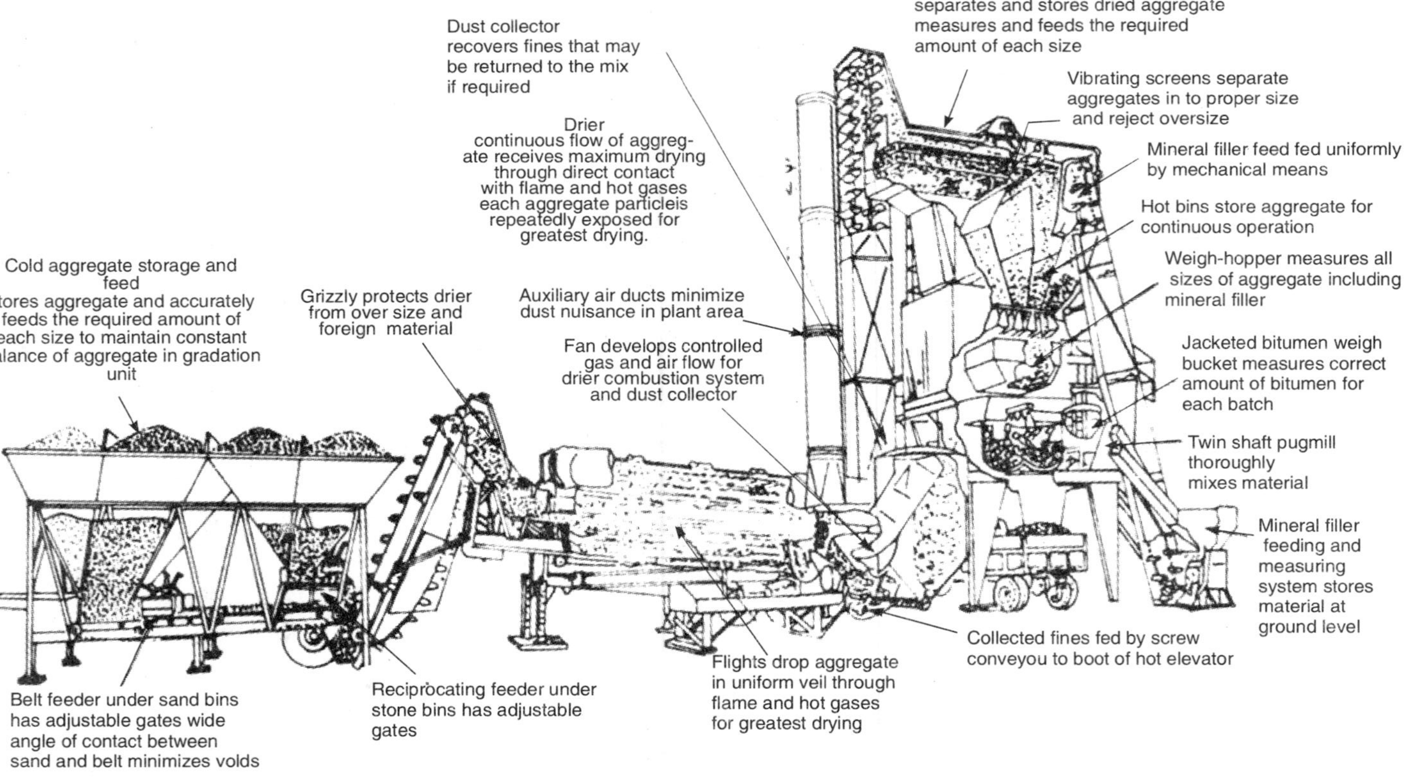

Figure 7.15 *Hot mixing plant*

8. There should be absolutely no leakage in the dryer unit as the flame is burning at a very high temperature as this may cause severe burn injuries to the operator.

9. Over size stone aggregates may be avoided in the hopper as they may cause damage to the moving parts.

10. The hot elevator column should be fully covered and the panels removed for inspection or repair should be immediately replaced.

11. Leakage or overheating of the bitumen should be avoided as it may result in burn injuries to the operator or even the bitumen may catch fire.

12. The bitumen tank should be properly covered and should only be opened at the time of filling the tank.

7.26 METHODS OF REDUCTION OF CONSUMPTION OF BITUMEN

Bitumen which is a bye-product of petroleum distillation is the primary material for the construction of black top road which are very common and popular in our country. With the current oil crisis facing the world it is likely that the shortage of bitumen may pose a difficult situation to maintain the roads in our country. Owing to improved refining process for crude tar as such as well as its usage increase in lieu of furnace oil, road tar is not available for road construction and the entire burden of road construction and maintenance has to be borne by asphaltic bitumen only.

The total consumption of bitumen in the country in the year 1975–76 was about 7,00,000 tonnes and in the year 1985–86 it has gone upto 72,00,000 tonnes. By the end of the eighth plan the rate of consumption will be difficult to be sustained. Since more than 95% of the total bitumen is consumed in the road making industry because of higher rate of bitumen specifications for the construction of National and State Highways, widening and strengthening of existing roads.

The price of bitumen in mid fiftees was Rs. 600 per tonne and in 1988 the price of bitumen has shot upto Rs. 3200 per tonne (approximately). Looking to the increased construction and so increased consumption of bitumen and its heavy cost it is an imperative need to find out ways and means for the reduction in the consumption of bitumen. There are two approaches for achieving the target. The first one is the direct approaching of attacking the specifications and the other approach an indirect approach of finding alternatives to asphaltic bitumen.

1. Direct Approach. One of the ways of reducing the consumption of bitumen could be in altering the specifications and choosing suitable construction methods and construction specifications so as to suit the requirements and minimise consumption. Although bitumenous courses can be quite satisfactorily used for the wearing course as well as the base courses, in case of shortage of bitumen and its high cost it should be used only on the surface layer. Now we will concentrate our discussion in Table 7.4 :

TABLE 7.4

Type of pavement surface	*Specification as laid down by IRC*	*Weight of bitumen required per 1 km for 3.5 m wide road*
1. Single coat surface dressing IRC : 17–1963.	170 to 19.5 kg per 10 m^2 of surface.	6.5 tonnes
2. Two coat surface dressing IRC : 23–1966 3. Surface dressing with precoated chips IRC : 48–1972	170–19 5 kg for the first coat and 9.8–12.2 kg per 1 cm^2 for the second coat.	111 tonnes
(i) Single coat	170 to 190 kg + 1% for coating per 10 m^2	7.5 tonnes
(ii) Two coat	17–19 kg for 1st coat and 10-12 kg/10 m^2 for second + 1% for a coating per 10 m^2	12.2 tonnes
4. Premixed carpet with seal coat IRC : 14–1970	For tack coat 7.3–9.8 kg per 10 m^2 for mixing 146 kg, and 68 kg/10 m^2 for seal coat	10.9 tonnes
5. 25 mm asphaltic concrete IRC : 29–1968	IV @ 5.70% bitumen by weight of mix + tack coat @ 7 kg per 10 m^2	14.2 tonnes
6. 40 mm asphaltic concrete IRC : 29–1968	IV b @ 5.7% by wt. of mix + tack coat @ 7 kg per 10 m^2	21.2 tonnes
7. 50 mm asphaltic concrete IRC : 29–1968	IC-b @ 5.7% bitumen by weight + tack coat @ 7 kg per 10 m^2	25.9 tonnes
(a) 25 semi dense asphaltic concrete	III-a @ 4.8% bitumen by wt. of mix + tack coat 7 kg/ 10 m^2	12.5 tonnes
(b) 40 mm semi dense asphaltic concrete	III-b @ 4.3% bitumen by weight of mix + 7 kg per 10 m^2 for tack coat	17.5 tonnes

Contd.

Type of pavement surface	*Specification as laid down by IRC*	*Weight of bitumen required per 1 km for 3.5 m wide road*
(c) 50 mm semi dense asphaltic concrete	III-c @ 4.3% bitumen by weight of mix + 12 kg per 10 m^2 for tack coat	20.1 tonnes
8. 50 mm full grouted macadam IRC : 20–1967	50 kg for grouting and 12 kg per 10 m^2 for seal coat	21.7 tonnes
9. 50 mm bitumenous macadam IRC : 27–1972	Mixing @ 3.8% by weight of mix + 10 kg per 10 m^2 for tack coat	17.7 tonnes
10. 75 mm built up spray grout IRC : 47–1972	Tack coat @ 10 kg and 15 kg/10 m^2 for each in two layers	14.0 tonnes
11. 75 mm full grout macadam IRC : 20–1966	Grouting @ 68 kg per 10 m^2 + seal coat 12 kg per 10 m^2	24.5 tonnes
12. 75 mm bituminous macadam IRC : 27–1967	Mixing @ 3.4% by weight of mix + 7 kg per 10 m^2 for tack coat	22.9 tonnes

From Table 7.4 the surface dressing is one which demands minimum quantity of bitumen and should be used for light trafficked roads. For high rainfall area where chances of rain water to wet the uncoated aggregates and result in loss of adhesion is more, a surface paint with precoated chips is most suited and economical too. For busy state or national highways a 20 mm premixed carpet with seal coat may be adopted. Asphaltic concrete which is no doubt a superior type of pavement with longer durability demands comparatively large quantity of bitumen and hence should be sparingly used for heavy traffic only.

Comparing quantitatively the two coat surface dressing with seal coat and premixed carpet with seal coat consumes 12% less bitumen than surface dressing. Hence premixed carpet should be preferred over two coat surface dressing with pre-coated chips. For light trafficked roads, only single coat surface dressing should be used. The use of built-up spray grout involves minimal quantity of bitumen. The quantity of bitumen required lor 50 mm bituminous macadam is nearly 17.7 tonne per kilometre for 3.5 m width of lane and for a structurally equivalent layer of 75 mm thick built-up spray grout layer is only 14.9 tonne. The saving in bitumen is nearly 26 percent.

In some countries crushed stone base with or without treatment with lime and cement and surface dressing carpet by mechanical sprayer and gritter provided a very good surface comparable with asphaltic concrete.

Thus the cost and consumption of bitumen can be reduced. Asphaltic concrete surface is no doubt a very superior surface but is very costly and its use should be restricted for heavy trafficked trunk routes only. For low rainfall area semi-dense carpet should be preferred. This will save nearly 26% of bitumen.

Apart from above saving measures, processes and handling of bitumen should also be done carefully to minimise losses. For heavy rainfall areas, anti-stripping agents should be used specially if leaner specification has been used. This will increase the durability of the surface.

(ii) *Indirect Method.* If bitumen emulsion which have low viscosity is used, will result in the saving of bitumen as the emulsion fibre is thinner and more uniform thickness can be obtained. In emulsions the quantity of solid bitumen varies from 50–60%. But the cost of emulsion is much more than the cost of bitumen. In western countries emulsion are preferred in view of environmental conditions and rainy season or wet climate unsuitable for hot mix. The Indian Road Congress has drafted tentative specifications for single and double coat surface dressing and premixed carpet using bitumen emulsion. The amount of bitumen and the amount of emulsion being the same, there is a saving of nearly 40% of bitumen.

The Central Road Research Institute has conducted intensive research on the road tar which is available in our country nearly 50,000 tonnes per year. New specifications based on the concept of pitch/anthracene oil raio were prepared to make a road tar comparable in performance to asphaltic bitumen. If refined tar as it is called, is used, it will save nearly 5–6% of asphaltic bitumen.

Low temperature tar is a bye-product obtained during carborization of inferior quality of coal to produce domestic smokeless fuel. Two pilot projects one at Regional Research Laboratory, Hyderabad and the other at Central Fuel Research Institute, Dhanbad were set up for the production of low temperature tar but they could not achieve the estimated target. Road Research Institute undertook a project to make low temperature tar suitable for road construction compared to a high temperature tar. The low temperature tar is non-sticky and high toxic and have given satisfactory performance on test-tracks. The long run success of this material will make our country self-sufficient in bitumen.

Tar-bitumen bends are being tried for durability and cost reduction. In many countries of the world like France, UK, West Germany tar/bitumen or pilch/bitumen blends have been satisfactorily used. Addition of tar to asphaltic bitumen not only improves the adhesive characteristics but also helps in producing more skid resistance surface. The conclusions of the detailed study and researches of Central Road Research Institute, are that by reducing the pitch/anthracene oil ratio of tar to 2, tar/bitumen blends with 10/20 bitumen can be produced in any proportion. Similarly use of rubberized pitch bitumen blends can be encouraged to save asphaltic bitumen by nearly 20–30%. With the above discussion the following conclusions are drawn :

1. In all type of road construction bitumen should be used judciously. The asphaltic concrete should be laid only on heavy duty roads and airfields. Secondary and light trafficked roads should be surfaced by premixed carpet.

2. Manufacture of bitumen emulsion on large scale should be promoted to reduce its cost.

3. Suitable pitch/bitumen blends should be developed and manufactured on large scale.

4. Use of low temperature tar should be encouraged.

5. Synthetic resins should be popularized.

PROBLEMS

7.1. Describe the process of laying two coats of surface dressing with stone chips and bitumen on a new surface of stone metal. Calculate the quantities of materials with specific mention of size and grade required for two coats of surface dressing for 100 sq. m of a road surface.

[Hint : Stone chipping of 12 to 20 mm size 1.8 to 2 m^3 will be required nearly for 1st coat and 9 m size chippings will be required 1.5 m^3 for the second coat. For 1st coat 200 kg of binder will be needed while for the second coat 150 kg of binder.]

7.2. Describe fully the process of laying the single coat of paint with bitumen (by hot process) on a water bound macadam road as a surface dressing. Give quantities of grit and paint (i.e., binder) you would require for one km of road.

7.3. What are emulsions and cut-backs ? Compare the merits and demerits of emulsions and bitumen.

7.4. Under what circumstances would you provide a surface painted road ? What are the advantages of a surface painted road over the waterbound macadam road ?

7.5. Write down the detailed specification for the first coat of surface painting. Give quantities of materials required for 4 m wide carriage way 1 km long.

7.6. Describe in brief the construction of premixed carpet road. Give detailed specification for a 4 cm thick premixed carpet.

7.7. Describe in detail, how you would lay a bitumen semi-grouted road surface 4 m wide over an old macadam full grouted road.

7.8. Describe in detail the construction of a bituminous macadam full grouted road.

7.9. Explain the various works involved in the maintenance of bituminous roads.

7.10. Write a short note on the possible defects in the asphaltic concrete mix.

7.11. What do you understand by the causes of disintegration of bituminous roads ? Describe them.

7.12. Write short note on the aggregate gradation for bituminous roads.

7.13. Calculate the rate of spread of tar in kg/10 m^2 and aggregates in cubic metres per 10 m^2 for surface dressing course 12 mm thick. The initial voids may be taken to be 50% and voids after final compaction may be assumed to 20%. The whip-off due to traffic is 10%. Binder should form 60% of the total voids. Sp gravity of tar is 1.2.

7.14. A specimen of asphaltic concrete has a height of 6.20 cm and a diametre of 10.16 cm. The weight of the compacted specimen in air is 1174.7 gm and in water the weight is 668.4 gm. When coated with paraffin its weight in air is 1220.9 gm and in water 664.4 gm. The specific gravity of paraffin is 0.90.

Material	*Specific gravity*	*Mix. composition % by weight*	*Aggregate composition % by weight*
1. Asphaltic cement	1.02	6.0	—
2. Coarse aggregate	2.58	52.0	55.3
3. Fine aggregate	2.72	34.6	36.8
4. Mineral Filler	2.70	7.4	7.9

Calculate

(i) Build density

(ii) Average specific gravity of aggregates

(iii) Percentage voids in compacted man

(iv) Percentage of volume occupied by asphalt

(v) Percentage of voids filled with asphalt of the aggregates

7.15. Compare the properties of tar, asphalt and emulsion as a binder for road construction.

7.16. List the different type of cutbacks. When are cutbakes used ? Discuss in brief the various tests conducted on bitumen.

7.17. Discuss briefly Marshal's Method of bitumen mix design. State its salient features.

7.18. State briefly the desirable properties of bitumen mix and what tests should be conducted on it.

7.19. What are the different bitumenous materials generally used for road construction ? Under what circumstances each of those materials are used?

7.20. Discuss the desirable properties of bitumen to look for highway construction of superior quality.

BIBLIOGRAPHY

1. The AASHO–Road test-Highway Research Board No. 22–1953.
2. Croney D and J. A. Loe, Flexible Pvements Institution of Civil Engineers 1965–30, 225–270.
3. Highway A.C. and N.W. Lister – Asphalt pavement – Ann Arbor, Michigan 1962.
4. Mallock H.R.A. Construction and wear of Road Institution of Civil Engineers 1909, 178, 111–181.
5. Steel D.J., Discussion on Classification of Highway sub-grade materials – Highway Research Board 1945, 25, 388–392.
6. Michigan State Highway Department. Field Manual of Soil Engineering, Lansing Machigan 1952.
7. Materials and Research Department – Material Manual of Testing and Control Procedures, Sacrometo, California Division of Highways 1960.
8. Lambe T.N. and Research Department – Material Manual of Testing and Control Procedures, Sacrometo, California Division of Highways 1960.
9. Tressider J.O.A., Review of Exiting Methods of Road Construction –Road Research Technical Paper No. 40–1958.
10. Road Research Laboratory-Protection of Sub-grades and granular bases by surface dressing – Road Nate No. 17–1953.
11. C.P. 2006–1960, Traffic Paving Structures.
12. I.R.C–20–1966, I.R.C–27–1972, I.R.C–47–1972, I.R.C–20–1967, I.R.C–27–1967, I.R.C–8–1980, I.R.C–10–1987.
13. I.R.C. Highway Research Board – 15–1981.
14. I.R.C. Highway Research Board –23–1984.
15. I.R.C. Highway Research Board–38–1989.
16. Specification for Roads and Bridges – M.O.S.T. (Roads wing), I.R.C. 1978.
17. Bituminous Materials in Road Construction Road Research Laboratory, London 1962.

18. Tentative Specifications for Priming of base course with bitumenous primers I.R.C. – 16–1965.
19. Road Tar, Indian Standard Institution. I.S. : 215–1961.
20. Specifications for surfacing with Road tar Shalimar Tar Products –1935.
21. Specifications for Paving Bitumen I.S.: 73–1961, Indian Standard.
22. Guidelines for the design of flexible pavements I.R.C. – 37–1970.
23. Mix Design Methods for hot mix asphalt paving. Manual series No. 2 – The Asphalt Institute – 1956.

8

Cement Concrete Roads

In this Chapter you will study,

• Advantages of Concrete Roads • Kinds of Concrete Roads • Construction of Premixed Concrete Roads • Specification for Concrete Materials • Design of Conrete Mix • Constructional Details • Joints in Concrete Road • Requirements of Concrete Pavement Joints • Cement Bound Macadam Roads • Crete-Ways • Tools and Plants for the Construction of Concrete Pavements • Fillers and Seal • Length and Width of Pavement Slab • Design of Dowel Bar • Prevention of Mud Pumping • R.C.C. Pavements • Prestressed Concrete Pavements • Admixtures

GENERAL

Cement concrete roads or concrete roads as they are generally called, had become very popular in this country and being used on a large scale. They are superior to many other roads especially bituminous roads. Their advantages are manifold.

8.1 ADVANTAGES AND DISADVANTAGES OF CONCRETE ROADS

Advantages

1. The concrete roads, if constructed properly, last longer and the maintenance charges are very low.
2. They provide a safe riding surface under all conditions.
3. They are dustless.
4. They can be laid on any sub-grade.
5. They do not develop corrugations.
6. They can be easily reinforced when it is so desired to resist high stresses.
7. The tractive resistance of such roads is very low due to non-slipperiness.

Disadvantages

1. The initial cost of construction is very high.
2. The construction of such roads require skilled labour and skilled supervision.
3. They cause noise under iron wheeled traffic.
4. They develop cracks due to variation of temperature in the atmosphere.
5. They cause glare due to reflected sunlight.
6. Resiliency is less than bituminous roads.

8.2 KINDS OF CONCRETE ROADS

The concrete roads may be divided into :

1. Premixed concrete roads.
2. Cement bound macadam roads.

The premixed concrete roads are constructed by mixing the various ingredients of concrete viz., cement, sand, coarse aggregates (stone metals) and water in proper proportions in a concrete mixer or central mixing plant and laid on the prepared bed of sub-grade or base as the case may be. It is consolidated and finished to a slab of desired thickness. In this case rolling is not required.

In the case of cement bound macadam roads, the road metal is laid on the prepared bed of subgrade or base and rolled to a desired thickness. Over the rolled surface, thin cement mortar or cement slurry is spread and the surface is finished.

8.3 CONSTRUCTION OF PREMIXED CONCRETE ROADS

The construction of these roads is carried in two different methods :

(i) Alternate Bay Method, and

(ii) Continuous Construction Method.

(i) **Alternate Bay Method.** In this method of construction, concrete is placed in bays of 4 to 5 m length and width equal to the width of the carriage

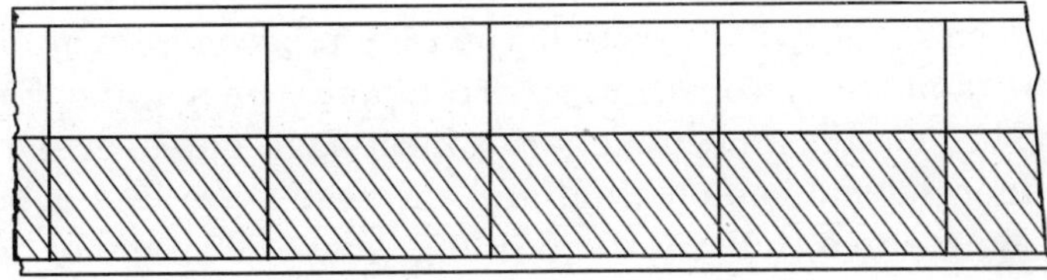

(a) *Successive bay method*

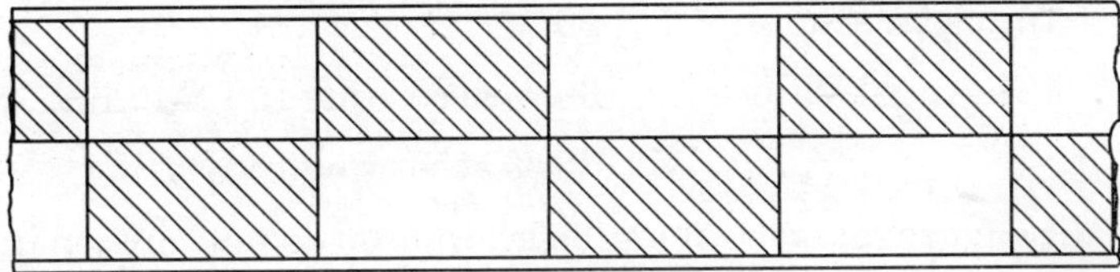

(b) *Alternate bay mothod*

Figure 8.1

way. The whole length of road is divided into rectangles or bays and concrete is filled in alternate bays and finished. The intermediate bays are filled after at least a week of the construction of adjoining bays. This method is only suitable when the carriage way is less than 4.5 m in width. This method is generally unsuitable for hot countries.

(ii) **Continuous Construction Method.** When the width of traffic way is more than 4.5 m the slab is laid in longitudinal strips, having a width of 2.5 m to 3.5 m. (This is also called strip method.) The slab along each strip may be laid in successive bays or in alternate bays as is shown in Fig. 8.1 (a) and (b). Usually the construction in alternate bays is preferred.

8.4 SPECIFICATION FOR CONCRETE MATERIALS

The ingredients of concrete are cement, aggregate and water, ordinary Portland cement is the binding material of aggregates. Rapid hardening cement is also used for the roads to be opened to traffic earlier than usual.

As the cement is essentially a mixture of clay and limestone, the properties of cement will vary depending on the source of raw material and the method of manufacture.

The coarse aggregate consists of crushed stone. The maximum size of coarse aggregates for cement concrete slab should not exceed one fourth the thickness of the concrete slab. The gradation may vary from 40 or 50 – 4.75 mm. The aggregates should be collected in two sizes one below 20 mm size and the other above 20 mm, where the gradation is from 50 – 4.75 the two groups should be below 25 mm and above 25 mm. Since the mineral aggregates form roughly 80 percent of the weight of concrete, it follows that the aggregate quality is of considerable in concrete road making. Fortunately, however good concrete can be made with most aggregates, even they are of different mineralogical composition and the result are less dependent on the type of aggregates than on the grading, size and shape of aggregates. Good quality of aggregate should be free from clay, dust, organic matter etc. and should have the following desirable limits :

Crushing Value	30%
Impact Value	50%
Los Angel Value	30 % according to I.S.I and 35 % according to IRC.

Good quality aggregates are non-absorbant and do not contain ingredients that are liable to decompose or undergo any change in volume.

The ingredients of concrete should be so proportioned that the concrete obtained should have a minimum of 40 kg/cm^2 modulus of rupture after 28 days of curing or to develop a minimum of 280 kg/cm^2 crushing strength after 28 days of curing. If a wet concrete or raw concrete as is being said generally, is placed in the roadway, the coarse aggregates are liable to settle, with the result that a layer of mortar may be produced at the surface during compaction. The process is known as bleeding results in the surface layer mixture having a very high water cement ratio which has poor resistance to abrasion. If an air-entraining agent is used in the concrete making process either directly or indirectly, the deleterious effects of the concrete can be largely prevented. While the advantages associated with air-entraining by far outweigh the disadvantages, it must be kept in mind that inclusion of air in concrete does cause a reduction in strength which is proportional to the amount of air-entrained. Owing to the increases in workability of concrete by entraining 4 to 5 percent of air, this loss of strength can be offset by reducing the water cement ratio by about 0.05 percent and decreasing the aggregate : cement by about 0.5 the proportion of sand can be reduced by 3 percent, as the tinny bubbles of water behave like sand particles.

8.5 DESIGN OF CONCRETE MIX

The concrete consists of aggregates (fine and coarse), cement and water. It is necessary to determine what proportion of each ingredients should be mixed in the concrete. This is known as concrete mix design. The ultimate aim of a concrete mix design is to obtain a concrete of desired strength and workability at the lowest cost. Another characteristic which all the methods have in common is that, even though the selection of the initial combination may be based on well-tried data obtained from previous experience, the ultimate design is always a result of extensive trial mixes which are tested at the site.

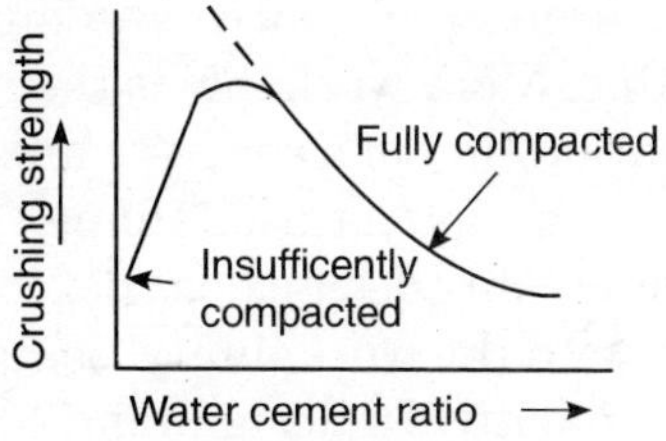

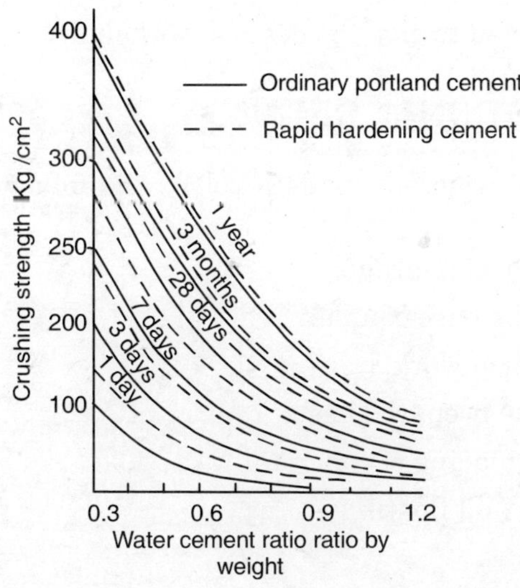

Figure 8.2

Water Cement Ratio Charts. Before discussing the concrete mix design procedure it will be interesting to know probably the most important aspect of concrete mix design is the relationship of water cement and strength. The water cement ratio law was formulated by Dr. Duff Abrams nearly 50 years ago. Essentially Dr. Abram stated that for a combination of cement and conventional aggregates in workable mixtures under similar conditions of curing, placement and test, the strength of concrete is solely a function of the ratio of the cement and the free water in the plastic mixture as shown in the graphs. When the water cement ratio is too low the mixture is not workable and difficult to compact, resulting in very low concrete strength. The peak point on the curve can be visualized on occurring where the concrete mix is just sufficiently workable to obtain near complete compaction with the maximum energy (or strength). As water : cement ratio is increased beyond the peak value, the strength begins to fall. But it is important to mention that it is not the sole criterion of strength for the concrete there are many other factors which control the strength of concrete. Along with strength concrete mix design be also workable. More workable is the mix less work is required to compact it fully. The workability of a concrete mix depends upon the water : cement ratio, cement : aggregate ratio, shape, size and grading of the aggregates. So a good concrete mix design does not concern only with strength or water cement ratio but with the workability, compactability with ease and by available means and economy. In designing a concrete mix these factors can be considered separately, the water : cement being choosen a basis for the desired strength, while the cement : aggregate ratio and the grading of aggregates are selected to give the desired workability.

8.6 CONSTRUCTIONAL DETAILS

The construction of concrete roads involves the following constructional stages :

1. Preparing the sub-grade.
2. Preparing the base course.
3. Placing of form work.
4. Watering the prepared base.
5. Mixing and placing of concrete.
6. Compaction and floating.
7. Belting.
8. Brooming.
9. Checking the finished surface.
10. Curing.
11. Edging.

1. Preparing the Sub-grade. The sub-grade is properly compacted by rolling, if necessary, and is brought to the required camber and gradient. The surface should be checked by means of a scratch template. The template should be kept at right angles to the centre line of the road. All elevations or depressions should be brought to the true profile.

2. Preparing the Base Course. Over the sub-grade, a base or foundation course is sometimes provided. In certain cases base is not provided. The function of base course is to provide levelled and smooth surface for the slab and hence to provide uniformity of support to the slab. The base for a concrete road may be :

(i) Water bound macadam,

(ii) Granular material, or

(iii) Stabilized soil.

(i) In the case of W.B.M. base the thickness should not be less than 15 cm. The base width should be atleast 50 cm more than the carriage way, projecting 25 cm on each side. The surface of the W.B.M. should be thoroughly compacted by rolling. The surface should not be made smooth but it should be left rough so that there should be proper anchorage between W.B.M. surface and concrete slab.

(ii) Granular material base provides high intrinsic strength and should be so graded that all particles will not knit together into a tightly packed mesh when compacted. The reason for this, is that there must be proper anchorage between the base and the cement concrete surfacing and hence it is desirable that the top of the base course should remain rough.

(iii) Stabilized earth base is also sometimes provided. The type of stabilization will depend upon the type of material available at the site and the type of soil.

3. Placing of Form Work. After the base course is completed, form work for the concrete slab is laid. The forms, to be set on the edges, can be of steel or timber. If of timber, they should be at least 5 cm thick capped with 5 cm angle-iron along the upper edge from the inside. The depth of the form work should be equal to the thickness of the slab. Channels are sometimes used for steel-form-works. The timber or steel-form-work should be fixed rigidly well in advance of concrete work to be started. It should be oiled and checked to line and grade.

4. Watering. After the form work has been laid, the surface of base or subgrade, as the case may be, is wetted with as much water as it can readily absorb. Water should not be allowed to stand on the surface. The surface should be kept wet for about 12 hours before concreting. The object of watering is to saturate the sub-grade or base so that it should not absorb any water from the concrete when it is laid over it.

5. Mixing and Placing of Concrete. The ingredients of concrete are mixed in proper proportion in a dry state. Generally 1 : 2 :4 mix is adopted. Mixing should preferably be done in mechanical mixers. Measured quantity of water should be added to have the designed *water cement ratio.*

The concrete is placed in the form works by manual labour, and it should be laid in layers of thickness not more than 5 to 8 cm or two or three times the size of aggregates. The concrete should be laid in the entire width of the form work and proceeded lengthwise. Each layer should be sliced and spaced with special tools, immediately it is laid. The top-most layer should be laid 6 to 12 mm higher than the actual profile for further tamping. The top layer should be laid to the required camber and gradient. Necessary transverse and longitudinal joints should be provided.

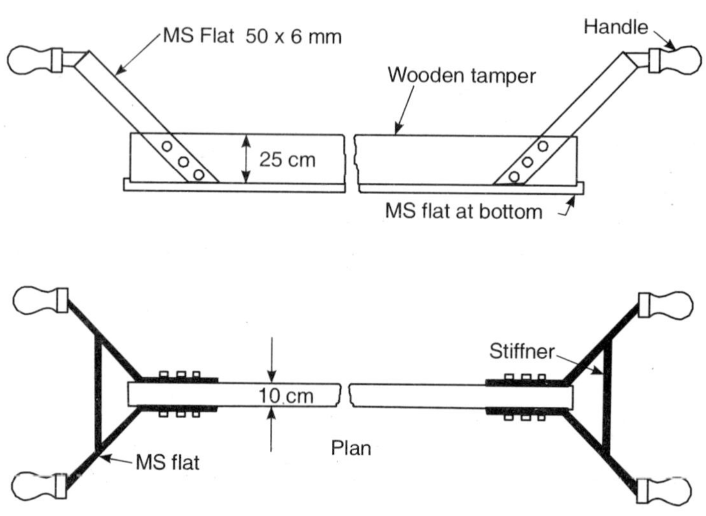

Figure 8.3 *Hand tamper*

6. Compaction and Floating. The compaction and consolidation of concrete is done by manual labour or by means of hand tampers as shown in Fig. 8.3. Consolidation may also be done by mechanical vibrations. The hand tamper is a wooden beam 10 cm thick and 25 cm deep, having a length equal to the width of the bay plus 30 cm. To the under-side of this beam a metal plate 5 mm thick is fixed. The tamper is used across the bay and tamping is done along the length of the bay. The surface is then tamped longitudinally. The irregularities caused due to tamping are rectified. The surface is then finished by hand floats.

The purpose of floating is to produce a uniform and even surface of concrete, free from transverse waves or corrugations. Floats are made of wooden boards 20 cm wide and 5 cm thick provided with suitable handles. The floating is carried out in longitudinal direction. The float is drawn in slow back and forth motion and slowly shifted towards the other edge. This will produce even surface free from corrugations.

7. Belting. The purpose of belting is to finish the surface of the concrete. The belt is 15 to 30 cm wide strip of canvas or rubber, fitted with handles at both ends. The belt is moved longitudinally by two persons. The process is rarely used in this country.

8. Brooming. Brooming is resorted to, if a rough or gritty surface is desired. The broom is gently pulled over the surface perpendicular to the centre line of the road from edge to edge. The brooming is done immediately after belting.

9. Checking. The finished surface is now checked. The road surface should conform to the required grade and camber. The surface is also checked against unevenness. The checking is frequently done during the floating process with a wooden straight edge 3 m long. A tolerance of 2 mm per metre can be given. If the difference is more than this tolerance, then the portion of the road should be corrected.

10. Curing. The finished surface, after 12 hours, is covered with gunny bags which are kept wet for about 24 hours. After 24 hours the gunny bags are removed and the surface is covered with a layer of sand which is kept wet for about 14 days. Sometimes the whole surface is divided into small bays by forming earthen banks or ridges 5 cm high. The bays are now filled with water to a depth of nearly 4 cm for 14 days. After curing, the surface is cleaned and washed.

11. Edging. Before the road is opened to traffic, brick edging is constructed to protect the slab. Earth is then spread on the berms upto the top of the brick edging.

8.7 JOINTS IN CONCRETE ROADS

The main difficulty with the concrete pavements is, that due to temperature variations, cracks are developed in the slab. If sufficient care is not taken at the time of construction, these cracks will make the road unserviceable as these cracks can never be repaired satisfactorily and will become a perpetual source of headache. To overcome this difficulty *joints* are provided. If the joints are properly spaced, cracks are reduced and if at all, only minor cracks will develop.

Joints can broadly be divided into two types :

(1) Expansion joints.

(2) Contraction joints.

The first provides for the expansion of concrete and the second to prevent irregular cracking due to warping or contraction. The expansion and contraction joints can also be grouped as :

(iii) *Warping joints.* These joints like the contraction joints are simple breaks in the continuity of the concrete slab. The opening of these joints is prevented by tie bars.

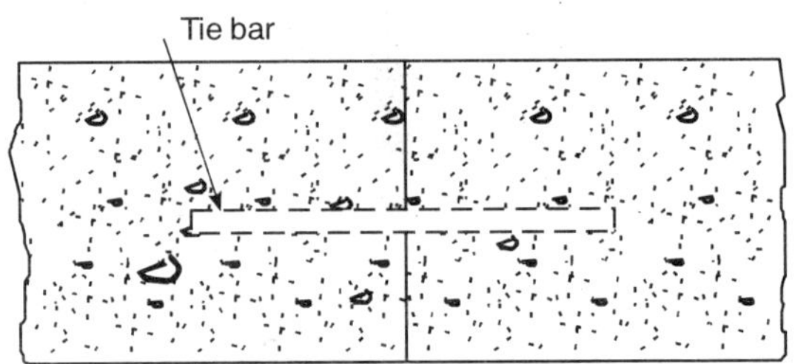

Figure 8.6 *Warping joint*

(iv) *Construction joint.* The object of construction joint is to facilitate the construction. They are provided when it is necessary to stop the work of concreting at any point other than the expansion or contraction joint. Tie bars of 10 to 12 mm diameter and 1 m long are provided at 60 cm centre to centre, at a depth of nearly 3 to 5 cm from the top of the concrete slab.

8.8 REQUIREMENTS OF CONCRETE PAVEMENT JOINTS

As has been discussed in Article 8.7 slabs can be divided into longitudinal and transverse joints. Longitudinal joints are parallel to the centreline of the road and transverse joints are perpendicular to the centre line. Irrespective of their position in the highway highest standard of care in their design and construction must be strongly emphasized. Following are the chief functional requirements of pavement joints :

1. A Joint must be Water-proof at all Times. This is perhaps the most important critarian. If surface water is allowed to enter through a joint softening of the sub-grade will take place and the uniformity of its supporting power will decrease. If the sub-grade is clayey, pumping may be promoted.

2. Free Movement of the Slabs must be Permitted at all Times. Both the filler and the sealing materials must be capable of withstanding the repeated expansion and contraction of concrete. There is a particular danger with expansion joints that if grit or other foreign material gets between the joints the expansion of concrete will be prevented, which will result spalling of edges. This can be overcome by taking adequate care in the selection and pouring of sealing material.

3. Joints should not Detract from the Riding Quality of a Pavement. Improper design and construction can result in excessive relative deflection of adjacent slabs. Not only this structurally undesirable but it also results in a most uncomfortable riding surface. Similarly, if proper care is not taken in selecting and placing the seal material, the transverse joints will cause uncomfortable riding surface.

4. A Joint should not be the Cause of an Unexpected and Undesired Structural Weakness. The transverse joints on cither side of the longitudinal joint should not be staggered from each other as transverse cracks will be induced in the slabs in line with the staggered joints. Moreover no joint should be constructed at an angle less than 90° to an adjacent joint or edge of the slab unless it is provided intentionally. The importance of not having acute angle corners can be gathered by considering a corner completely unsupported and then subjected to a load. The stresses induced in the opposite corner will be 20% more of 80° angle, 75% more for 60° angle and 175% more for 30° angle.

5. Joints should not Interfere with the Placing and Curing of Concrete. This is possible and can be best accomplished by having as few joints as is possible. When they must be used, then easy and economical construction can be facilitated by using the simplest type of joint consistant with the structural design of the pavement.

8.9 CEMENT BOUND MACADAM ROADS

On the reconditioned surface of W.B.M. road, a 12 cm thick layer of road metal of 36 to 48 mm size is spread and rolled tightly with 8 tons roller to get a partially consolidated layer of 10 cm thickness. Cement and sand are taken in the ratio of 1 : 2. About 5 bags of cement per cubic metre of road metal and 34 litres of water per bag of cement is used to form a fluid cement mortar or grout. This grout is poured on the prepared surface of macadam. After grouting, stone chippings of 15 to 18 mm size are spread on the surface and forced into the mortar on the surface by longitudinal tamping or light rollers. The grout penetrates into the surface layer which gets consolidated also. Longitudinal and transverse joints are provided as in other concrete roads. Such roads are not common in India.

8.10 CRETE-WAYS

India is a country of villages. Village roads are very important and it should be tried to have durable roads under iron-wheeled traffic and at the same time they should be cheap in construction and in maintenance. Generally the following types of roads are constructed in villages :

(1) Earth roads

(2) Moorum roads

(3) Gravel roads

(4) W.B.M. roads

These roads give way easily under the bullock cart traffic and hence it becomes necessary to repair them periodically or to recondition the whole surface. This becomes a very costly affair in the long run. The present economic condition of the country do not afford to construct other durable type of roads such as bituminous roads or concrete roads.

A type of *trackway* or *wheeler* has been devised for the bullock carts. This type of trackway made of concrete is called *crete-way*. The trackway can be made of stone pavings or slabs or bituminous surfacings. These trackways are quite durable and cheap and are becoming common and popular especially in those places from where heavy bullock cart loads move to nearby industrial areas or sugar mills, etc.

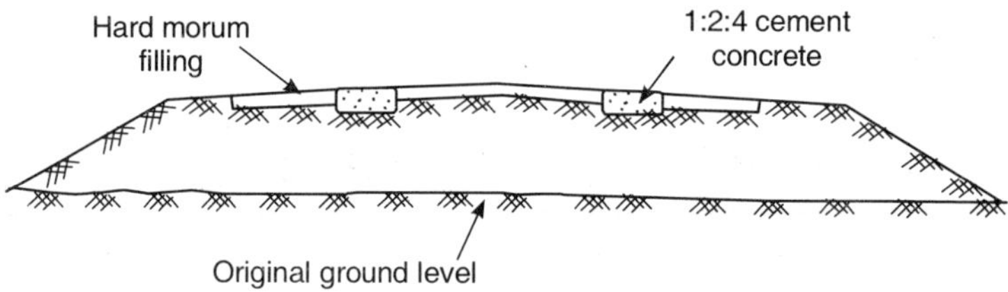

Figure 8.7 *Crete-ways*

Crete-way consists of two narrow longitudinal strips of road surface on which the two wheels of the bullock cart or any other vehicle can move. The strips are generally 60 cm to 75 cm wide, 10 to 15 cm thick and are 1.30 m to 1.5 m centre to centre apart.

The crete-way slabs may be constructed in situ or precast. Generally precast slabs of 2 m length and 10 cm thickness are placed in the trenches excavated for the purpose. The beds of trenches are properly watered and compacted. Tongue and groove joint is provided between the two adjoining slabs.

8.11 TOOLS AND PLANTS FOR THE CONSTRUCTION OF CONCRETE PAVEMENTS

Following Tools and plants are used for the construction of concrete pavements :

1. Concrete Mixer. Concrete mixers of adequate capacity of the batch type is provided. It has a rated capacity of not less than 0.2 m^3 of mixed concrete. The mixer is equipped with a water measuring device for measuring water required per batch. Some mixers are provided with counters and turning devices and automatic device for locking and discharging the lever.

2. Vibrating Screeds. Vibrating screeds comprise of wooden or mild steel screed with suitable handles driven by vibrating units mounted thereon.

3. Internal Vibrator. It comprises of vibrating head with suitable motive power either of compressed air or of a petrol engine for proper control and compaction of the mass concrete. It is used to ensure compaction of the concrete along with the forms and also to avoid any tendency of honey-combing of the edges of the slab.

The hand tools used are :

Float, straight edge, bell, fibre brush, edging tools etc.

8.12 FILLERS AND SEAL

Regardless of their function all joints of a cement road must be water proof at the time of construction and shall maintain like that throughout its life.

Further more any cracks which develop in the slab throughout their life should be made water proof as soon as they are detected. Failure to do this will result in the infiltration of grit and water with consequent failing of joints/edges and creation of non-uniform foundation/base/subgrade. The materials which are used to make the joints water proof are known as fillers and seal. Both filler and seal materials are used in expansion joints whereas for other joints only seal in used.

Filler. Fillers are used to provide the gaps for expansion joints, at the time of construction, and to provide support for the sealing compound. Materials which are to be used as filler materials should have the following characteristics as per IRC :

1. It should be able to be compressed without extrusion.
2. It should be sufficiently elastic and return to its original position after the compression force is released.
3. It should be durable so that it could retain its characteristics throughout the designed life of the pavement.

Filler materials used in joints which satisfy these criteria are soft wood, fibre board, impregnated with cut-back and emulsion, sometimes, cork is also used for this work. A filler material which has excellent compressibility and recovery properties is cellular rubber but because of its pliability it is difficult to maintain the material during the time of constructions hence not recommended for use.

Sealer or Seal. Expansion joints which are formed with filler materials only are neither watertight nor do they prevent the ingress of grit. Thus not only are contraction, warping and construction joints scaled with a water tight compound but so also are expansion joints. Following are the requirements of a seal material as per IRC :

1. It should be capable of adhering firmly to concrete under all weather condition.

2. It should be sufficiently ductile in cold weather to accommodate itself to the widening of joints without cracking and be capable of withstanding repeated expansion and contraction over long periods without disintegrating.

3. It should not melt and flow down the joints during hot weather.

4. It should, under all weather condition effectively resist the ingress of grit and water.

5. It should be durable and not harden or soften with the passage of time.

6. Its colour should not be contrast the colour of concrete.

There are many materials satisfying the above conditions and are used as sealing compound, it is important to note that no single material can satisfy all the characteristics as some of the requirements contradiet each other.

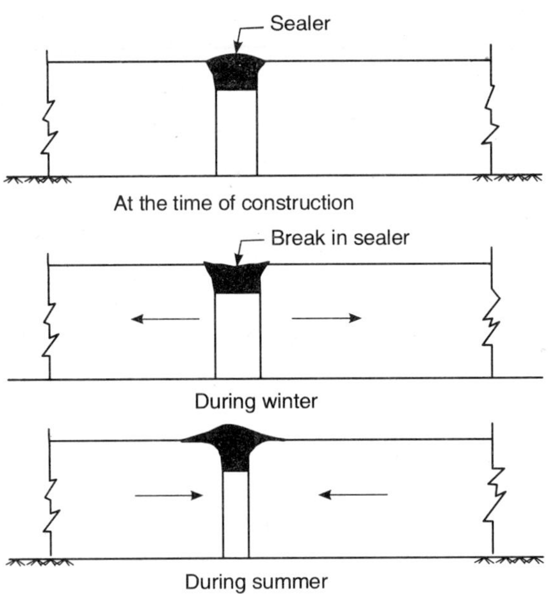

Figure 8.8 *Sealer*

Bitumen is used either alone or with mineral filler as a sealing compound. Kubber compounds have shown excellent results. The sealing joints should be 1 cm wide and 2 cm deep (normally). The amount of material needed to seal the joints is dependent on the functions of the joints.

8.13 LENGTH AND WIDTH OF PAVEMENT SLAB

The geometric features of a concrete pavement decide the length and width of a pavement slab and the shape of the pavement cross section.

Length. By length of a pavement slab is meant the distance between the two transverse joints or distance of construction between deliberately installed construction joints . There is a wide spread disagreement amongst the engineers of different countries as to how much the length of the pavement slab be. The length can vary from 20 m to 300 m. In India the normal length varies from 3 m to 50 m. According to IRC the maximum distance between two expansion joints should not exceed 45 m if the thickness of slab is 200 mm and the contraction joints are 15 m apart.

Width. Constructional requirements and practical needs rather than theoretical consideration dictate the width of slab on most concrete roads in our country. The width of slab should not be more than 5 m for constructional reasons as the more wider slabs will develop irregular longitudinal cracks in approximately mid width as a result of excessive transverse bending stresses due to the development of temperature and load stresses. Not only are these cracks unsightly but difficult to seal also. Now world wide it has become a standard practice to construct the concrete pavement equal to the width of carriage way which is 4 m.

Thickness. The stresses developed at the corners and edges of the slab are considerably more than on the interior of the slab. For more economical constructions it has become a standard practice to reduce the thickness of the slab at the interior parts. This is of course a very difficult work and involves lot of labour in shaping the sub grade and base course for reducing the thickness of the slab in the interior parts. In our country uniform slab thickness is constructed.

8.14 DESIGN OF DOWEL BAR

Dowel bars are used for load transfer from one pavement slab to the other. The dowel bar should be strong enough to withstand shearing stresses, bending stresses and bending of concrete. If p is the load transfer capacity of a dowel bar of diameter d and length l, then

$$p \text{ in shear} = 0.785\, d^2\, fs' \qquad \text{(i)}$$

$$p \text{ in bending} = \frac{2d^2 f_s}{l \quad 8.8\, w} \qquad \text{(ii)}$$

$$\text{and in bearing on concrete} = \frac{fe\, l^3 d}{12.5\,(l \quad 1.5\, w)} \qquad \text{(iii)}$$

where

fe = permissible bearing stress in concrete in kg/cm^2

fs = permissible flexural stress in dowel bar in kg/cm^2

fs' = permissible shear stress in the dowel bar kg/cm^2

w = width of joint

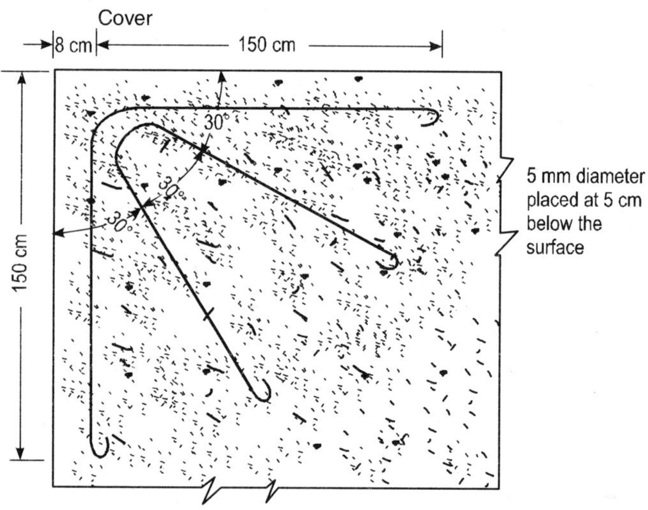

Figure 8.9 *Corner reinforcement*

For bar to be equally strong in bending and shear equate Eqs. (i) and (iii) to find out the length of bar for assumed diameter and standard width of joint. The distance on either side of the load position upto which the dowel bar is effective in load transfer is equal to 1.8 l. The corners are sometimes reinforced as shown in Fig. 8.9.

Example 8.1 *Design a dowel bar for a cement concrete pavement if permissible shear stress in dowel bar is 1000 kg/cm² and permissible flexural stress is 1400 kg/cm². Bearing stress in concrete is 100 kg/cm².*

Solution

$$p \text{ in shear} = 0.785 \; d^2 \, fs' \qquad \text{(i)}$$

$$p \text{ in bending} = \frac{2\,d^2\,fs}{l \;\; 8.8} \qquad \text{(ii)}$$

Assume diameter of bar = 2.5 cm and width of joint as 2 cm (ω)

Equating Eqs. (i) and (ii)

$$0.785 \times (2.5)^2 \times 1000 = \frac{2 \quad (2.5)^3 \quad 1400}{l + 8.8 \times 2}$$

Solving the equation we get the value of l, the length of dowel bar to be 41.26 cm.

The length of the dowel bar can be obtained from the formula

$$l = 5d \quad \frac{fs}{fc} \quad \frac{l + 1.500}{l + 8.800}^{1/2}$$

where fs is flexural stress and fc bearing stress of concrete substituting the value for fs and fc, we have

$$l = 5 \quad 2.5 \; \frac{1400}{100} \quad \frac{l \;\; 1.5}{l \;\; 8.8}^{1/2}$$

Solving $\qquad w = 2$

$$l = 40.5 \text{ cm}$$

Total length of dowel bar will be

$$40.5 + 2 = \mathbf{42.5 \text{ cm}}$$

8.15 PREVENTION OF MUD PUMPING

By mud pumping is meant the forceful ejection of water and soil from below the cement concrete slab when it is laid directly on the sub-grade of fine grained plastic soils. Pumping occurs with most severity on soils with high clay contents. It generally does not occur, even under the heaviest wheel loads, on non-plastic soils of granular nature. Pumping will not occur if free water under the slab is not available if the frequency of heavy load is low and the soil beneath is not in suspense. Frequency of movement of heavy traffic on the road is the most important factor responsible for mud pumping, when the pavement slab deflects under the wheel load and make way for the soil water mixture to eject. Mud pumping can be minimised by limiting the expansion joints to as few as is absolutely essential and substituting closely spaced and dowelled contraction joints.

Water which infiltrates through the pavement joints, cracks and edges is invariably the source of free water which causes pumping. If the sub-grade is not free draining the soil goes into suspension where pockets of water are collected at the void spaces in the sub-grade and it is this suspension which is being ejected when the slab deflects under the wheel load. Continuous passage of heavy traffic leads to pavement deflection and the ejection of sub-grade soil water mixture. This can be prevented by adequately sealing the joints and cracks so as to prevent entry of water into the sub-grade and by efficiently drawing off surface water as quickly as possible.

Finally pumping can occur if the sub-grade soil goes into suspension in water. This can be prevented by either chemically stabilising the soil to some adequated depth or by providing a sub-base of dense graded or open graded material. A dense graded sub-base is provided in an undrained trench which is some what under than the pavement. It is compacted to such an extend so as to make it practically impervious to water. If the gradation of the sub-base is poorly designed, it will have a high moisture holding capacity, while lack of compaction will lead to pavement cracking at joints as a result of traffic causing further compaction of the sub-base. Open-graded sub-base should also be compacted but here the coarseness of the gradation must be such as to permit adequate drainage without itself becoming clogged by the intrusion of soil from the sub-grade. An open graded sub-base is more expensive to construct since it must be provided with a subsidiary drainage system which must be regularly maintained.

8.16 R.C.C. PAVEMENTS

When the sub-grade is poor and there are large temperature variation, the cement concrete pavements are reinforced by welded wire mesh. The function of the reinforcement is to hold the cracked slab together and prevent widening of crack gaps. Sufficient load transfer is effected by the reinforcement. The reinforcement does not provide any additional flexural strength to the pavement. The function of mesh reinforcement is more or less the same as that of the dowel bars in contraction joints and the tie bars to withstand tensile forces due to the movement of the slab and the frictional resistance between the bottom fibre of the pavement and the sub-grade.

Hair cracks generally appear on the surface of the pavement but they do not open up. For reinforcement to be more effective, more reinforcement should be provided in the longitudinal direction and the reinforcement should be placed in mid-depth of the slab. Steel wire fabric should conform to I.S. 432–1960. The reinforcement generally provided varies from 3.5 kg to 7.0 kg of wire mesh per square metre of the pavement surface.

8.17 PRESTRESSED CONCRETE PAVEMENTS

Prestressed concrete pavements are not very common and popular. Prestressed roads can be compared economically with other road pavements but still not very popular. It is argued that if the constructional techniques further improve, prestressed concrete pavements may prove cheaper than other pavements.

The purpose of prestressing a concrete structure is to augment its tensile strength by inducing an initial compressive stress in the concrete before it is subjected to loading. Prestressing is carried by incorporating steel tendons, stretching and subsequently fixing them to the concrete in such a way that induced tension in the steel is balanced by compression in concrete. Stretching may take place before the concrete is hardened (pretensioning) or after the concrete is hardened (post tensioning). The second method of prestress is by casting the concrete between two rigid abutments which are of sufficient strength to receive the prestressing thrust. Jacks are then placed between the abutments and the concrete and operated to compress the concrete. Prestressed pavements can be built in continuous length upto 120 m without joints. Elimination of joints without cracks in the pavements could be considered advantageous, in view of the maintenance problem of joints.

Prestressed concrete pavements can be classified according to whether they are of the individual slab type or of the continuous type. The individual slab type prestressed pavements are only used.

Prestressed concrete pavements are particularly suitable when the sub-grades are weak and the traffic is heavy as the thickness of prestress slab is less than the R.C.C. Slab. Thicker slabs have more temperature differentials and liable to cracks easily. The other main advantage of prestressing is the less number of joints.

The principal disadvantage associated with prestressed concrete slabs are the construction difficulty particularly on curves, vertical and horizontal. In urban areas where the pavements have to be cut for laying cables, sewers and water pipes, the prestressed is lost as soon a cut is made in the slab. Strict safety precautions are to be implemented during construction.

Following points should receive due consideration for designing prestressed concrete pavements:

(i) 120 m length of slab can be prestressed.

(ii) 3.6 m width lane for prestressed pavement is desirable.

(iii) A minimum thickness of 15 cm is recommended for the desired cover of tandoms.

(iv) A minimum of 22 kg/cm^2 of prestress is recommended for 120 metres long prestress pavement slab with 10 percent increase for losses.

(v) A tandom of 7.0 mm of diameter of tensile strength of 142–173 kg/cm^2 are employed.

(vi) Only standard methods of prestressing viz., Freyssinet Magnel-Blaton, Le-Mc Cali and Gifford Udal, should be used. Freyssinet method is very common.

8.18 ADMIXTURES

Admixtures for concrete are materials that are added to concrete at some stage in its making to give it new properties. The use of admixtures has increased substantially during the recent years. There is a regional variation in the use of concrete admixtures. For colder region accelerators and air entrant admixtures will be preferred whereas for warmer region retarders are preferred. The general attitude towards admixtures was one of disbelieve and disinterest But engineers and builders have started taking interest in the use of admixtures.

Admixtures are of various types and can be broadly classified as Primary Admixtures and Miscellaneous admixtures. The primary admixtures are organic materials and act as plasticizers, air entrants and water reducing agents. The miscellaneous admixtures are finely divided, pozzalanic and cementitious materials.

Water Reducing Admixtures. The water reducing admixtures are the primary admixtures which possess as their primary ability to produce a given workability at a lower water cement ratio than that of controlled concrete without any admixture. These admixtures are used in three different ways.

1. For attaining the same strength and workability at lower cement content than a controlled concrete without adversally effecting its durability.
2. For achieving greater strength with the same workability at lower water cement ratio (cement content being same).
3. For achieving same strength but higher workability particularly with coarse aggregates.

Retarders. The retarders are chemical substances used to retard the setting process. Their presence in the water reduces the solubility of the anhydrous constituents of the cement and their rate of solution is delayed. Some of the retarders cover the cement particles with a very thin film and cause retardation of the process of cementation. Retarders are preferred in frost prone and arid zones just to nullify their effect on the hardened concrete surfaces.

PROBLEMS

8.1. Under what circumstances you would recommend the construction of a cement concrete road ? What are its advantages over the other types of road surfaces ?

8.2. What are the stages of construction of a cement concrete road ? Describe in brief.

8.3. What do you understand by the term 'crete-way' ? Where they are used and what are their advantages ? Give specifications for a crete-way and describe in brief the construction of a precast crete-way.

8.4. Why are joints provided in a cement concrete pavement ? What are the various types of joints ? Describe the construction of any one joint.

8.5. A water bound macadam road in a city area is to be improved either by adopting a bituminous pavement or a cement concrete pavement. Which of the two, do you prefer ? Give reasons for your choice.

8.6. State and discuss the requirements of concrete pavement joints.

8.7. State the procedure of designing a concrete mix for highway construction. Only salient points should be discussed.

8.8. List the various tools and plants required for the construction of cement concrete pavements.

8.9. Discuss the objects of providing the following type of joints in a concrete pavement:

(i) Longitudinal joints

(ii) Warping joint

(iii) Construction joint

8.10. State the procedure of designing a dowel bar for a contraction joint. What assumption are made in its designing ?

8.11. Compare the advantages and disadvantages of a cement concrete pavement and R.C.C. pavement.

8.12. State in brief the method of construction of a prestressed concrete pavement. Why prestressed concrete roads have not become popular ?

8.13. Design a tie-bar system for a cement concrete road with the following data :

Slab thickness = 22 cm

Slab width = 3.6 metres

No. of lanes = 2
Weight of slab = 480 kg/m^2
Bond stress = 17.5 kg/cm^2
Allowable working stress in steel = 1400 kg/m^2

[Hint.

$$\text{Area of steel required} = \frac{b \quad \mu \quad W}{fs}$$

here b = 3.6 m, fs = 1400 kg, W = 480 kg and μ coefficienl of friction between the slab and the sub-grade may be taken as 1.5.

$$\text{Length of lie bar} = \frac{2fs\ As}{B}$$

Here B is bond stress, p is the perimetre of the tie bar)

8.14. Discuss the importance of fillers and seal materials to be used in the joints of a cement concrete road.

8.15. What is mud pumping in a cement concrete road and how can it be prevented ?

BIBLIOGRAPHY

1. Standard specification and code of practice for the construction of cement concrete Roads IRC : 58 – 1974.
2. Concrete Pavement Design — Portland Cement Association, 1951.
3. Guidelines for the design of concrete pavements for Highways IRC : 58–1974.
4. Ministry of Transport and Roads Research Laboratory — A guide to the structural design of Rigid Pavements, Road Note No. 29, London.
5. Portland Cement Association sub-base, subgrade, and shoulders for cement concrete pavements, 1960.
6. Portland Cement Association — Cement concrete pavement, 1951.
7. Road Research Institute Laboratory — Concrete pavement manual, 1953.
8. Abrahams D.A. Design of Concrete Mix Bull No. 1 Structural Material Research Laboratory.
9. Road Research Laboratory. Concrete Roads London, 1953.
10. Troxell G.E. and Hammer E. Davis composition and properties of concrete. McGraw Hills, 1956.

11. Cement Concrete Journals, 1974–88.
12. Teller L.W. and E.C. Sutherland – Concrete pavement – Design and construction and experimental study, HMSO, 1955.
13. Glanwille W.H., AR collins and D.D. Mathews. The grading of aggregates and workability of concrete, Road Research Technical paper No. 5, London.
14. Blake L.S. and K.M. Brook. The construction of Major Concrete Roads 1955–60. Technical Report No. 363, Cement and Concrete Association, 1962.
15. Bradbeny. R.D. cement concrete pavement Mix Design and construction, Washington D.C.

9

Hill Road Construction

In this Chapter you will study,

• Alignment of Hill Roads • Formation • Planning of Hill Roads • Survey • Gradients • Economics of Hill Roads • Cross Drainage Works • Effects of Poor Drainage • Drainage Criteria of Hill Roads • Hill Road Side Drains • Retaining and Breast Walls • Construction of Hill Roads • Land Slides • Geometrics of Hill Roads • Blasting • Handling and Transportation of Explosives • Blasting Procedure • Detonators • Tools and Accessories for Blasting • Requisites of Explosives • Precautions in the Storage of Explosives • Tunnels • History of Tunnels • Alignment and Grade of Tunnels • Location of Centre Line on the Ground • Transferring Centre-Line to Inside of Tunnels • Transferring Centre-Line to Inside of Tunnels• Methods of Location of Centre-Line of Curvilinear Tunnels • Method of Providing Grade • Shafts • Location of Shafts • Construction of Shafts • Methods of Tunnelling • Tunnelling in Soft Ground • Tunnelling Methods • Rotary Excavator • Compressed Air Tunnelling • Mucking and Hauling • Methods of Mucking • Loading Machines • Muck-Cars • Mucking in Steep Grade Tunnels • Car Changer • Tunnel Lining and Grouting • Concrete and R.C.C. Lining • Ventilation of Tunnels • Drainage of Tunnels • Ground Water Removal • Permanent Drainage • Safety Measures

GENERAL

Hill roads are also called *Ghat* roads in India. Hill roads present more difficulty in their design, construction and maintenance. These roads are very dangerous and sometimes fatal accidents occur on such roads. Hence very special care is taken in the design and construction of such roads. Surface drainage is very important in hill road construction. Hill roads are classified as :

(i) Motor roads, (ii) Bridle paths, and

(iii) Village paths or tracks.

Motor paths are important from the engineering point of view. They are generally meant for fast moving traffic.

Bridle paths are meant for pedestrians and other pack transports such as horses, ponies etc. They serve as feeders to the motor roads, thus connecting interior places with the main communication system. The width of the bridle paths generally varies from 2 to 2.5 m. The ruling gradients for bridle paths is 1 in 10. The maximum permissible gradient can be 1 in 7½. The bridle paths can have either a W.B.M. surface or earth or natural surface brought to the required camber and gradient. Side drains should be provided for bridle paths.

Village tracks or paths connect small villages and other working places over which cattle and people move. The width of village paths varies from 1 m to 1.5 m.

9.1 ALIGNMENT OF HILL ROADS

The success and utility of a ghat or hill road depends on the proper alignment of the road. During the alignment of these roads care should be taken to see that the road should be as short as possible and should have easy gradients. In the alignment of the road, a number of sharp curves such as hair pin bends and corner bends will have to be provided (Fig. 9.1). As far as possible the minimum radius of the curves should be 15 m. The desirable radius is of course 30 m. The aim of the alignment should be to establish the easiest,

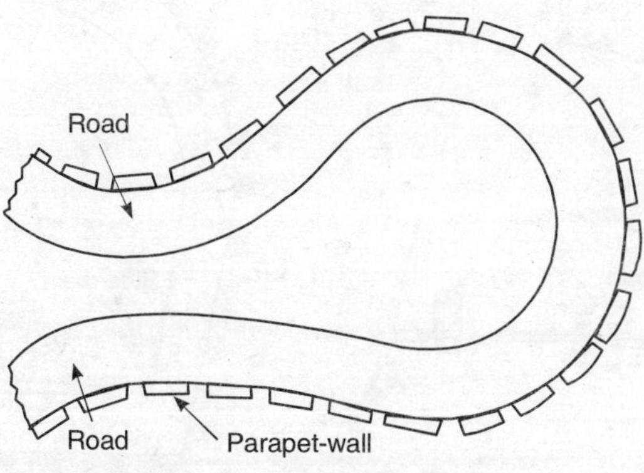

Figure 9.1 *Hair-pin curve*

shortest and the most economical line of communication between the obligatory points in consideration, physical features of the country and the traffic need of the area served. The principle which the engineer should bear in mind is to align the road so that the vehicles can travel with ease and in safety and the expenditure of motive power and wear and tear of the vehicles should be reduced to a minimum, as also the cost of construction and maintenance of the road should be as low as possible.

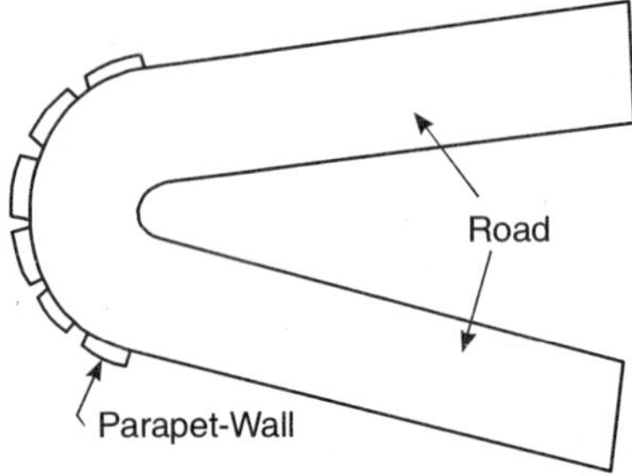

Figure 9.2 *Corner-Bend*

9.2 FORMATION

The formation of a hill or ghat road may be :

(i) Wholly in cutting.

(ii) Partly in cutting and partly in embankment.

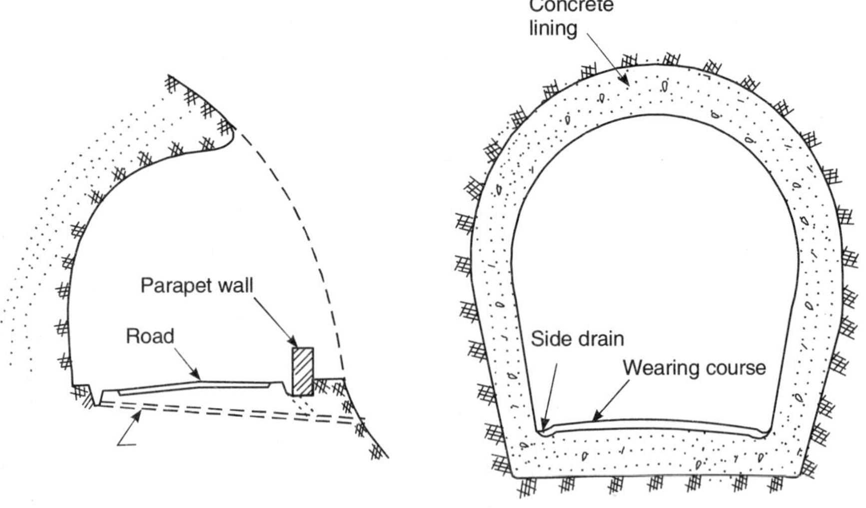

Figure 9.3 Road in half tunneling

Figure 9.4 Road in tunneling

(iii) Wholly in embankment.

(i) Road in Cutting. The road in cutting is taken when the hill side slope is very steep. The roads in cutting may be in half tunneling or in full tunnelling. Half tunnel is used where the strata of rock, dips away from the road side and when the rock is sound and hard. The roads through full tunnels are rarely constructed as they are costly.

(ii) Road in Cutting and Embankment. Hill roads are generally constructed partly in cutting and partly in embankment. Road sections may be 2/3rd in cutting and $\frac{1}{3}$rd in embankment. Road in cutting and embankment is constructed when the side slope is not very steep and the cost of cutting through rock is too much.

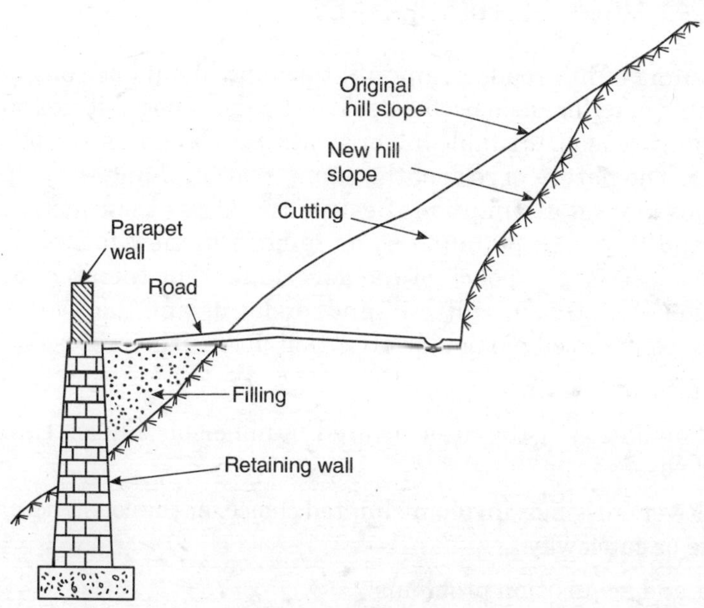

Figure 9.5 *Road in embankment and cutting*

(iii) Road in Banking. Sometimes the entire road will have to be taken in embankment. This happens when a depression comes in the alignment.

In this case two retaining walls one on each side of the road will have to be constructed. This type of construction is also very costly and hence avoided as far as possible.

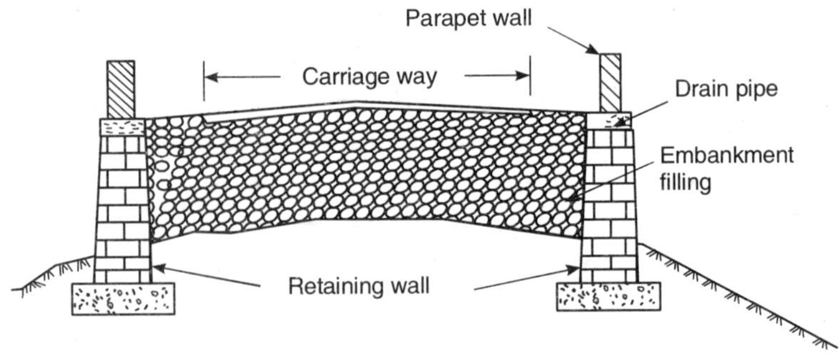

Figure 9.6 *Road in banking*

9.3 PLANNING OF HILL ROADS

The planning of hill roads is a uphill task and should be considered in the background of unforeseen problems. This includes possibilities of land slides during rainy season, unstable hill or strata with boulders and other complex problems. The nature of rock of the same location changes so abruptly that it becomes beyond assumption. Besides the above the construction needs careful and thorough planning so as to keep in view future development schemes based on the policy of the government for the area, avoidance of acquisition of costly agricultural land, orchards and thick forests etc. The following points need particular attention during the process of planning :

(i) Type of terrain

(ii) Population of the area covered, number of villages through which the road passes.

(iii) River crossings, involving limited choice for the construction of culvert or bridge or causeway.

(iv) Land acquisition problems

(v) Land slides

(vi) Economics of construction of roads.

(vii) Future and anticipated traffic

(viii) Gradients

9.4 SURVEY

Thorough and proper survey plays a vital role in the selection of an appropriate alignment. This also goes a long way to ensure economy in cost.

It is seldom desirable to compromise on geometric standards, safety requirements, drainage and cost reduction on construction and maintenance of road. Sharp curves and steep gradients should be avoided because their subsequent improvement would involve heavy expenditure. The cutting and embankment or walling length should balance as far as possible to ensure economic construction and stability. During the process of survey the following considerations should be kept in mind :

1. The alignment should not pass through steep rocks or soft unstable hill sides.

2. The road should serve the purpose to its fullest even with longer length and costlier alignment.

3. Future development plans of the area should be the part of the road planning and survey.

4. The survey, planning and alignment should take care of the integrated development of the area particularly rural, backward and inaccessible areas of the locality.

5. Survey and alignment should aim at selecting such locations so as to avoid unstable hill side to minimize undue maintenance cost.

6. Such locations should also be avoided which have strategic military importance, having rich mineral and forest value.

9.5 GRADIENTS

While selecting alignment, consideration of gradients to be choosen should be such as to cater the needs of ultimate traffic which has to use the road. Jeepable roads and bridle paths with steep gradients and smaller radii of curves for short distances could be adopted. For important roads steep gradients and smaller radius curve can pose operational problems and become uneconomical in the long run. It has to borne in mind that improvement on steep gradients and sharp curves at a subsequent time will become difficult and a very costly affair subsequent changes in grade or curves may lead to dislocation of traffic, very costly diversion etc. Under stiff rocky strata, sometimes limiting gradients have to be accepted on overall considerations of traffic needs and cost. At the alignment stage, consideration of future traffic expected to develop are necessary for the selection of gradient is a very important aspect. Jeepable route and bridle paths may be given steeper gradients but for heavier traffic like trucks and buses steep gradients and sharp curves cannot be given or provided otherwise it will lead to operational problems.

9.6 ECONOMICS OF HILL ROADS

The following points are worth considering in the case of hill roads for economics and future development of the area.

Future Traffic. (i) Consideration for future traffic at the alignment stage should be made so that future development of the road is possible and feasible at a reasonable cost and in reasonable time.

(ii) Dry stone pitching on loose and unstable hill side face should be provided/done to avoid heavy slips. Dry stone pitching is comparatively more economical than permanent construction like retaining and breast wall.

(iii) Heavy breast walling or even dry stone pitching should be avoided where slip clearance could be a cheaper solution. Heavy breast walls should not be constructed in slip zones even if widening is not to be done because breast walls cannot retain heavy slips. Breast walls in such cases even do not reduce slips.

(iv) Unlined drains should be constructed where possible particularly where future widening at an early stage, is envisaged.

(v) Culverts and scuppers should be constructed for the full width of the expected future width of the road.

(vi) Allowance for grade changes or re-alignment should also be made for an economic and viable future planning.

9.7 CROSS DRAINAGE WORKS

While aligning a hill road, care should be taken to see that natural drainage is not obstructed and costly and un-manageable structures are avoided. To achieve this objective the following measures are taken.

(i) Natural water sources should not fall on the road or its alignment but instead cross drainage should be provided from hill side to the *khud* side.

(ii) Scuppers should be provided on alignment through thick forests.

(iii) Precast and monolithic R.C.C. culverts should be avoided as the abutments get damaged and the slabs do not remain intact. Hence precast R.C.C. deck slab of 30 – 45 cm width should be preferred, so that in case abutments are damaged the deck slab could be removed without much inconvenience to the traffic.

(iv) Causeways or Irish bridges should be discouraged wherever possible. Causeways are to be provided where construction of culverts or bridge is very very costly and obstruction to traffic is insignificant. The depth of flow of water in the causeway should not exceed 60 cm and the duration of flow is not more than $1\frac{1}{2}$ hours at a time in a day.

(v) Beds of culverts and scuppers should be provided with cross slopes not flatter than 1 : 15 so that self clearing of discharge is possible. At the same time bed slope should not be so steep as increase erosion tendency on the up stream or down stream side. 20 cm – 30 cm pitching or *kharanji* will be helpful in reducing erosion or scouring. Sometimes curtain walls are to be provided.

9.8 EFFECTS OF POOR DRAINAGE

Drainage needs arise from the accumulation of rain water or any other water getting accumulated on the road or its shoulders. The accumulation of water is injurious to the life of the road and safety of traffic. The effect of accumulated water on the road surface or its shoulders may result in one or more of the following :

(i) Reducing bearing capacity of the sub-grade

(ii) Eroding soil near the roadway or its shoulders, thus making the road risky.

(iii) Disturbing the hill zone near the road causing slip and reducing strength.

9.9 DRAINAGE CRITERIA OF HILL ROADS

To ensure speedy and smooth flow of discharge of water accumulated on or along the road side, efficient drains should be provided along one side or both sides of the road. The provision of road drainage, therefore, includes not only ensuring flow along or across the road surface but it also covers drainage of its sides including hill faces above the road, along the road and *khud* side. Drainage schemes, therefore require provision for effective flow and discharge to a suitable outfall from or along the roads of all the discharge likely to get accumulated through suitably designed drainage system.

A carefully planned provision of road drainage is, therefore, of vital importance and needs of any road construction and subsequent maintenance. The drainage system of a hill road comprise of the following studies.

(i) Sources of drainage discharge.

(ii) Volume and velocity of discharge and the time period.

(ii) Type of surface over which the drainage flows prior to its reaching the point of drainage work.

(iv) Quantity of water flow likely to be absorbed in the road surface, shoulders, or any exposed surface.

(v) Any other special feature of the drainage problem based on previous knowledge and experience, such as steepness of hill face, stability of hill

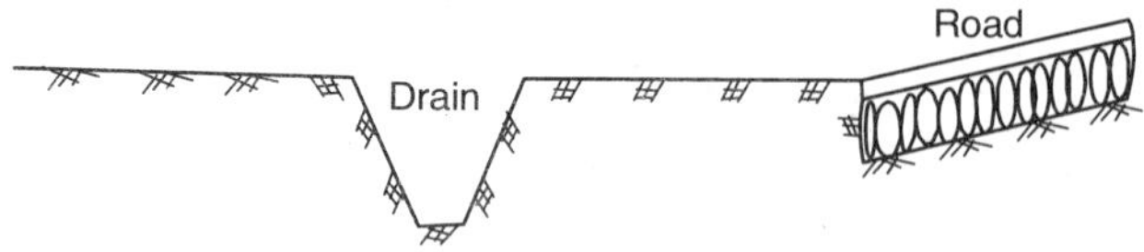

Figure 9.7 *Hill side drain*

face, gradient of drainage on upstream side, likely gradient on the down stream side.

Inadequate drainage or poor drainage will have the following adverse effects:

(i) In submergence or saturation, the road base and sub-grade will become soft and slushy.

(ii) Pot holes, excessive patches and sometimes baggy areas are developed.

(iii) Subgrade is depressed which causes overall failure of road surface.

(iv) Surface of road develops corrugations and undulations due to upheaval and creeping action in the subgrade, the sub-base and the base course materials.

(v) Road shoulders get damaged and road edges are eroded.

Efficient hill road drainage should have the following aspects :

(i) The road surface should have adequate camber.

(ii) The adjoining hill side face above the road should not be such as to collect and allow percolation of water from the hill face under the sub-base or road structure.

(iii) There should be no accumulation of water on the sides of the road.

(iv) Seepage of water from the road surface and from the sides should be prevented or should be minimised. Seepage water should be drained through sub-surface drains.

(v) Obstruction or hindrance in the drainage system should be removed at the earliest.

(vi) Minor cross drainage system should be provided at regular intervals.

(vii) For un-metalled or water bound macadam roads rain cuts are provided to reduce percolation of water into the sub-grade. Rain cuts should lead to the side drains on the hill side and on the *khud* side and are connected in line with the top of retaining walls or parapet tops.

9.10 HILL ROAD SIDE DRAINS

For better service and low maintenance, efficient surface drainage should be provided for hill roads. Actually speaking drainage in hill roads is of prime importance. The drainage in such roads may be surface drainage and cross drainage.

In hill or ghat roads, drains are generally provided on one side of the road, usually on the hill side. Catch water drains are provided higher up in the hill to intercept the run off from the hill slope and divert it towards a nearby nallah. The catch water drains should not be less than 4.5 m from the edge of the road.

The shape of drains near the road side is (i) Angle type drains, (ii) Kerb and Channel type, and (iii) Saucer type. These drains are shown in Fig. 9 7.

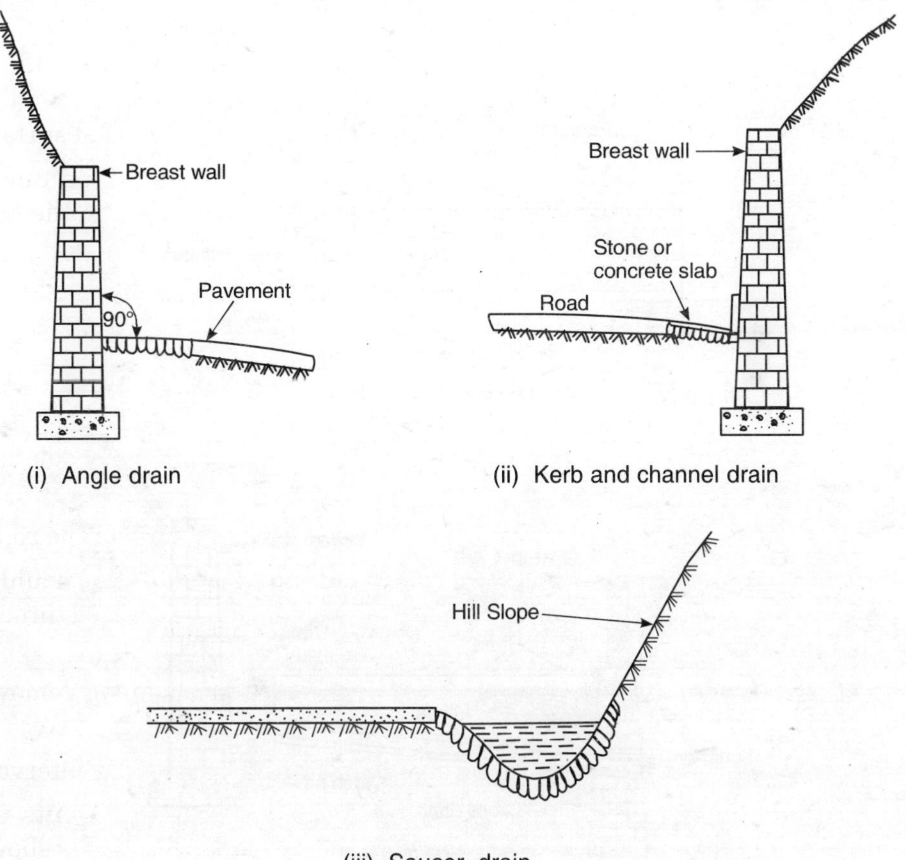

Figure 9.7

To prevent the side drains from overloading and thereby causing the road surface flooding, cross drainage should be provided at frequent intervals. Cross drains also help in reducing the size of side drains. Cross drains are provided by means of small *under drains, scuppers, Irish bridges,* etc.

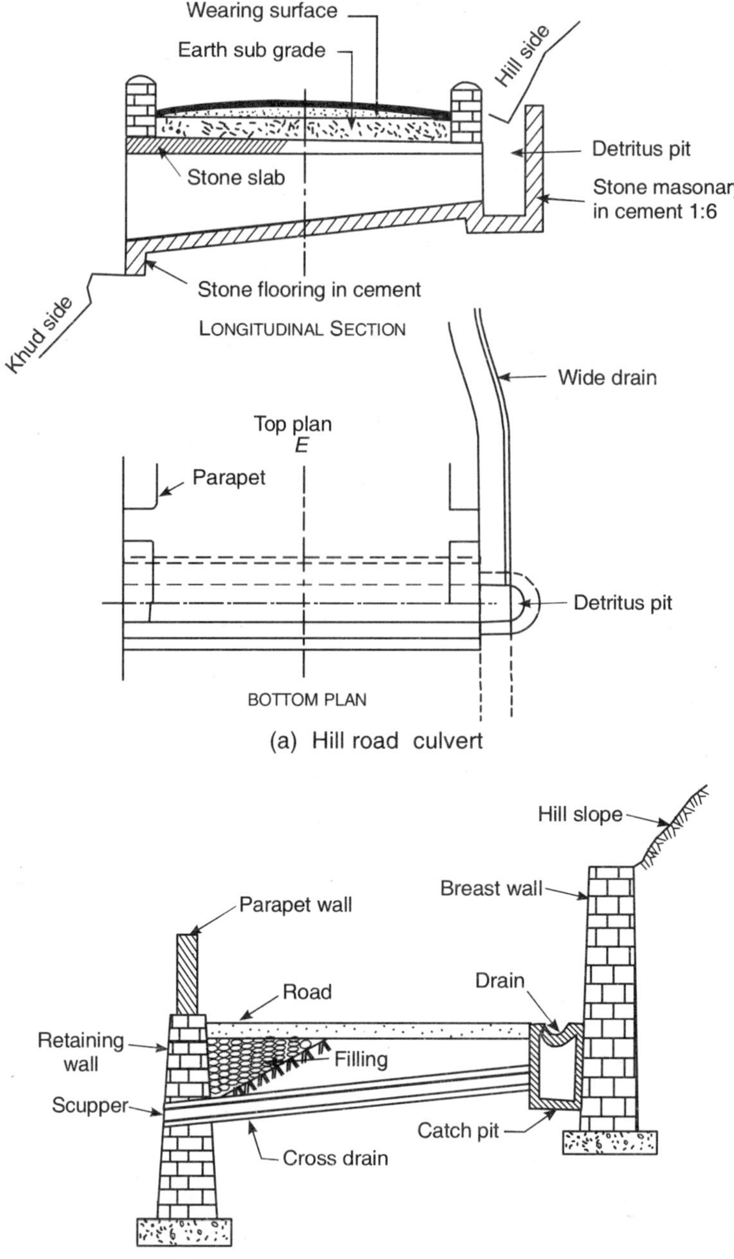

(a) Hill road culvert

(b) Section through cross drain and catch pit

Figure 9.8

9.11 RETAINING AND BREAST WALLS

Retaining Wall. It may be defined as a wall built to resist the earth pressure of filling and the traffic load of the road. It is commonly used in hill roads when the road goes in embankment, partly cuttings and partly embankment. Retaining walls can be constructed in dry stone masonry, stone masonry, brick masonry or cement concrete.

Dry stone masonry retaining walls are the simplest and are mostly used in hill road construction. The stability of such walls does not depend upon the quality of materials used in the masonry but on the workmanship, that is, the arrangement of stones in the wall. The stones to be used should be of a large size and should be roughly hammered and dressed. The wall should have a minimum top width of 60 cm and the front face should have a batter varying from 1 in 4 to 1 in 3. In principle the height of dry stone retaining wall should be restricted to 6 m only. For higher walls, the lower portion of the wall should be made in dry masonry and the upper portion in mortar.

Breast Walls. These are stone walls provided to protect the slope of cutting in natural ground from the action of weather. The section of a wall to be adopted depends upon the height of the wall, the nature of the backing, and the slope of cutting. The front and the back batter varies from 1 in 2,with a minimum top width of 60 cm.

9.12 CONSTRUCTION OF HILL ROADS

The principles of construction being same as that of ordinary plain roads, the construction of roads in hilly terrain is costly, time consuming and hazardous task. The cost of construction of a 6 m wide road comes out to be nearly Rs. 60–80 lacs per kilometre depending on the altitude. It takes 3–4 years to construct 100 km length of road using mechanised methods. In our country nearly 3,00,000 sq km hilly terrain is to be covered by roads. Most of the area thinly populated and above an altitude of 5000 m above the mean sea level. A terrain can be divided into four groups depending on the cross slope

(i) Level (L)	0 to 99%
(ii) Rolling (R)	10 to 24.9%
(ii) Mountainous (M)	25 to 60%
(iv) Steep (S)	above 60%

Hill roads will have cross slope more than 25% (cross slope means slope of a line approximately perpendicular to the alignment). A hilly mountainous area is characterized by a highly broken relief with widely differing elevations, steep slopes, deep gorges and a large number of water courses. Owing to complex topography the route has to be ineffectively increased.

The geometric design standards in respect of motor roads are more or less similar to that of plain roads. The bridle paths are meant for pedestrians and animal pack transport. The width of a bridle path vary from 2.0 m to 3.5 m with as high a gradient as 1 in 10. The minimum width specific for various type of hill roads is

National and state highway	9.0 m – 11.5 m
Major and other district roads	7.5 m
Village roads	6.5 m
Minimum radius of curvature	14.0 m
Minimum length of transition curve	15.0 m
Maximum gradient	1 in 40
Maximum super elevation	1 in 10

Stages of Construction. The important stages of construction of a hill road are

(i) Reconnaissance

(ii) Trace cut

(iii) Detailed survey

(iv) Formation cutting

(v) Drainage and protective works

(vi) Pavement construction

(i) *Reconnaissance.* During the process of reconnaissance a route is selected for alignment. Reconnaissance consists of study of topographical sheets, aerial survey sheet, meteriological maps etc. Sometimes aerial survey has to be performed for inaccessable routes. After reconnaissance the proposed alignment is pegged for trace cut and detailed survey.

(ii) *Trace cut.* It is defined as a path 1.0 m to 1.5 m wide prepared along the proposed alignment for the purpose of detailed survey. It is prepared along the natural configuration of spurs and re-entrants etc. Sometimes a detour or go on a steeper rise and fall till a comfortable or easy terrain is reached. Sometimes access in valley is achieved by dry rubble filling, particularly in case of hard rock terrain. Places where there are steep rocky terrain, it is economical to provide timber planks over a bally framework resting on ledges, instead of trace cut. The trace cut is done at a ruling gradient of 1 in 25. This gives sufficient margin to obtain a ruling gradient of 1 in 20 when final formation is completed. At zig-zags this fracture leads to difficulties because it is not always possible to obtain correct curve compensation to gradient. The ruling gradient is changed to almost level at the hair pin bends or sharp curves to ensure that the finished gradient is correct.

(iii) *Detailed survey.* After the trace cut deluded survey is conducted along the trace cut nearly 15 m on either side on straight route and 30 m on curves and hair pin bends. Reduced levels are recorded at an interval of 10 m and cross sections at an interval of 20 m. Contour survey plan at all the curves should be prepared with a contour interval of 2 m.

(iv) *Formation cutting.* Initially a 3 m wide formation cut by widening the trace cut, is prepared. The initial cut also generally follows the natural configuration of spurs and re-entrants. After the initial cut the road is widened to its full width. While the final formation is being prepared, the initial cut is being filled and blocked. Later the curves are improved to correct radius and formation is graded to final grade. The hill sides should be dressed to the following recommended slopes :

Ordinary soil, moourm or hard clay	$1 : 1$ to $1 : \frac{1}{2}$
Disintegrated rock	$1 :$ to $\frac{1}{2} : \frac{1}{4}$
Soft rock and shale	$1 : \frac{1}{4}$ to $1 : \frac{7}{8}$
Medium rock	$1 : \frac{1}{12} \; 1 : \frac{1}{6}$
Hard rock	Nearly vertical or half tunnelling if the height is more than 7.5 m.

As briefly indicated in Art. 9.3 the formation can be in partial cutting and filling, full or half tunnelling or in embankment. For tunnelling blasting is sometimes desired. The detailed process of blasting is dealt with in subsequent articles.

(v) *Drainage and protective works.* The drainage in hill roads forms an important part of hill road construction. Similarly the protective works such as breast walls, parapets, retaining walls etc. should be properly planned and constructed before the actual construction of pavement starts.

(vi) *Pavement construction.* Pavement construction of hill roads is definitely more hazardous than in the plains Dozers, rollers etc., cannot work on very steep gradients so sometimes ramps are prepared for the movement of dozers etc. Hill roads are subjected to high intensity of rainfall, hence W.B.M. surfaces will require high degree of maintenance because of its permeable nature. Impermeable pavements like bituminous and cement concrete are prepared. Generally bituminous surfacing over W.B.M. base is used as cement concrete pavements are costly and require more time for construction. Indian Road Congress has recommended 175/225 penetration grade bitumen for the construction of hill roads in cold regions. Cut back with slight heating is also used. RC3 cutback has been found to be useful and economical in regions subjected to freezing temperature. Mastic asphalt surfacing should be used in strategic hill roads carrying very heavy traffic like tanks.

9.13 LAND SLIDES

It is defined as a displacement of a big mass of rock rapidly or all of a sudden without giving any warning. It is also defined as the displacement of residual soil or settlement adjoining to a slope in which the centre of gravity of the moving mass advances in downward direction. It is the failure of the rockmass in shear. The landslides can be classified into four categories depending on the mechanics of its failure viz. falls, slides, flow and complex.

Falls. These landslides occur due to tension failures. A portion of the rock mass rolls down the rest of the rock without interaction.

Slides. This is caused purely due to failure of shear resistance. The shear cracks are visible before the slide takes place.

Flow. This type of landslide is just like the flow of a viscous material having practically no shear resistance.

Complex. It is the combination of two or more of the above reasons of failure.

Causes of Landslides. Following is the list of various causes of landslides:

(i) Earthquake and nearby blasting shocks.
(ii) Nearby cxavation and errosion
(iii) Increase in pore water pressure
(iv) Seepage of water from artificial sources
(v) Deforestation of the area
(vi) Heavy and continuous rain
(vii) Snow melting
(viii) Friability of soil
(ix) Water logging of the area
(x) Spontaneous liquification

Prevention of Landslide. Landslides are one of the most serious problems of the maintenance of hill roads. Landslides can be prevented by any one or more of the following methods :

(i) By preventing pavement seepage

(ii) By providing adequate surface and sub-surface drainage

(iii) By reducing earth slopes and

(iv) By providing adequate protective works.

9.14 GEOMETRICS OF HILL ROADS

On account of the topography of the land the geometric standards of hill roads are different than those of plains.

Camber. On straight reaches of hill roads camber is provided which is normally 10–20% higher than the camber provided for the roads in plain. Following values of camber are recommended for different surfacing of hill roads:

Earth or Moourm roads	1 : 25 to 1 : 33
W.B.M. and Gravel roads	1 : 30 to 1 :40
Bituminous roads	1 : 40
Mastic asphalt and cement concrete roads	1 :50

Sight distance. The stopping sight distance is calculated by the following formulae :

$$\text{Stopping sight distance} = 0.278\ V.t + \frac{V^2}{254f}$$

where

V = speed of vehicle in km/ph

i = reaction time

f = coefficient of resistance

Normally the value of t is taken to be 3 secs and the value of f to be 0.4–0.5 On this basis the I.R.C. recommends the following safe stopping sight distances for various design speeds.

Speed		*Sight distance*
20 km/ph	=	20 m
25 km/ph	=	30 m
30 km/ph	=	35 m
40 km/ph	=	50 m
50 km/ph	=	70 m

The overtaking sight distance is calculated by the following formula :

Overtaking sight distance = $0.556\ V_b + 2S + 0.278\ T\ V_b + 0.278\ V.T$

where

S = space between the vehicles

V_b = speed of vehicle to be overtaken

V = speed of overtaking vehicle

T = overtaking time

In this formula the following assumptions are made:

V_b is taken to be $(V - 16)$ kmph

S is taken to be $(0.2\ V_b + 6)$ m

and $T = \sqrt{\dfrac{14.4}{A}}$ sees where A = acceleration in kmph.

The minimum overtaking sight distance for various speeds is

Speed	*Overtaking sight distance*
30 kmph	90 m
40 kmph	145 m
50 kmph	210 m

Radius on horizontal curves. The radius on horizontal curves can be calculated by the following formula

$$R = \frac{0.008V^2}{e\quad f}$$

where

V = design speed in kmph

e = superelevation

f = coefficient friction

As per I.R.C. recommendations, the value of e varies from 1 in 15 to 1 in 10 and the value of f is 0.15.

Transion Curves. The length of a transion curve is calculated by the following formula

$$L = \frac{0.0125V^3}{CR}$$

where

L = length of transition curve

V = speed in kmph

R = Radius of curve

C = Constant and is equal to $\frac{80}{V \quad 75}$. The value of C is taken to be 0.76

I.R.C. recommends a minimum length of 10 metres upto 40 kmph and 20 m for design speed of 40–50 km/h.

9.15 BLASTING

Blasting of rocks is necessitated where cutting by hand tools becomes ineffective, slow and uneconomical. Blasting consists of drilling holes filling the holes with explosives and firings. The explosives consist of combustible materials, such as sulphur, charcoal etc. and oxygen which is obtained from sodium or potassium nitrate. Following rock blasting explosives are generally used:

(a) Blasting Powder. It consists of 75% potassium nitrate, 15% charcoal and 10% sulphur.

(b) Dynamite or Nitroglycerine. It is a very strong explosive consisting of 75% Nitroglycerine and 25% silicate earth.

(c) Gun Cotton or Nitro-cellulose. It is made by saturating cotton with coc. Nitric acid. It is even more powerful than dynamite but it is highly inflammable in dry air, therefore it is always kept damp.

(d) Cordite. It is used in the form of a stick or cartridge and is a combination of gun cotton and Nitro glycerine. It is as good as gun cotton or Dynamite but cheaper.

(e) Rock-a-Rock. It is composed of 79% potassium chlorite $KClO_3$ and 21% Nitrobenzoil.

(f) Blasting Gelatine. It consists of 93% nitroglycerine and 7% guncotton.

(g) Liquid Oxygen. It is also sometimes used as an explosive.

9.16 HANDLING AND TRANSPORTATION OF EXPLOSIVES

Explosives are hazardous materials and demand great caution and care in handling and use. Before transporting the explosives from one place to another, it must be ascertained that the explosives are well packed in suitable containers and are being tested for their stability and sensitivity. No other inflammable material should be placed near the explosives or carried along with them. Different types of explosives should not be kept or carried together.

Often the explosives are to be stored in a place before they are brought for use. Great care should be taken in their storage so as to prevent accidental explosion. Explosives are stored in a specially built stores known as *Hazardous Buildings* away from dwelling places. Different type of explosives should be stored in different apartments.

9.17 BLASTING PROCEDURE

For blasting, holes are driven either manually or mechanically, to a depth to which the rock is to be blasted. The holes should be cleaned and dried. After the hole has been properly cleaned a charge of blasting powder is deposited in the lower portion of the hole and is lightly tampted with a wooden tamping road. A primming needle, coated with greas is inserted and then dry earth or moorum is filled in the hole in layers of 4 cm. Each layer is lightly tempted. After the hole has been filled and tempted, the primming needle is removed, thus a hole is formed. Each time when the hole is filled with earth, the needle is being rotated so as to facilitated its removal. The hole is now filled with gunpowder and a fuse is introduced. A fuse is a cotton thread impregnated with gunpowder. Wickfords safety water proof fuse is the best among the various fuses available. It burns at the rate ol 60 cm per minute. One end of the fuse is kept in the explosive and the other end outside atleast one metre from the hole. The fuse is then fired.

When the blasting is required on a large scale, a series of holes are driven and each of them is charged with the blasting powder and fired simultaneously. Series of blasting holes are fired by an electric fuse wire by an electric spark by remote control.

To facilitate blasting, line of bedding is located along which the rock can be broken easily. The explosion develops a tremendous force acting in all directions. The explosive gases escape the way they will find the least resistance. These lines of bedding or fissures along which the rock will break

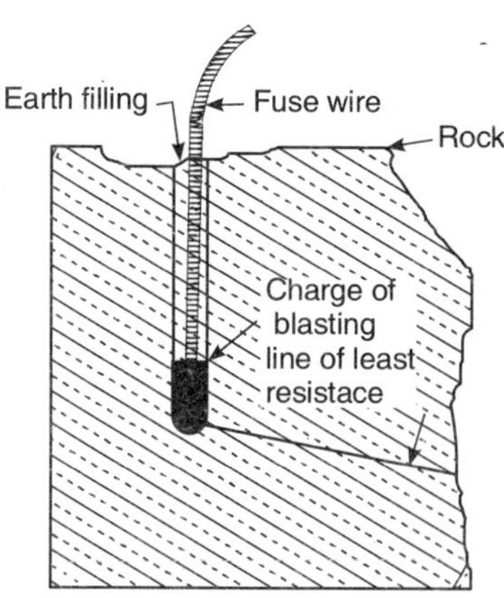

Figure 9.9 *Blasting*

or detach is known as line of least resistance. If the blasting charge is not tempted properly, the explosive gases will escape along the line of blast hole and the rock will not break. If the tamping is too hard it will still not explode. Hence tamping should be done very carefully. The mis-fire may cause accidents and the following precautions may be taken :

(i) No hasty attempt should be made to find out the cause of mis-fire. Let a good deal of time pass before approaching the blast hole.

(ii) The mis-fire hole should never be re-drilled. A second hole should be drilled at least 1 metre away from the mis-firehole.

The quantity of blasting powder to be used will depend upon the nature of the explosive, toughness of rock, depth to which the rock is to be broken and the length of the line of least resistance. As a general rule, $\frac{1}{2}$ kg of blasting powder will explode 4.5 tonnes or 17 cu. metre of soft rock and 2.75 tonnes or 10 cu. metre of hard rock. The blasting hole diameter will vary from 2.5 cm to 7.5 cu m and are spased 1.5 times the depth of holes. Depth of hole will depend on the depth of line of least resistance. Following bore diametres and holes are recommended:

Depth of hole	*Diameter*
1 m to 2 m	2.5 cm
2 m to 3.5 m	4.0 cm
3.5 m to 5.0 m	6.0 to 7.5 cm

9.18 DETONATORS

These are small metallic caps of very sensitive type of explosives in small quantity. By the effect of fire, friction and hammering the detonators explode instaneously. Normally TNT is used in aluminium containers for detonators. The detonators are of two types viz. electric detonator and plain detonators. The electric detonators are fired by electric current supplied by a battery while a plain detonator is fired by fire through a fuse wire or fuse coil. Fuse coils are manufactured with insensitive explosive materials.

It is to be noted carefully that the amount of charge, capacity of detonator, spacing of holes, depth of holes etc., should be carefully planned and decided so that the explosion is not ineffective.

9.19 TOOLS AND ACCESSORIES FOR BLASTING

Following tools and accessories are required for blasting.

(i) **Prickers.** These are sticks made of hard wood 15 cm long and 6 mm in diameter and are used for making holes in the cartridges for inserting detonators.

(ii) **Crimper.** These are used for placing plain detonators on safety fuses.

(iii) **Cap sealing compound.** These are required when holes are likely to contain water. The junction between fuse and detonator is properly coated with this compound to make it water tight.

9.20 REQUISITES OF EXPLOSIVES

Following are the requisites of explosives :

(i) **Power.** Power of explosive means the intensity and strength of the explosive and may be compared with their blasting capacity or capability. Blasting gelatine is one of the most powerful explosives and its strength is taken as 100%. The power of all other explosives will be compared with the power of blasting gelatine.

(ii) **Velocity of Detonators.** It is the rate at which the detonation passes through a column of explosives. Explosion effective increases as the detonation velocity increases.

(iii) **Density.** Greater the density of the explosion material, greater will be the explosion effect.

(iv) **Resistance to Water.** Some explosives deteriorate rapidly under moisture and wet conditions while some explosives withstand for longer duration under moist and wet conditions.

(v) **Sensitivity.** An explosive should be insensitive to normal handling shocks and frictions, yet it should be sufficiently sensitive to detonation.

9.21 PRECAUTIONS IN THE STORAGE OF EXPLOSIVES

Following precautions should be taken while storing explosives :

1. Explosives should not be brought near the fire, lighted lamps or candles or a burning cigarette.

2. The explosive should be stored in dry clean and well ventilated magazines.

3. The explosives should be handled by an experienced and license holder.

4. Explosives and detonators should be kept in separate boxes and should be carried by separate persons.

Precautions during use

Following precautions should be taken while using explosives :

1. Iron and steel tools should not be used for opening cases. Only wooden or copper impliments should be used.

2. Never store explosives in the sun.

3. Replace the cover of cases immediately after taking out the explosives.

4. Explosives should not be carried in loose packets.

5. Never insert anything else other than fuse coil in the detonators.

6. Do not remain near the explosives during storm or thunder storm, nor they should be transported.

7. Never use damaged explosives or tools.

8. The explosives should be handled by a licensee.

Precautions while drilling and charging

1. Never start drilling before it is sure that the rock face contains no unfired explosive.

2. Check the condition of the hole with stemming rod before inserting the cartridge.

3. Force should not be used while inserting the cartridge in the hole.

4. Never keep large unwanted stock of explosive near the shot holes.

5. Cut detonating fuse from the rear immediately after the primer has reached the bottom of the hole. Insert only freshly cut fuse in the detonator.

6. Never force the detonator excessively into the cartridge.

7. Never try to soften cartridge by heating or rolling on the ground.

Precautions while firing

1. Never use short length fuses.

2. Always use approved crimpers for fixing detonators on a fuse.

3. Never use cartridges for lighting a fuse.

4. Always use fuse lighters. Use of Match box should be avoided as far as possible.

9.22 TUNNELS

An underground engineering structure or artificial gallery used for transporting traffic, sewage, water, oils and minerals etc. is known as *Tunnel.*

Following are the advantages of tunnels :

1. Tunnels protect the pavements from the weathering action of rains, snow and other natural influences, hence reduce their maintenance cost.

2. Tunnels protect the cities and other conveyance vehicles, passages which are inside them, from destruction during war bombardings.

3. At some places tunnels are cheaper for transporting water, gas, laying railway lines, roads across rivers or mountains than bridges or open cuts.

4. In most congested cities having no space for the construction of railways or roads, tunnels are constructed to provide subways for removing traffic congestions.

5. When the hill consists of soft soil and it is uneconomical to maintain the open cut, due to large number of slips, tunnels are constructed.

Following are the disadvantages of the tunnels :

1. Tunnels require special equipments and modern techniques for their construction, which are not available everywhere.

2. Tunnels require much longer time for their construction as compared to bridges and open cuts.

Generally when the depth of cut is more than about 170 m, tunnels are provided.

9.23 HISTORY OF TUNNELS

Tunnels were constructed in the very early age of civilization. It is said that the first tunnel was built about 4,000 years ago between two buildings by the Egyptians in Babylonia. The first tunnel under Eupharates river lined with brick masonry was constructed in Egypt connecting the Royal Palace to Temple of Jove.

In 54 A.D. Emperor Claudius built about 5.9 km long tunnel for carrying water of a spring under Appennine Mountains. The internal dimensions were about 3.0 m × 1.8 m and it was completed in twelve years by 30,000 labourers. The slaves and the prisoners were given the job of tunnel construction in the shape of severest punishment. Romans also constructed a 900 m long road tunnel about 2000 years ago under Posilipo hill.

One of the oldest Greek tunnels was built about 2600 years ago for carrying water on the island of Samos. This tunnel was about 1.8 m × 1.8 m and about 1500 metres long.

Upto seventeenth century, practically there was no improvement in tunnel construction methods. The use of gunpowder was started in France in 1679–81 for construction of Languedoc canal. In France during 1766–1777 Harecaslle tunnel was constructed having about 3 m × 4 m section, 2.4 km length. First railway tunnel was also constructed in France in 1826 on Roanne–Andressieax line. One of the longest tunnel used for navigation was constructed in 1927 known as Rov Tunnel in France.

Upto seventeenth century, practically there was no improvement in tunnel construction methods. The use of gunpowder was started in France in 1679–81 for construction of Languedoc canal. In France during 1766–1777 Harecastle tunnel was constructed having about 3 m × 4 m section, 2.4 km length. First railway tunnel was also constructed in France in 1826 on Roanne-Andressieax line. One of the longest tunnel used for navigation was constructed in 1927 known as Rov Tunnel in France.

Use of shields was introduced in 1825 by Sir Isam Bard Brunei for the construction of Thames river tunnel in Britain. In 1869 use of iron lining was started by Peter William Barlow while constructing tunnel under river Thames in London. The first railway tunnel was 3.2 km long about 9 m × 7.5 m cross-section, known as Box-Hill tunnel constructed around 1840 in Britain. In Britain use of shields and compressed air was done for the construction of Rotherhithe tunnel completed in 1908.

Connaught Tunnel 7.2 m × 80 m size, 8 km long is the first railway tunnel of Canada. Portage tunnel 60 m × 5.7 m size, 270 m long is the first railway tunnel of U.S.A., which was completed in 1833 in Pennsylvania.

60 m high × 7.2 m wide, 7.6 km long tunnel was constructed during 1857–1875 in U.S.A. which connects Boston with Albany. America's first vehicular tunnel was built in 1866 which is known as Washington S.T. Tunnel. It was constructed under Chicago river.

9.24 ALIGNMENT AND GRADE OF TUNNELS

While deciding the alignment and grade of tunnel the following points should be kept in mind :

1. As far as possible the alignment should be kept straight because it will be best and economical.

2. For providing good drainage minimum grade of 1 in 50 should be provided in drains. In the case of longer tunnels from the middle of tunnel grades may be provided on both sides to carry drain water towards both ends.

3. As far as possible minimum grade should be provided in tunnels and their approaches.

4. Proper precautions should be taken for efficient ventilation and lighting of the tunnels.

9.25 LOCATION OF CENTRE LINE ON THE GROUND

The correctness and economy of constructing tunnels entirely depends on the accuracy of the surveying. Therefore, survey work should be done with great accuracy and it should be checked several times during execution of the construction work also.

The location of centre line on the ground is done as following.

(1) When the length of the tunnel is small, the centre line is accurately located on the ground by means of a theodolite during calm and clear day time. The wooden pegs are driven in the ground at various short intervals and the centre-line is marked on the top of wooden pegs by means of nails. If it is not possible to drive the wooden pegs in hard stratum, the surface of the rock is painted with white paint and the centre line is marked with black paint over it. The centre line located so is checked several times and after thorough checking permanent monuments of stone or concrete are placed on the centre line at various places.

(2) When the length of the tunnel is more, accurate and elaborate survey methods are used for transferring the centre line on the ground. Large size Micrometer Transit Theodolite should be used and the centre line should be located by means of triangulation method. All the operations for locating the centre line on the ground must be performed with greatest accuracy and should be checked several times by independent methods, to avoid any error in the layout of centre line. After accurately locating the centre line on the ground, permanent stone or concrete monuments should be provided on the centre-line.

9.26 TRANSFERRING CENTRE-LINE TO INSIDE OF TUNNELS

(When the shaft is located directly over the centre-line)

Generally the centre line is transferred underground through the opening of a shaft. This is a very difficult work and requires more attention because the alignment will entirely depend on it. If any mistake is committed in it, the position of that part of the tunnel will be dislocated. This is done as follows :

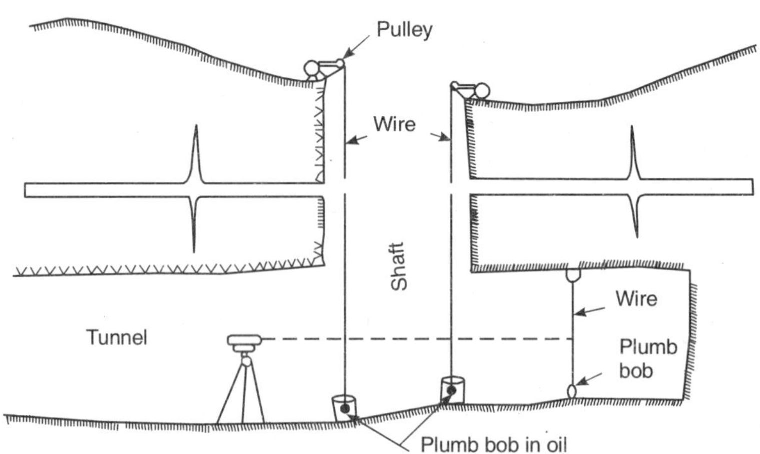

Figure 9.10 *Method of transferring centre-line from the top to the inside of tunnel*

Two plumb-bobs attached to fine wires are suspended in the shaft. The wires are wound on reels placed on the mouth of the shaft. By revolving reels plumb-bobs are immersed in water or oil buckets, so that they may not be affected by vibrations. With the help of a theodolite these wires are brought on the centre line at the ground.

When the wires at top are lying on centre line, they will automatically be on centre line in the shaft also. Now one theodolite is kept about 7 metres away from wires in such a way that, when observing through the telescope

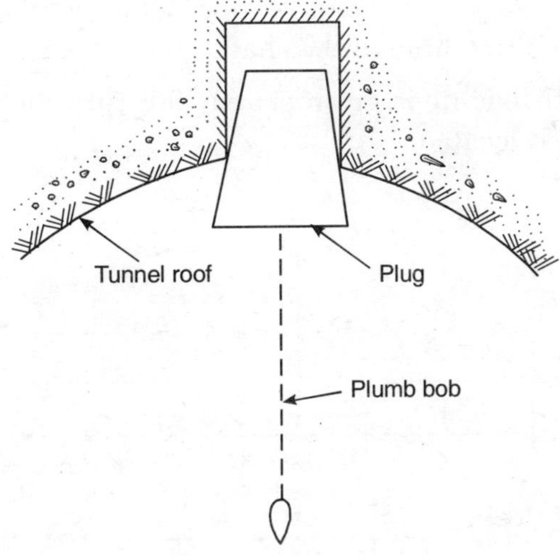

Figure 9.11 *Details of the pegs driven in the ceiling of the tunnel*

both wires are in one line. In this position the axis of telescope will lie in the centre line of the tunnel. Now the construction of the tunnel in one or both the directions is started along the centre-line, which has been fixed by driving pegs in the roof of the tunnel and suspending plumb-bobs from them as shown in Fig. 9.11.

9.27 TRANSFERRING CENTRE-LINE TO INSIDE OF TUNNELS

(When the shaft is located to one side of centre-line)

In some cases the shafts are placed on one side of the tunnel. In such cases the centre-line cannot be directly transferred to the inside of the tunnels. For transferring the centre-line from the ground to the underside following method is generally adopted. (Refer Fig. 9.12).

(a) Line *AB*, exactly perpendicular to the centreline of the main tunnel, is set out on the ground by means of a theodolite.

(b) Permanent stone or concrete pillars *P* and *R* are erected on the line *AB*, on both the sides of the shaft. Between the pillars a wire is stretched and from it two plumb-bobs are suspended as described above. With the help of these wires line *AB* is transferred inside the tunnel.

(c) Line *RS* is also established parallel to the centre-line of the main tunnel, intersecting line *AB* at *L*.

(d) Point *M*, the intersection of main shaft centre-line and line *AB* are located by means of a theodolite inside the tunnel.

(e) Keeping theodolite at point *M*, line *AB* is sighted by means of two wires suspended on *AB* through the shaft.

(f) Now the theodolite is given exactly 90° turn and the main centre-line of the tunnel is located.

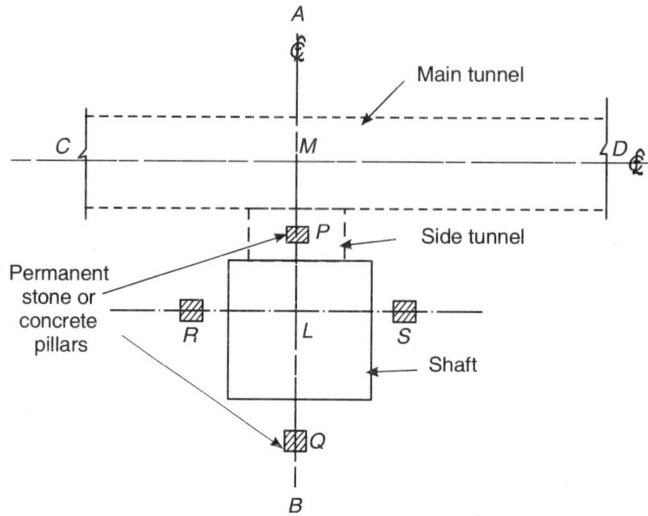

Figure 9.12 *Method of transferring centre-line to the tunnel through shaft in the side*

The centre line located so, should be checked by various independent methods. After ascertaining the accuracy the construction of the tunnel should be started on one or both the sides.

9.28 METHODS OF LOCATION OF CENTRE-LINE OF CURVILINEAR TUNNELS

Following two methods for the location of centre line of the curvilinear tunnels are usually adopted in practice :

(i) By chords and deflection angles

(ii) By tangent offsets.

(i) *By chords and deflection angles.* In this method the centre-line points *A, B* and *C,* of the tunnel are located as follows :

(a) For any chord length l, the angle subtended by the chord is first calculated by the formula

$$\sin \frac{\ }{2} = \frac{l}{2r}$$

where

l = length of the chord

r = radius of the curve

(b) The theodolite is set at the tangent point A and the back sight along Ax is taken. Now the telescope is turned 180° along the horizontal axis to x'. Now an angle of θ/2 is subtended and point B is located by measuring chord length l.

(c) The theodolite is shifted to station B and by the similar procedure point C is located.

(d) Now the theodolite is shifted to another point, C, *D, E,* ....etc., and the points on the centre-lines of the curves *D, E,*...etc., are located.

(e) All the poins *A, B, C, D* ,...etc. are joined together by a smooth curve, which is the required centre line of the curvilinear tunnel.

(ii) *By tangent offsets.* This is the common method used for locating the centre-line of the curvilinear tunnels.

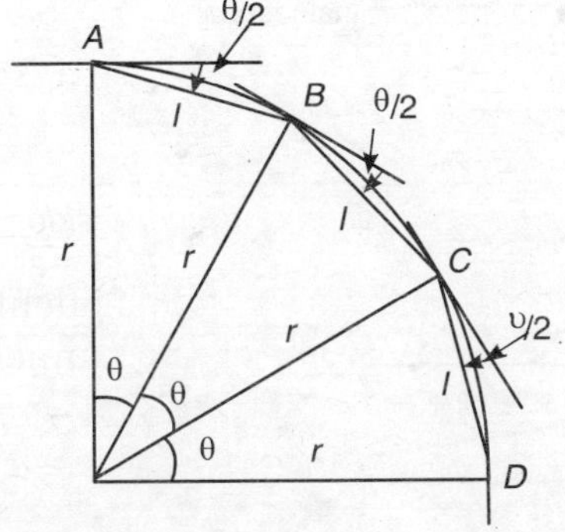

Figure 9.13

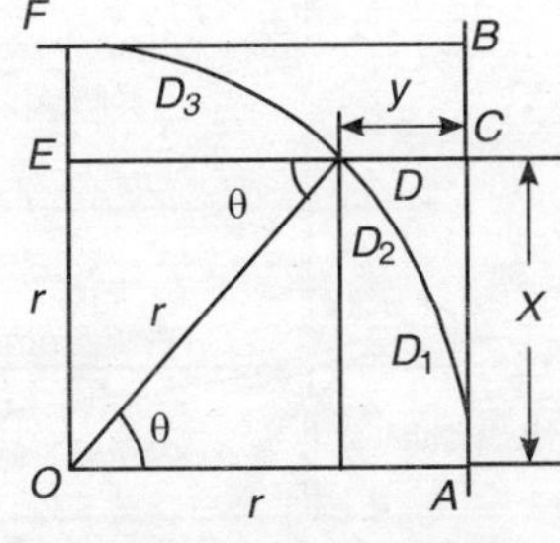

Figure 9.14

Referring to Fig. 9.14, if the radius of the curve of the tunnel is r, then

$$OE = AC = x$$

$$OA = OD = OF = r$$

$$AC = OE = x$$

$$DC = y$$

$$ED = \sqrt{OD^2 - OE^2}$$

$$= \sqrt{r^2 - x^2}$$

$\therefore$ Length of ordinate $DC = y = EC - ED$

$$= r - \sqrt{r^2 - x^2}$$

$\therefore$ $$y = r - r.\cos\theta$$

In this method for various lengths of the tangent x_1, x_2, x_3, ... etc. the ordinate lengths y_1, y_2, y_3, ... etc. are calculated. All the points D_1, D_2, D_3,...etc. corresponding to the various ordinate lengths are located and joined by means of a smooth curve, which is the required centre-line of the curvilinear tunnel.

9.29 METHOD OF PROVIDING GRADE

Following method is generally adopted for giving grades or slopes to the tunnel. (Fig. 9.15)

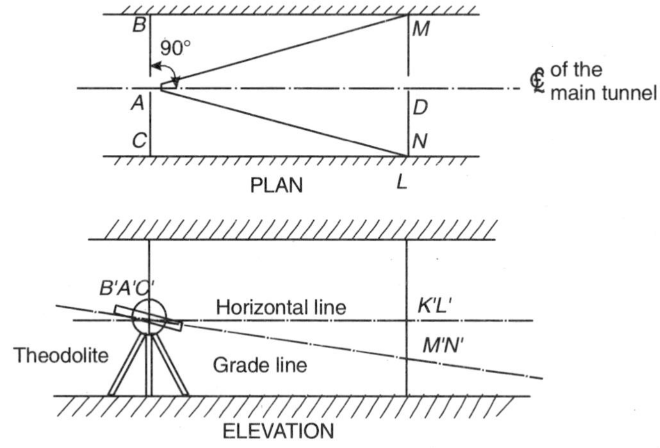

Figure 9.15 *Method of providing grade in tunnel*

1. The theodolite is kept on the centre-line of the main tunnel at *A* and then centred and levelled.

2. Now it is rotated exactly 90° horizontally and points *B* and *C* are located on the sides of the tunnel and plugs are inserted at these points, exactly to the level plane of the theodolite.

3. Keeping the telescope exactly in horizontal level it is directed along the tunnel and points *K* and *L* are marked on the sides of the tunnel as shown in Fig. 9.15.

4. The vertical distance *KM* = *NL* is calculated exactly to give the required grade for the measured horizontal distance *BK* = *CL*. Now plugs and spades are driven at points *M* and *N*.

Plumb-bobs are suspended through piano wires in shafts for transferring levels from the ground surface to the bottom of the shafts. Two marks are made on these wires, one at the top and the other at about 1.5 metre above the springing level of the roof arch.

With respect to the bench marks the corresponding reduced levels of the top and bottom of the tunnel are calculated. With the help of these reduced levels, reference points are fixed at the flow levels corresponding to the required grade of the tunnel. The grade of the tunnel is checked several times, to avoid any possible error.

9.30 SHAFTS

When the length of the tunnel is small, it can be constructed by doing excavation works from one side. But in case of considerable length and to complete the construction work in short time, the excavation can be done

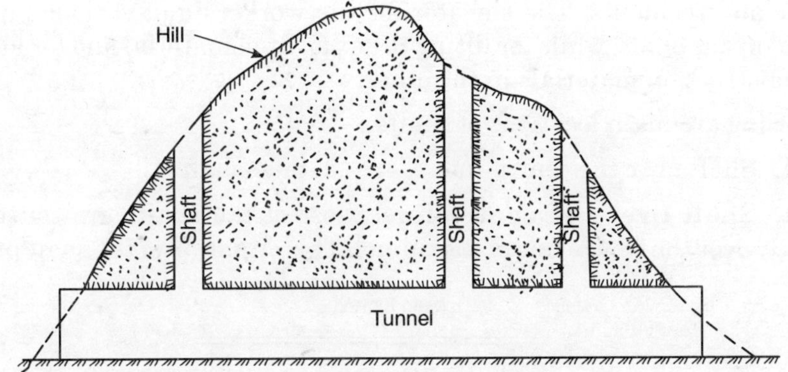

Figure 9.16 *Longitudinal section through a tunnel*

from both the ends of the tunnel simultaneously. But when the length of the tunnel is very long and the work is to be completed in short time, the vertical openings are made from the surface of the hill which meet the tunnel at various intermediate points. These vertical openings are called shafts and are constructed at suitable places to expedite the construction work.

Figure 9.16 illustrates the longitudinal section through a tunnel with three shafts. In this case the work can be started from both the ends and also from each shaft in both directions. After transferring the centre-line to the underground, in each shaft work is started in both the directions. The excavated materials are taken out from the shafts.

The shafts are also constructed in tunnels even when the work is started from one or both ends, to avoid delay in disposing of the materials, when the excavation has been done upto a considerable length, because when the tunnel has been constructed upto a long distance, it will be costly to dispose of the material from the mouth of the tunnel. In such cases it is always economical to dispose of the material through the shaft.

9.31 LOCATION OF SHAFTS

The site for the location of the shaft should be selected such that its construction cost is minimum with maximum advantage. If the site permits the advantage of the valley should be taken for the location of the shaft. If the shaft is located in the valley, a wall should be built around it to prevent the entry of water. While deciding the number of the shafts, the advantages gained from a large number of working faces should be compared with the additional cost of the shaft.

While locating the working shafts for sewers, care should be taken to see that it should cause minimum obstruction to the traffic and annoyance to public and residents. The elevator for the workers and various pipes are placed in one shaft, while for lifting the excavated material and for lowering the construction materials in the other.

Following are main locations of shafts :

1. Shaft over the centre-line.
2. Side-shafts

1. Shaft over the centre line. Most of the shafts are constructed directly over the centre line of the tunnel. This situation of the shaft provides

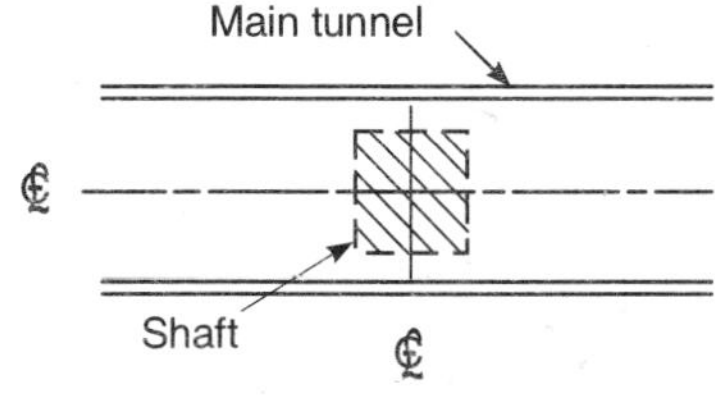

Figure 9.17 *Shaft over the centre-line*

maximum facility in affording for hoisting out the excavated materials, taking down the workers and construction materials from the surface in case of long tunnels.

2. Side-shafts. The shafts which are not directly constructed over the centre line of the tunnel, but are constructed in the side are called side-

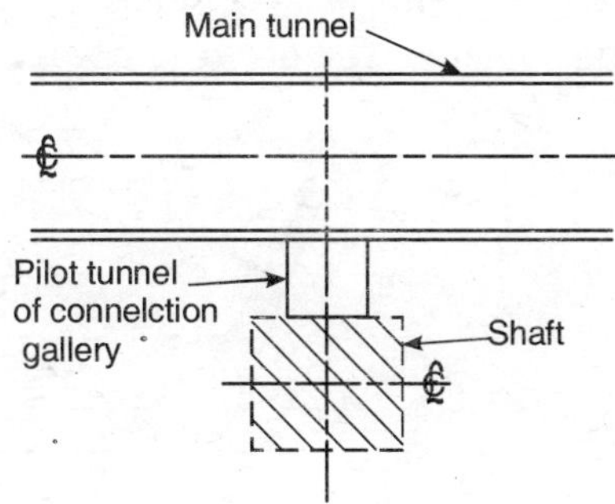

Figure 9.18 *Side shafts*

shafts. These shafts require a transverse gallery connecting with the tunnel. The side gallery should be of such size and shape that the construction equipment machines can be easily taken through the shafts to the tunnel.

9.32 CONSTRUCTION OF SHAFTS

The methods to be adopted for the construction of the shaft mainly depends on the nature of the ground. In common practice the shafts are usually sunk down from the top toward the tunnel. In special cases however the shafts can be constructed from the tunnel in upward direction. The construction of the shaft in the upward direction is cheaper as the muck is dropping down and can be directly trapped into the muck carrying cars.

Following are the main operations for the construction of the shafts in the rocks:

1. Drilling and blasting.
2. Mucking.
3. Timbering.
4. Pumping.

1. Drilling and Blasting. Holes are drilled in the rocks by various drilling machines and equipments, such as jackhammers and pneumatic compressed air operated equipments. The large size shafts are excavated by 'stepped down' technique for permitting the drilling operation and the taking out mucking outside simultaneously.

2. Mucking. It is the operation of removing the excavated material from the shaft and dumping it outside at a suitable place. The mucking operations can be done by manual labour or by cranes. Figure 9.19 shows the mucking operation by crane. In the mucking by crane two buckets are used, when one is being loaded, the other is being hoisted to the surface.

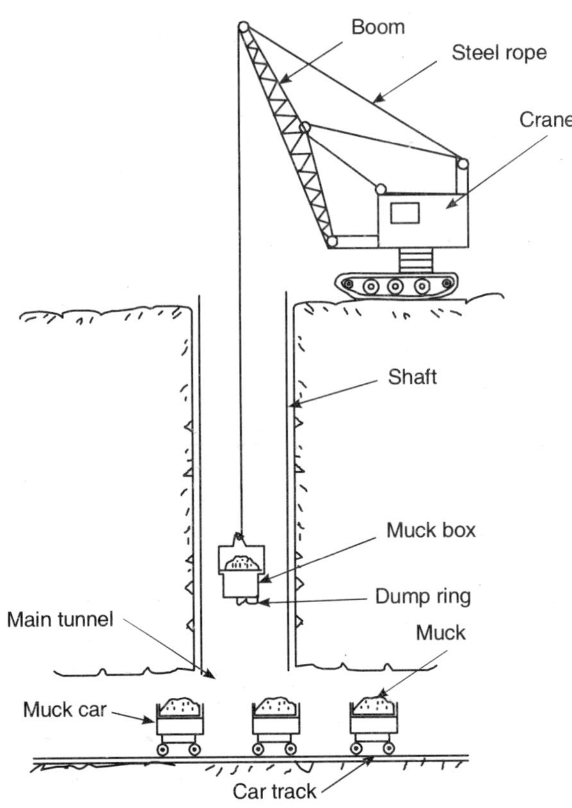

Figure 9.19 *Mucking operation by crane*

3. Timbering. It is process of providing temporary support to the cut soil sides against falling till lining is done or finally filled up after the completion of the work. Figures 9.20 and 9.21 show various methods of timbering.

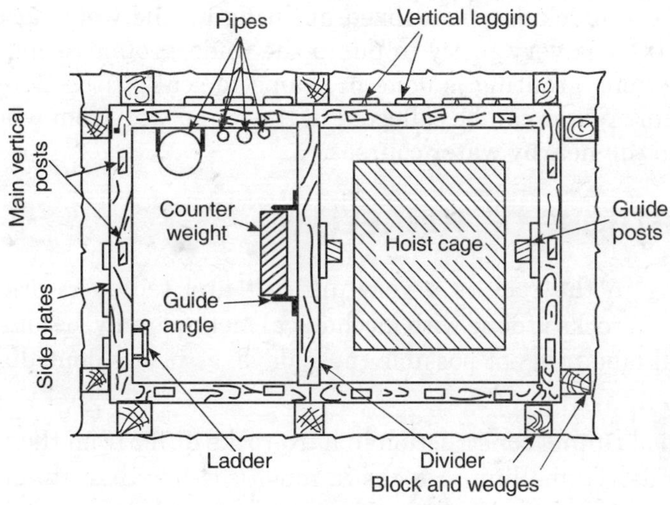

Figure 9.20 *Timbering for small shafts*

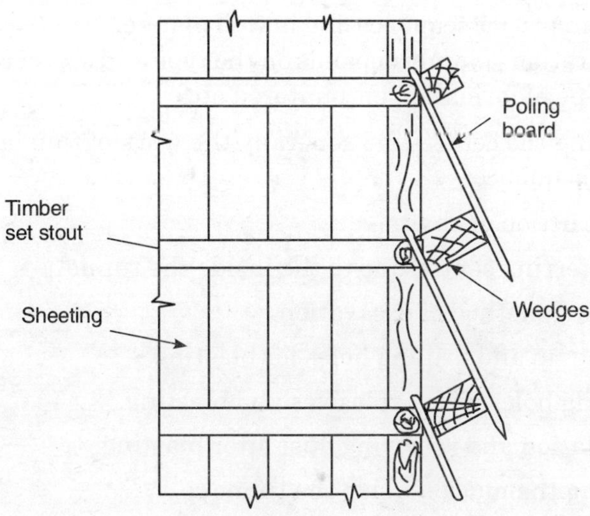

Figure 9.21 *Timbering for deep shafts in soft soils*

4. Pumping. Usually pump-sets are used for lifting the water from the shaft. This water may be due to seepage from underground or water used during construction operations. In the beginning the buckets used for removing the muck can be utilized in lifting up the water also. In case of deep shafts, it is very costly to pump the underground seepage water. In such situations grouting is done in seams to seal off the flow of water to some extent. Water can be collected in the sump well from where it can be pumped to the nearby water courses.

9.33 METHODS OF TUNNELLING

Tunnelling in the rock is a very difficult and tedious work. Nowadays tunnelling in rocks is done with mechanical methods only, because tunnelling by manual labour is not possible these days, as it was done during ancient times.

Method of tunnel construction in hard rocks differ from the tunnelling in soft grounds. Tunnelling in rocks in much costlier than tunnelling in soft ground. As rocks are self-supporting, they require less supporting. As slight deviation will involve huge wastage of money, greater care is required in tunnelling through rocks. In rocks the tunnelling work can be started in various sections for expediting the work.

Sequence of Operations. All the operations in the tunnel construction must be done in proper sequence and in well planned manner. The sequence of various operations mainly depends on the type and size of tunnel, method of blasting, type of formation encountered etc.

After locating the centre-line generally the work of tunnelling is done in the following sequence:

1. Construction of shafts.
2. Transferring centre-line to the inside the tunnel.
3. Deciding method of excavation.
4. Setting up and drilling holes for blasting.
5. Loading holes with explosives and blasting.
6. Ventilation and removing dust after blasting.
7. Carting the muck outside the tunnel.
8. Pumping and removing ground water from the inside of the tunnel (if any).
9. Providing supports inside the tunnel.
10. Lining of the tunnel.

9.34 TUNNELLING IN SOFT GROUND

Soft ground mainly means variety of materials of the soil. During doing tunnel excavation work some materials require immediate supports for the soil, whereas some materials can stand unsupported for some time. The various supports used at various places depend on various factors. Sometimes before timber supports were used, but nowadays these have been replaced by linear plates and shields.

Following types of grounds are commonly met with during tunnelling in soft ground:

(i) **Soft Ground.** In this type of ground the roof requires instant support after doing excavation. The side-walls can remain standing without any support for small duration of time only (Example clayey soils).

(ii) **Firm Ground.** In this type of ground the roof can also remain unsupported for few minutes, and the sides can remain without support for 1–2 hours. (Example dry earth, firm clay).

(iii) **Running Ground.** Sand, gravel and other cohesionless soil comes in this category of ground. These soils require instant support during excavation work and cannot remain unsupported in any case.

(iv) **Self-supporting Ground.** These are soft stones such as sand stones, cemented sands, mud stones etc. These grounds can remain unsupported for short lengths of 1.5 to 3.5 metres for short periods, but after that they require support.

The method to be adopted for tunnelling in the soft ground mainly depends on various factors, such as size of the ground, type of ground, equipments available for construction and sequence of the tunnelling etc. Large section tunnels require more supports which should be immediately placed in position than smaller section tunnels.

While tunnelling in soft grounds, explosives are not required, as the excavation can be done with hand tools such as pick axes, shovels etc. Due to heavy strutting and supports inside the tunnel, there will be more obstructions in the movement inside the tunnel, due to which the progress of the work is reduced. The tunnelling work is done with maximum possible precautions as soft ground may collapse and may cause accidents. All the supports and struts should be sufficiently strong to bear the load and pressure coming over them.

9.35 TUNNELLING METHODS

Following are the main operations in tunnelling :

1. Excavation.

2. Providing supports and strutting.
3. Carting excavated material outside the tunnel.
4. Lining.

Following are the main methods used for tunnelling in the soft grounds :

1. Forepoling method.
2. Needle-beam method.
3. Linear Plate method.
4. Shield method.
5. Compressed Air method.
6. English method.
7. Italian method.
8. Army method.
9. Belgium method.
10. American method.
11. German mehtod.
12. Austrian method.

Forepoling Method. In olden days this method was very common, but now it has been replaced by compressed air method. Very skilled labour is required for doing tunnelling in soft ground by this method. As this method

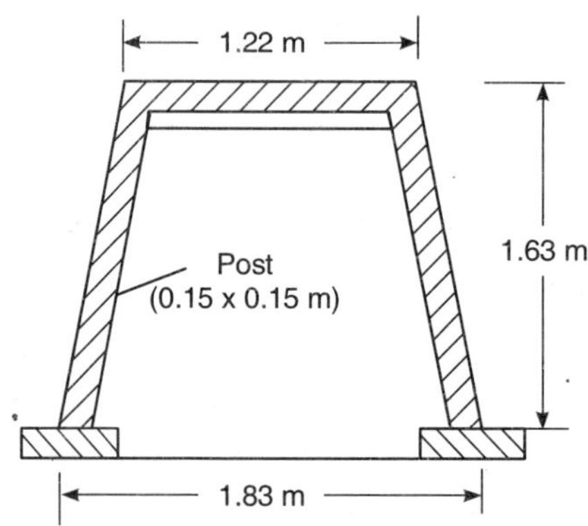

Figure 9.22 *Wooden bent*

requires manual labour, and timber in large quantity is required for supporting purpose the work of tunnelling in this method is to be done in the following sequence only :

1. At the designed location on the surface, the shaft is sunk from the ground surface to the grade level and the timber sheeting is provided to keep the soil from collapsing.

2. Wooden bent is set up a few centimetres from the sheeting and securely braced as shown in Fig. 9.22.

3. Now the holes are drilled in the timber sheeting as shown in Fig. 9.23, above and below the cap.

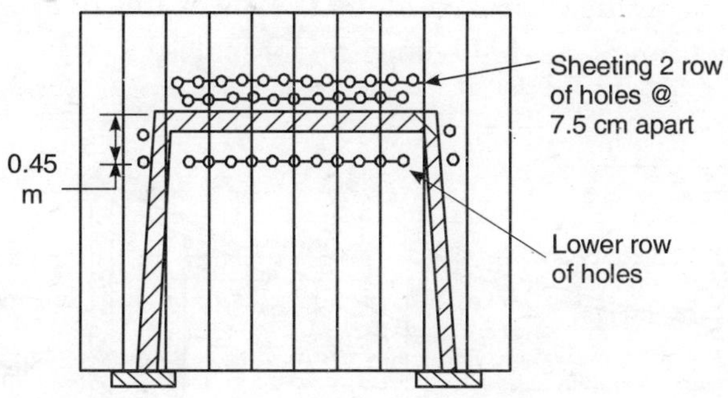

Figure 9.23 *Position of holes in the sheeting*

4. The sheeting above the cap is cut out along the top lines of the holes.

5. Now the wedge end shaped 'Fore-poles' having size about 170 × 15 × 5 cm are driven through the cut in the sheets, into the ground at an inclination as shown in Fig. 9.24 upto their half lengths.

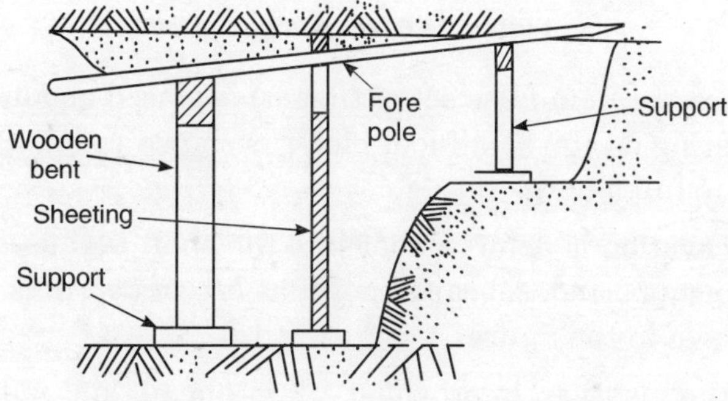

Figure 9.24 *Method of driving forepoles*

6. A timber plank is placed along the roof and wedges are placed to press the hanging sides of the forepoles from top as shown in Fig. 9.25.

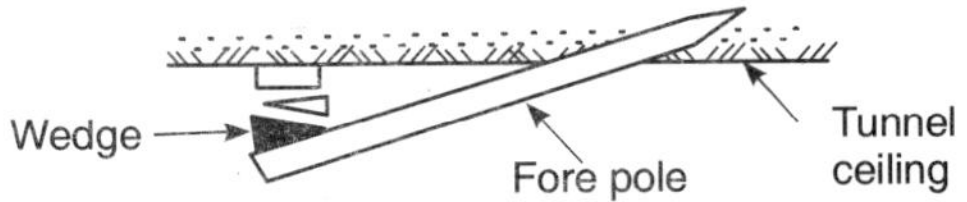

Figure 9.25 *Method of placing wedges*

7. The face sheeting is broken along the lower line of holes below the cap and the excavation is started below the forepoles.

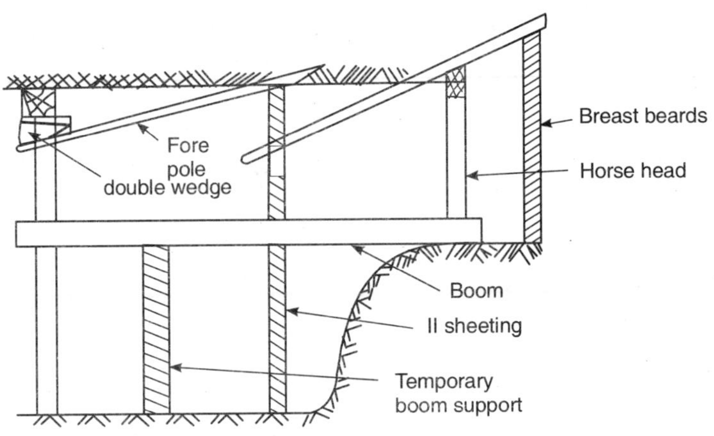

Figure 9.26 *Horse-head placing*

8. Now horse-head (a false set of timber) is placed about 60 cm from the sheeting, resting on the small foot block as shown in Fig. 9 26. Spikes are then driven to full length.

9. The excavation is done further into the shaft because the ends of the spikes are unsupported. A board known as *breast board,* is provided 45 cm ahead as shown in the figure.

10. Now the next cap is provided 1.2 m ahead, and is held in position temporarily by a single post set on the bench as shown in Fig. 9.27 (a) and (b).

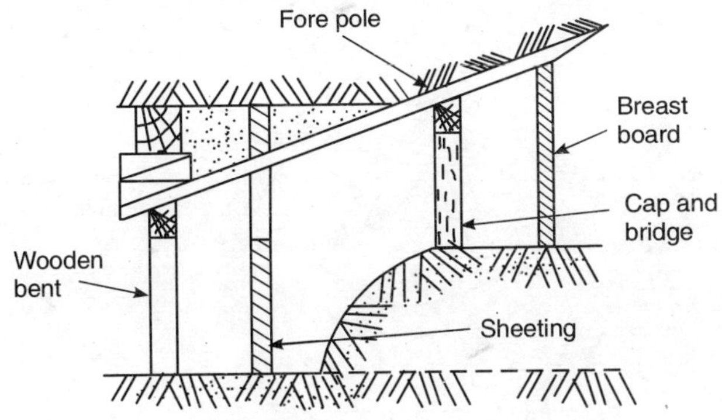

(a) Providing caps

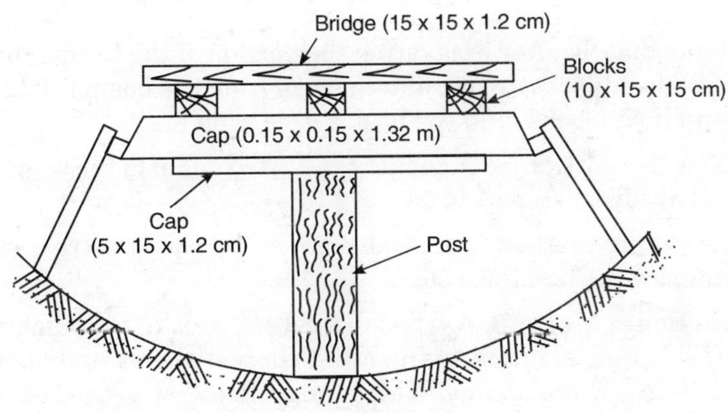

(b) Cap and post

Figure 9.27

11. The side spikes are driven to their full length.

12. A pair of needle-beams known as *boom* is provided to support the forward cap and the excavation in the lower portion is done as shown.

These needle-beams are placed on the sides of the tunnel with a post under the middle.

13. While providing the booms the breasting boards are removed one by one while doing the excavation. The breast boards are immediately reset one at a time at the front end.

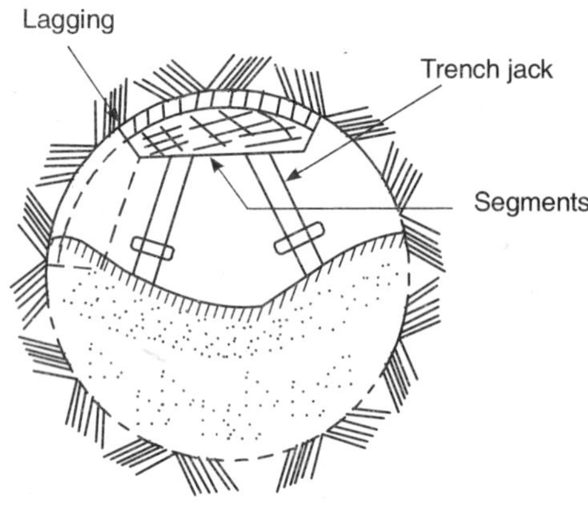

Figure 9.28 *Laggings*

14. Immediately after excavating the portion of the tunnel to the grade, the legs and foot blocks are set and the load from the boom is transferred to the legs and foot blocks, and the booms are taken out.

15. Now the work of the tunnelling is started again in the same sequences as given above from steps 1 to 14.

The forepoling method is a good method for the construction of small tunnels for sewers, pipelines etc.

Needle Beam Method. As this method requires 5 to 6 m long timber or R.S. Joist beam in addition to the other timber boards and struts, this is known as *Needle Beam Method.* This is an economical method of tunnelling. This method is only suitable for the soft grounds, whose roof soil can stand without support for few minutes and the sidewalls can stand for 1–2 hours without support.

The work of tunnelling in this method is generally done in the following sequences:

1. A drift known as *monkey-drift* is first made in size of about 1 metre on the working face of the tunnel.

2. Now the roof of the drift is supported by lagging carried on the wooden segments, which are supported on the trench jacks, as shown.

3. A needle beam is placed in position as shown in Fig. 9.29 with its front end resting horizontally on the planks resting on the floor of monkey drift. The rear end of the needle beam is supported on the strut post, resting on the lining of the tunnel.

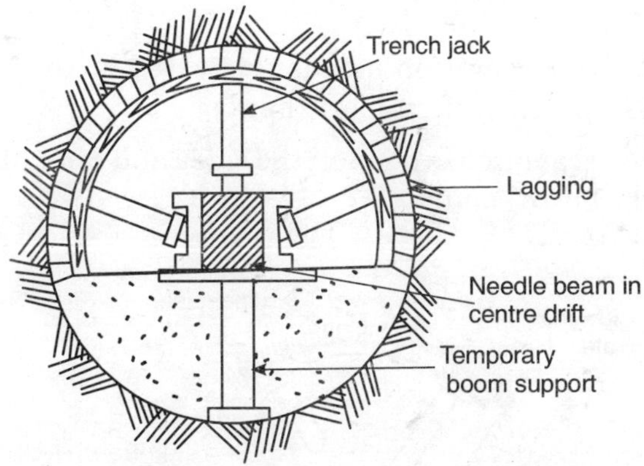

Figure 9.29 *Needle seam method*

4. Trench jack is placed on the top of the needle beam to support the roof through lagging.

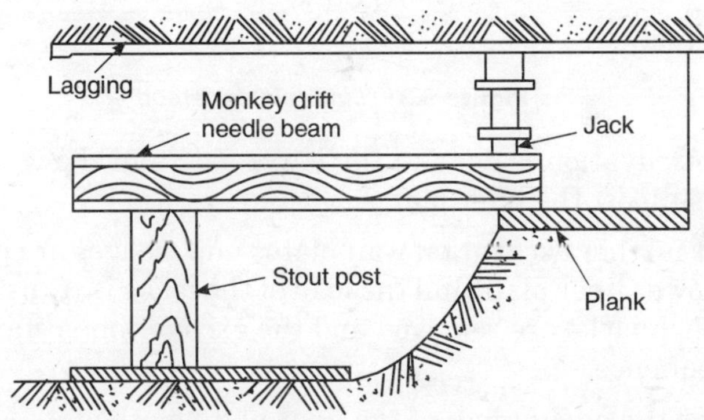

Figure 9.30 *Needle beam in monkey drift (longitudinal section)*

5. Now the drift is widened sideways and the roof is supported as above and the whole tunnel section is excavated.

6. The lining work follows the excavation and offer the removal of the excavated material.

The needle beam method requires large number of the jacks, which also cause obstructions in the efficient working of the workers.

Liner Plate Method. In this method pressed steel liner plates are used to support the soil during excavation work. The standard size of the plate is 91 cm × 41 cm. All the four sides of these plates are flanged 5 cm with holes for bolting with each other. Following is the sequence of operation in liner plate method:

1. First about 40 cm deep hole is excavated at the crown and a liner plate is placed in position to support the roof.

2. Now the excavation is done on the sides and liner plates are provided and bolted with central plate.

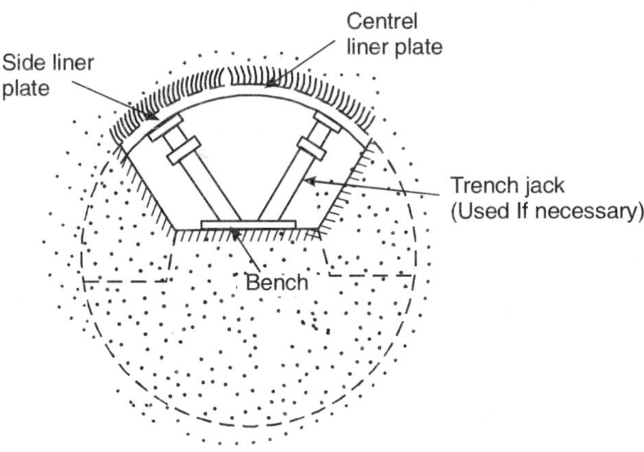

Figure 9.31 *Liner plate method*

3. The excavation is done in the lower side and the trench. Jacks are provided to support the liner plates.

4. After further excavation wall plates and wedges are provided at the ends of the lower liner plate and the load of the jacks is transferred to them. Now the trench jacks are removed and the excavation is started further in the above sequence.

Following are the advantages of the liner plate method :

(a) The excavation can be done quickly and economically.

(b) These are lighter in weight and easy to handle than timber planks and struts.

(c) They require less number of joints.

(d) They are more fire-resistive than timber.

(e) If the lining of R.C.C. is to be done, the plates can be cmbeded in the concrete and the cost of reinforcement can be reduced.

Shield Method. Nowadays tunnelling in soil is done with the help of a mechanical device known as *shield.* Shields are used for the construction of circular tunnels. Following are its main parts.

(a) The *skin* or *outer shell* which is constructed with steel plates, by riveting or welding them together.

(b) *Cutting-edge* which cuts the soft material.

(c) *Hydraulic-rams* which extend and push the shield forward. With the help of these hydraulic-rams the cutting edge is pushed forward to cut the material.

(d) *Tail.* It is the last portion of the shield. At this place lining is done so that the roof of tunnel may not collapse.

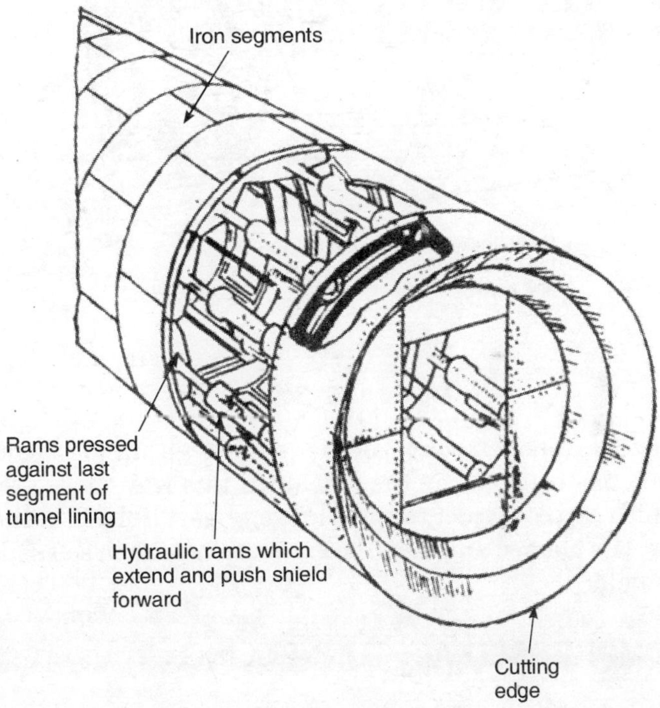

Figure 9.32 *Head shield*

Inter-structure is constructed in such a way that it can take all types of stresses which will come over it. Generally it consists of ring-girders, which are braced together.

9.36 ROTARY EXCAVATOR

This is also a type of shield in which cutting knives are provided in addition to cutting edge. This shield rotates while moving forward, therefore it is called *Rotary Excavator*. Figure 9.33 clearly shows all the component parts of

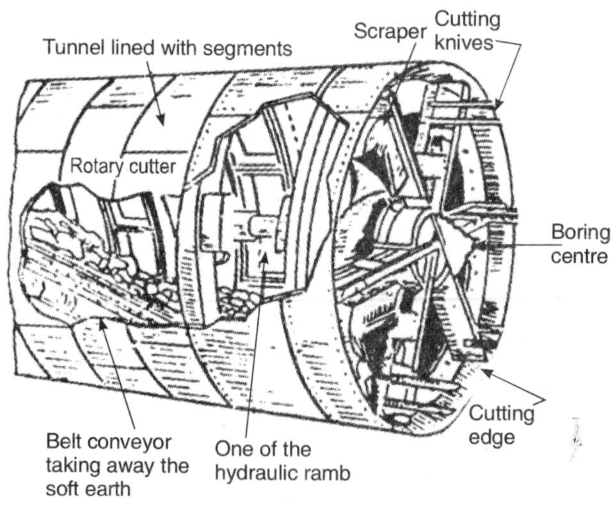

Figure 9.33 *Rotary shield*

a Rotary Excavator. Belt conveyor is provided within to take away the soft earth which has been cut by the shield. In this way these are mechanical devices which can construct tunnels in very short time. If in the way a hard rock comes, it is blasted and then the excavation work in soft soil is continued with the shield.

9.37 COMPRESSED AIR TUNNELLING

This method is commonly used these days for tunnelling in the soft ground having water bearing stratum. In this method compressed air is forced into the enclosed space, which prevents the collapse of the sides and the top of the tunnel. Compressed air is used with air-tight locks and in conjunction with the shield.

The air pressure forces back the percolating water or water mixed soil and keep the tunnel dry. In case of large diameter or size tunnel, the pressure of water or soil is not uniform from top to bottom. The water/soil pressure at the bottom is greater than that near the roof. In such cases the pressure of compressed air inside the tunnel is kept equal to the mean hydrostatic pressure along the centre-line of the tunnel.

As the compressed air escapes through the pores of the soil, it continuously decreases. Hence the air pressure should be varied from time to time to the actual required pressure inside the tunnel.

9.38 MUCKING AND HAULING

The operation of loading and removing the excavated or blasted materials from the construction sites and dumping it at the predecided site is called *mucking and hauling.*

The loading of the excavated material and its removal is an expensive and time-consuming operation during the construction operations of the tunnels. The removal of the excavated material is started as soon as possible after excavating the material by any suitable method.

Mucking and hauling is a major item of the tunnel construction. The time taken in this item is about one-third to one-half of the total time taken in the construction of the tunnel. The method to be adopted for mucking and hauling should be decided after thoroughly considering all the points.

9.39 METHODS OF MUCKING

Following two are the main methods of mucking :

1. Hand mucking
2. Machine mucking.

1. Hand Mucking. In olden days this method of mucking was used, but nowadays it has been totally replaced by the machine mucking:

However on certain small jobs, the hand mucking is adopted which is economical as well as unavoidable. Some such situations where hand mucking is adopted are as follows :

(a) In soft soils which can be easily excavated by hand tools.

(b) For making enough space and headroom for placing the construction equipment or machine at the site.

(c) For providing space to the belt conveyor.

(d) For the construction of the small tunnels, whose construction by the machine mucking is uneconomical.

(e) For excavating and trimming the sides of the tunnel and shafts.

(f) For the final cleaning and giving grade to the formation of the tunnel.

2. Machine Mucking. There are various methods for doing the operation of mucking by machines. All the machines used for the mucking are usually operated by the electric or compressed air.

Following types of machines are usually used in the tunnel excavation works

(a) Loading Machines.

(b) Muck cars.

9.40 LOADING MACHINES

Following loading machines are usually used during the construction of the tunnels :

1. Shovels.
2. Crawler shovels.
3. Revolving shovels.
4. Conway-digger.
5. Gathering arm loader.
6. Duck-bill loader.
7. Scraper.
8. Vibrating type loaders.
9. Excavators.
10. Mine-car loaders.

1. Shovels. The shovels are reliable and economical machines for doing the excavation and loading of the muck in the muck-cars. Usually bucket type tunnel shovels operated by lifting the loaded bucket to the attached arm directly over the machine which discharge the muck directly into the car or on the rope-way conveyer.

2. Crawler Shovels. Such shovels are most commonly used because they need no track and can work even on slopes. These shovels require greater working area, about 7.0 m wide tunnel. Electric or diesel engine operated shovels arc available. These shovels cannot be used for the construction of small narrow tunnels.

3. Revolving Shovels. Fully revolving type shovels can be used in the narrower tunnels and tracks. Such shovels are capable of full revolving. Usually these are operated by compressed air and can be used even in the construction of 2 m × 2 m cross sectional tunnel.

4. Conway-digger. These diggers are electrically driven excavators. The muck is loaded in the muck-car or conveyor belt or conveyor bucket by simply raising the bucket, tilting it backwards, and allowing the muck to slide down through the chute.

5. Gathering Arm Loader. It is also an electric operated crawler mounted loader. It collects the muck by means of two eccentrically operated arms. It essentially consists of a revolving chain conveyor which carries the muck backwards and dumps it into the muck cars or conveyor belt or trucks etc. The gathering arm loader has very high loading capacity and are commonly used in the tunnel construction.

6. Duck-bill Loader. This loader works on the sliding action developed by a rapidly accelerated plate moving into the muck to load. The main advantage of this loader are that it permits face operation, such as drilling, cleaning, setting of roof arches etc. It also proceeds with safety while loading the muck. As a matter of fact it is an improvement over the vibrating type loader.

7. Scraper. In this machine a sloping steel chute at the front of the machine is used as a cutting edge of the scraper. The cut material is filled in the dump type bucket. The scraped material is dumped at the required place or waiting skip.

8. Vibrating Type Loader. It essentially consists of a trough having a shovels-shaped front end which can be given required vibrations into the heap of the muck. Due to vibrations the muck materials starts moving in the slope of the trough and into the skips waiting at the rear end of the loader. This machine has high loading capacity but requires small working space. This loader is not suitable for loading of hard rocks. It has the disadvantage of not working satisfactorily on the uneven floor surfaces or on curved surfaces.

9. Excavators. Excavators are used for excavation of large size tunnels about 10.0 m or more. Electric or Diesel operated excavators fitted with shortened booms are usually used for excavation and loading purposes. They require large working area and high head room. They have high loading efficiency.

10. Mine Car Loaders. The mine car loader machines move at a very high speed. These are operated by the compressed air. These have high efficiency in fast moving and require small working area.

9.41 MUCK-CARS

Various types of muck-cars are used for hauling muck from the tunnel. Following types of muck-cars are commonly used :

1. Muck Boxes. These boxes are made of hard wood. Hoop iron on flat iron reinforcements are provided at suitable places to increase the strength of the boxes. Suitable rings are provided in the bottom for removing the bottom and dumping the muck.

2. Battle-ship. It is a simple large size box, which is carried into the tunnel for carrying out the muck on flat cars or trolley chasis.

3. Side-dump Car. This type of car is used for hauling of muck of rock tunnels. These cars have compact design and have large hauling capacity. These cars dump the muck on one side of it, by opening the side door or revolving the car along the horizontal axis.

4. U-bottom Cars. These muck-cars are commonly used for taking out the muck of soft excavation. The muck can be dumped by these cars by two workers.

5. Quarry Cars. These are commonly used for hauling of the muck of hard rocks obtained by the blasting. As the rock pieces of large size get excavated by blasting which cannot be taken out by small cars. Quarry cars are of large size. These are not of self-dumping type.

9.42 MUCKING IN STEEP GRADE TUNNELS

Figure 9.34 shows the method of mucking in steep gradient of tunnel where the gradient is 1 in 50 or more. In such cases the mucking of the tunnels is made self-mucking by providing an inclination of about 40° or more to the horizontal. In the beginning the tunnel is excavated upto about 30 m. The cross-section of the tunnel is closed with wooden bulk-head. Two openings are provided in the wooden bulk-head, one for discharge of the muck and another for entry of workers as shown in Fig. 9.34.

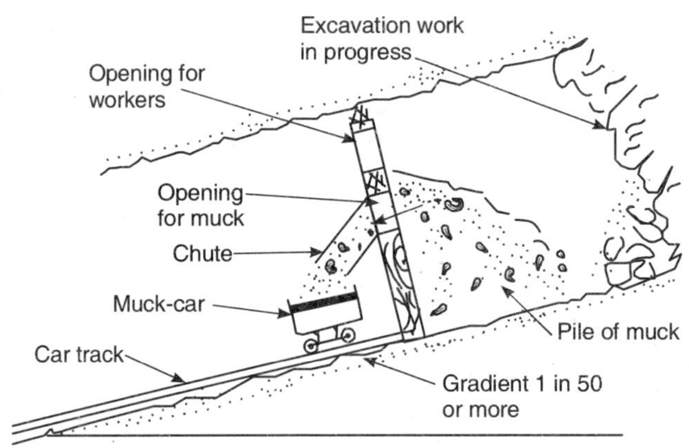

Figure 9.34 *Method of mucking in steep grade tunnels*

The muck is allowed to accumulate behind the bulk-head and is allowed to discharge through the chute on the muck-cars. As the work proceeds the bulk-head is moved forward to the successive new positions.

9.44 CAR CHANGER

For the rapid transportation and minimum loss of time, it is most essential that the muck-cars should be quickly changed from one track to another. It is most important that as soon as one muck-car is loaded, it should be immediately replaced by the empty car to the muck loading machine.

Following are the common methods of layouts of the muck-car tracks :

1. The Grasshopper Method. Figure 9.35 shows the grasshopper method. It is used for narrower tunnels. In this method the empty cars are stored in an overhead track on the large truss frame. The loaded cars are

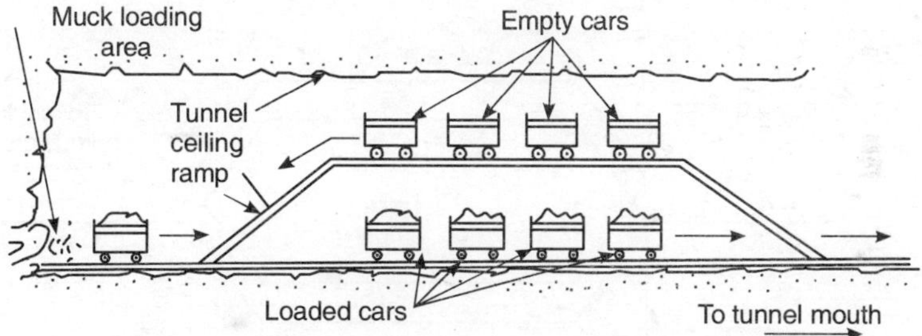

Figure 9.35 *Grasshopper method of car-changing*

allowed to pass underneath. Immediately after loading and moving the muck-car, the empty car is allowed to come down through the ramp to occupy the position of the loaded car and the loading machine starts loading it.

2. Passing Track. In this method a side track is provided for moving the empty cars toward the loading area. Figure 9.36 shows the method of passing

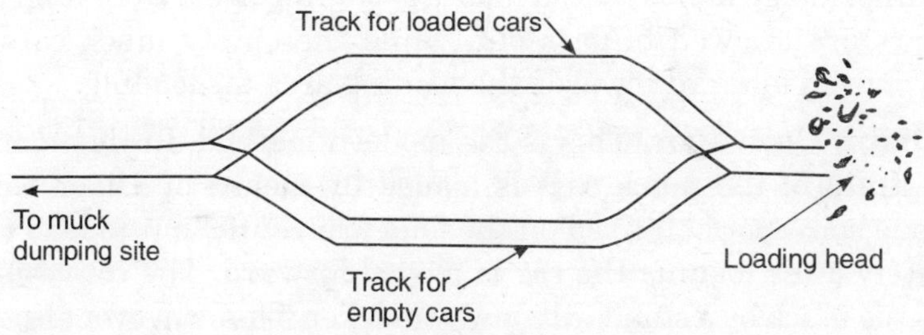

Figure 9.36 *Passing track layout*

track. This method is possible only in case of wider tunnels where two tracks can be laid side by side. In this method it becomes more and more difficult and delay in moving the passing track forward as the tunnel work is advanced. But this method is more commonly adopted.

3. The Cherry Picker. It is an old device which was used in the early days for interchanging of cars. In this method the empty car is lifted off the track and is moved sideways directly on the flow so that the loaded car can pass through the track. After passing the loaded car, the empty car is again placed on the track and moved. In this method only one track is used, therefore it is cheaper but it is not efficient and is time consuming method.

4. California Crossings. It is also known as California Switch. Figure 9.37 shows the layout of this switch. In this method short double track assembly completely welded into a unit with frog and switches is laid in position as shown in Fig. 9.37. This assembly slides along the top of the

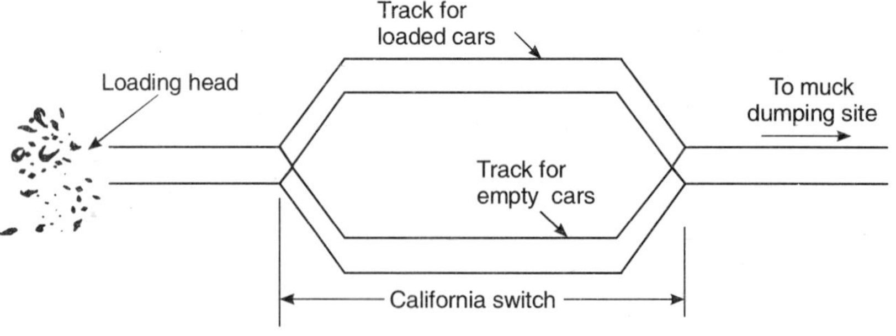

Figure 9.37 *California crossings*

single main track and the car is allowed to change from one track to another by means of switches. The ends of the rails of the track are made tapered to fit the main track so that the car can move over the California crossing, to and from the main track without any difficulty.

The California switch is placed near the loading area, so that on one side loaded cars are allowed to pass out, while the empty muck cars can be promptly moved forward towards the loading area for loading.

5. Dixon Conveyor. This is the modern method. In this method the complete train of the muck cars is loaded by means of 'Dixon Conveyor' having long conveyor belt fixed on the long arm, sufficient to load the cars. Immediately after loading the car is moved forward. The conveyor belt is loaded by the muck by a standard conway mucker. This conveyor also requires an extra set of tracks for moving the muck cars or their trains.

9.44 TUNNEL LINING AND GROUTING

Tunnel lining is done in the tunnels to give the finishing touch to the cross-section and preventing the collapse of side or roof ground soils. Lining serves many purposes which depend upon the type of the ground and the nature of the permanency of the tunnel. Temporary or primary lining is the temporary support of the roof and the side walls during tunnel construction. The permanent lining is the lining which gives the actual shape and size of the tunnel for the purpose for which the tunnel is constructed. Only permanent lining shall be described in this chapter. For the tunnel lining bricks, stone or concrete can be used. But in modern days only R.C.C. or Plain Cement Concrete with temperature reinforcement are used for lining purpose.

Objects of Lining. Following are the main objects of the permanent lining of the tunnels :

1. It provides the correct shape of the tunnel.

2. It withstands the soil, pressure and prevent the collapse of soils in case of soft ground.

3. It keeps the inside of the tunnel free from water percolation as suitable outlets are provided.

4. It binds and keeps the loose rock pieces in position and keeps the tunnel safe.

5. It prevents the rock from air slake.

6. It reduces the maintenance cost of the tunnel, because if lining is not done even in case of hard rocks, the loose pieces may fall from time to time and make the tunnel unsafe for traffic.

7. In case of water and sewer tunnels the lining will reduce the friction against the flow and will also reduce the turbulence of flow which may be caused due to projecting pieces of stone.

8. In soft ground the lining gives strength to the structure and makes it safe.

9.45 CONCRETE AND R.C.C. LINING

This type of lining is the most common method used in soft grounds as well as hard rocks. The thickness of the concrete lining mainly depends on the following :

1. Size and shape of tunnel section.

2. Relative amount of the vertical and horizontal pressures on the lining at various places.

3. In case of water tunnel, the amount of maximum internal pressure.

4. Final use of the tunnel.

5. Condition of the ground around the tunnel.

6. Conditions under which the tunnel has been constructed.

The thickness of the concrete lining should be as thin as practical for economy reasons. As a matter of fact there is no fomula for the design of the concrete lining of the tunnels through rocks. Every tunnel will have its own conditions of design, analysis and the circumstances of loads coming over the lining. As a thumb rule the approximate thickness of the tunnel lining is kept 2.5 cm for each 30 cm of the bore diameter of the tunnel.

Sometimes the thickness of the lining is determined by

$$T = 82\,D$$

where

T = thickness of lining in mm

D = diameter of tunnel in metres.

R.C.C. lining is required in the tunnels, when the tensile stresses in the lining exceed the permissible tensile limit of the tunnel lining. The placing of concrete in R.C.C. is more difficult than cement concrete. In case of R.C.C. lining the thickness of the lining at the crown of the tunnel is generally more than 25 cm. Ribbed or round steel bars may be used for the reinforcement. The reinforcement bars placed close to the inner surface at the top and close to the outer surface at the springing points.

Advantages of Concrete and R.C.C. Lining. Following are the main advantages of the concrete and R.C.C. lining :

1. It makes the tunnel water tight.
2. It provides smooth surface.
3. The maintenance cost is the lowest possible.
4. It can be easily moulded and casted in any desired shape.
5. It can be casted continuously in the shape of a shell without joint.
6. Its thickness can be controlled during casting.
7. Steel can take all types of stress and make the concrete safe against all types of stresses, which may come over it.

Form-Work for Lining. Concrete lining is done by pouring the concrete in the form-work. The form-work should be very accurately constructed and should show the true outline of the finished tunnel section.

The concreting of the tunnel is done in three operations and therefore the form-works are designed and constructed in three operations which are:

1. Ground mould form-work for the concreting in the flow or invert of the tunnel.
2. Leading frame for the lining work of the side walls.
3. Trusses form-work for the lining of the roof arch of the tunnels.

1. *Ground Moulds.* These moulds are used for casting the invert of the tunnel. These are generally made in two parts, which are joined at the middle by means of fish-plates and bolts for easiness in handling. While doing the

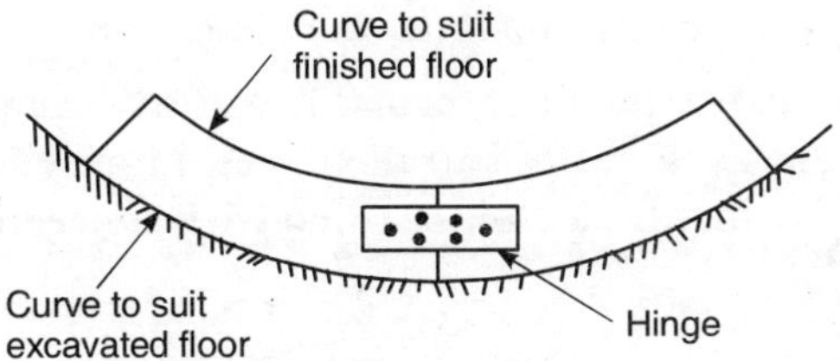

Figire 9.38 *Typical ground mould in position*

concreting two moulds are used at a convenient distance apart and chords are stretched across them, to obtain the true profile of the flow surface. The levelling and the centring of the mould before pouring the concrete, is done by means of a theodolite. Figure 9.38 shows typical ground mould in position.

2. *Leading frame.* Figure 9.39 shows a leading frame in position for casting the concrete of the side walls of the tunnel. These may be made of

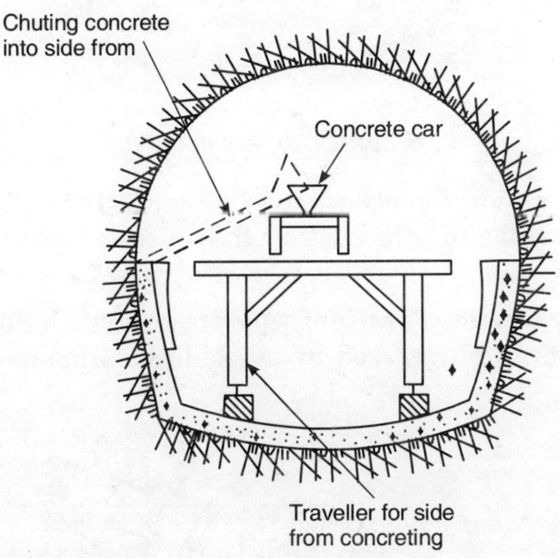

Figure 9.39 *Leading frame in position*

good thickness planks, one edge of which is cut to the curve of the side excavation and the other edge to conform to the inner face of the finished side wall. These leading frames are set with their lower end resting on the finished invert. The tops of the frames are correctly checked by means of theodolite.

3. *Trusses form-work.* The construction of this form-work is similar to those of arches. Sufficient head room is provided in the centre for convenience of working place. Centre form-works have a support roof pressures in addition to the weight of the roof lining, therefore these should be designed carefully. Figure 9.38 shows the centre form-work in position.

The lining is done by travelling forms. These are first placed at one section and after the lining, are shifted to next section and so on. Generally one small track is laid over which trollies filled with concrete move and take the

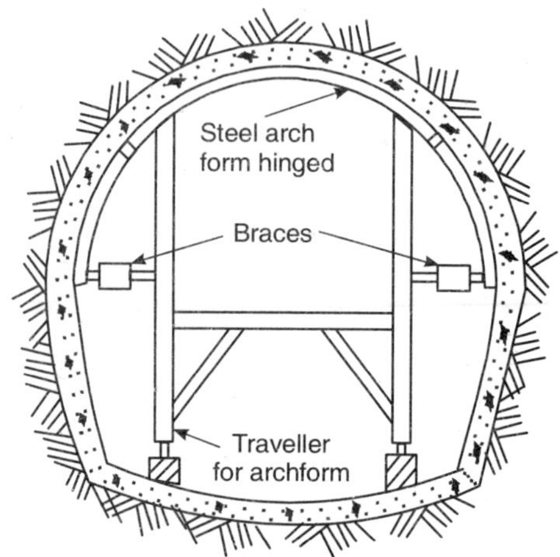

Figure 9.40 *Lining of Roof*

concrete to the required place easily. The concrete is placed in position by means of pump and gun, which throw the concrete inside, to the tunnel side and form space.

Telescoping Forms. The use of separate form-works for invert, side walls and the roof has been replaced by single unit forms nowadays. For large

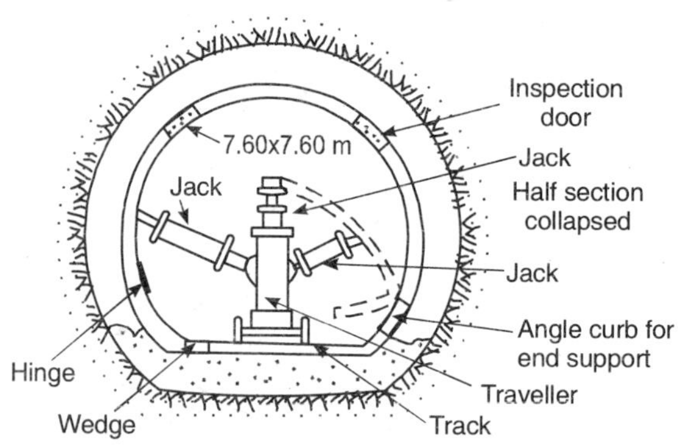

Figure 9.41 *Telescoping steel form*

tunnels separate forms are used, but for small tunnels single unit forms known as *telescoping forms* are used. Figure 9.41 shows this type of form in position.

The main ribs in this form are made up of sections hinged together, so that the whole unit would be collapsed after hardening the concrete, and moved forward. Due to this folding of the form at the hinge, these forms can be taken out without disturbing the concreting. Telescoping forms are suitable for lining of tunnels upto 8.8 m diameter. These forms carry jacks for erecting and collapsing the rib sections.

9.46 VENTILATION OF TUNNELS

At the time of constructing tunnels, different operations such as blasting, drilling holes, mucking etc. are done due to which poisonous gases are produced in the tunnel. To safeguard the workers, it becomes compulsory to remove these gases and bring inside the fresh air. The ventilating system must have the following qualities :

1. The poisonous gases produced by blasting must be taken out immediately, so that work is not held up even for shortest period.
2. The whole tunnel must be free from the dangerous gas fumes, it should not leave any deads pots in the tunnel.
3. The dust should also be taken out along with the dangerous gases.
4. At the face of the tunnel, comfortable atmosphere must be developed, so that workers can work easily.

Methods of Ventilation. The following are the various methods of ventilation :

(A) Mechanical Method. The mechanical ventilation is provided by:

1. *Blowing air inside the tunnel.* This is also called propulsion method. In this method fresh air is forced by electric fans or blowers or with the both.

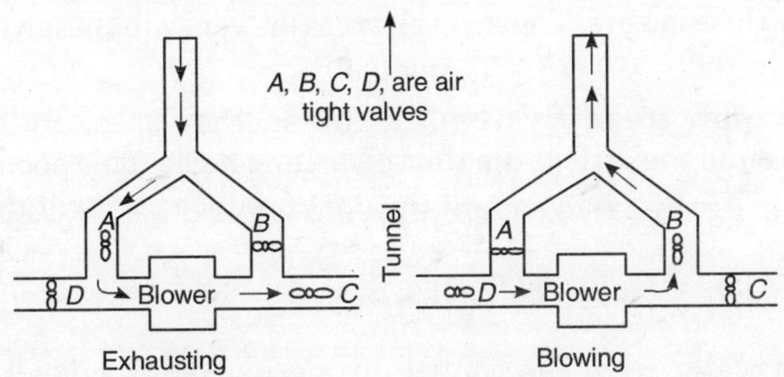

Figure 9.42 *Combined method of blowing and exhausting for ventilation*

In blowing, fresh air is blown by blower fans, mounted on one or more input shafts. The vitiated air is being forced out naturally due to pressure build up by blowing. This method is suitable for short length tunnels only.

2. *By exhausting the air from the tunnel.* Sometimes vitiated air is taken out by means of exhaust fans. The fresh air is being drawn inside through portals or inlet shafts. This method is good in removing dust and foul air quickly from the tunnel.

3. *By combination of blowing and exhausting the air.* Sometimes combined blowing and exhausting method is used. This is called balanced method. In this method fresh air is forced inside the tunnel by blowing fans and the vitiated air is sucked out by exhaust fans. This method is very good and is mostly used in tunnelling.

In the combined method of blowing and exhausting, blowers and exhaust fans are installed for forcing fresh air in the tunnel. Sometimes the blower and the exhaust fans are installed in the shafts connected to the tunnel and meant for this purpose.

(B) Natural Ventilation. Due to difference in temperature inside and outside of the tunnel, flow of air starts. This natural flow can be increased by constructing shafts at various places over the tunnel. But this method does not provide sufficient ventilation, which is required in practice.

9.47 DUST CONTROL

At the time of drilling holes and blasting sufficient quantity of dust is mixed in air which is harmful to the health as it causes uncurable lung disease. The following precautions are generally adopted for controlling the dust:

1. **Wet-drilling.** At the time of drilling holes, water is added in the hole which prevents dust. Modern drilling machines are fitted with wetting arrangements, the water also keeps the drill bit cool.

2. **Respiration Method.** In this method special types of respirators are used by the workers. These respirators prevent inhalation of dust by the workers. But these are not very common.

3. **Vacuum Hood System.** In this method specially built hood is provided around the drill at the time of drilling holes. This hood is connected to a suction hose and removes all the dust produced by drilling.

9.48 DRAINAGE OF TUNNELS

Removal of water from the tunnel during its construction and after the construction is very essential. Generally at the time of construction the water which is used for wet drilling and which comes through seepage is removed by pumps. Drainage ditches or sump wells are provided at suitable intervals in the tunnels in which the water from both sides is collected and from where it is pumped out at suitable intervals.

After the completion of the tunnel the water is taken out by means of open drains provided in the centre or both sides of the tunnel by gravitational force. Sufficient slope is provided in the drains to take out water from the tunnel.

Sources of Water. Mainly water comes in the tunnel from the following two sources :

1. Wash water which is used during drilling holes and used to wash the cuttings inside the tunnel.

2. Sub-soil water, which percolates inside the tunnel from the pores of the soil or through the natural fissures in the rocks or caused during blasting of the rocks. The quantity of water entering from the ground is sometimes very large and causes water to flow inside the tunnel.

The quantity of water from the first source, described above, can be easily determined, but the calculation of quantity of water from the second source is not easy. Exploratory holes are to be drilled at various places, in order to determine the quantity of water, which would come from the second source.

Water Handling. During the construction of the tunnel, if the ground water is trickling continuously from the roof and is causing obstruction in the construction, it can be diverted to the side-drains (constructed along the both sides of the tunnel) by providing a false roof under the roof by corrugated G.I. sheets. These side-drains will carry the water outside the tunnel.

If the heading is driven on level ground, the water should be removed by pumping it through shafts. If the heading is driven on upgrade side of the tunnel, the water can be removed through slopy drains. On the other hand if the heading has been driven through shafts, all the water is to be pumped through the shaft only.

9.49 GROUND WATER REMOVAL

The quantity of the ground water entering the tunnel can be removed from the tunnel by the following two methods :

1. By draining through the open-ditch constructed along the tunnel on both sides at the invert.

2. By pumping.

1. Open Drainage System. This is the simplest method for the removal of water. This system is suitable in hard rock bases and water resistant soils.

Following are the disadvantages of this system :

(a) Valuable working space is wasted for in the construction of open drains.

(b) It is not possible to drain-off the water at all grades, because on flat grades, self-cleansing velocity will not develop.

(c) If the muck or debris block the flow of water, it will create objectionable ditches and pools inside the tunnel.

(d) The construction of open drains in the sides weakens the section of the tunnel as the continuity of the shell is broken.

Due to the above disadvantages, in modern tunnels, open drains are not preferred for the removal of the water from inside of the tunnel.

2. Pumping Drainage System. In modern tunnelling methods, the removal of water from the tunnel by pumping system is preferred. Two types of pumps are commonly used. First is the piston-type pumps, which are available in vertical or horizontal models, and are usually operated by compressed air. Second type is centrifugal pumps which can be operated by electricity or compressed air. Being smaller in size than first type, these can be easily installed or even submersed inside the water in the sump-wells.

For the removal of water by the pumps, sump-wells are constructed at regular intervals from 300–500 metres apart or at wet spots where water is percolating in the tunnel. A series of pumps are installed in each sump-well. The pumps remove the water of their sump to the next sump, from where it is pumped to the next. In this way the water is taken out in stages to the outside of.the tunnel. The capacity of the pumps, size of the sump well and the pipes go on increasing from the middle of the tunnel to the mouth sides. To overcome this difficulty sometimes one pipe line is laid all the way inside the tunnel, and all the pumps are connected to it through check valves. In this case all the pumps of uniform capacity can be used.

For handling the water during the construction of the shafts, centrifugal pumps are hung from the cable work more satisfactorily. In case of deep shafts, booster pumps are installed in the sumps constructed into the side of the shaft. If the site conditions require high capacity pump should be installed in the bottom of the shaft to pump the water from the tunnel.

9.50 PERMANENT DRAINAGE

At the time of completing the tunnel, permanent drainage arrangement is provided , to take out the water entering the tunnel continuously, to save the road or railway track.

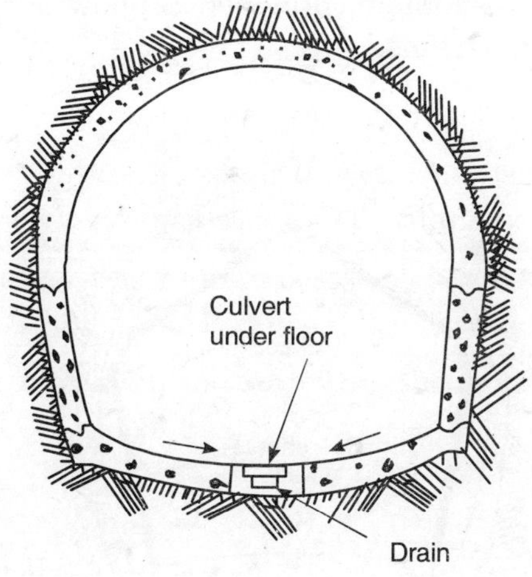

Figure 9.43 *Central drainage system*

If the tunnel carries two railway tracks, one open drain is provided in oetween both the railway lines. The section of the drain should be sufficient to carry the water without submerging the floor of the tunnel at self-cleaning velocity. The central drain can also be provided covered with facility for the inspection and maintenance as shown in Fig. 9.34.

If the tunnel carry single lane traffic, one or two side-drains can be provided as shown in Fig. 9.44.

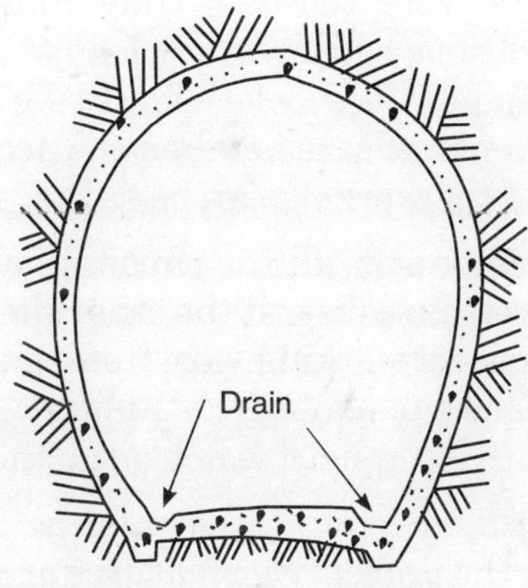

Figure 9.44 *Tunnel with two side-drains*

In case the water continuously trickles from the roof of the tunnel, false roof of the G.I. sheets should be provided as shown in Fig. 9.45 to divert the roof water to the side drains.

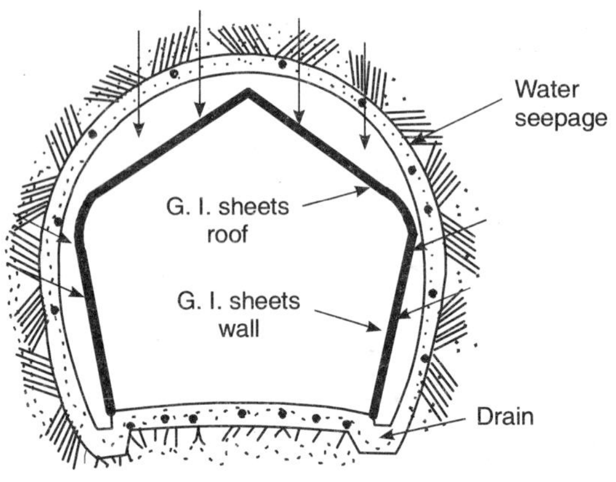

Figure 9.45 *False-roof of G.I. sheet in the tunnel*

If the water is also trickling from the side-walls and obstruct in the free movement of the traffic, it is prevented by erecting the side-walls also with G.I. corrugated sheets to prevent the splashing of water on the pavement or the track.

9.51 SAFETY MEASURES

Tunnel construction is very tedious, time consuming and hazardous operation. There are various causes which lead to accidents. At the time of tunnelling maximum possible care should be taken to prevent any mishappening or accidents. If care is taken the accidents can be prevented in time and the lives of the workers can be saved.

It is impossible to think over all the causes of accidents. At one place if accident occurs due to certain cause, at the other site it may be due to another cause. The site conditions of working vary from one place to another, hence one safety measure at one site may fail at the other site. Hence the executive persons should constantly keep strict watch on all the operations of tunnelling to prevent the accidents.

If during tunnelling the preventive measures are rigidly followed, most of the accidents can be eliminated. Following are some of the preventive measures which should be followed during tunnel constructions :

1. The design of the planks and vertical supports should be checked

carefully, because most of the accidents occur due to rockfall. The walls and the roof should be frequently checked to prevent rock-falls.

2. Defective or weak timber should not be used under any circumstances, as it may give way under load.

3. Safety rules, regulations and the preventive measures should be taught to every worker, working on the site.

4. All safety rules should be followed strictly without any violation. Labourers' and staff's regular meetings should be held from time to time and they should be helped for understandings and the application of the safety rules.

5. All electric light and power lines should be properly installed and all connections should be well insulated as per prevailing codes of practice and rules.

6. Drilling of holes, loading of holes with explosives and the firing should be done with great care, keeping all the precautions in view.

7. If any charged hole mis-fires, following precautions should be taken:

(a) If the leg wires are still intact, they should be connected to the electric ignition equipment again and refired.

(b) If the hole is not exploded, the stemming should be washed out by water jet down upto the explosive. New primer should be inserted and fired.

(c) Another method is to drill a new hole about 60 cm from the unfired-hole, loaded and fired.

8. Extra light should be provided where essential materials are stored.

9. All the tools and equipments should be kept in best working conditions, as bad tools may lead to the accidents.

10. Water should not be allowed to remain inside the tunnel, as it may cause obstruction in the walking and may lead to skiding, rusting of the steel and other tools etc.

11. All the working sites should be kept well lighted, as poor lighting may lead to accident.

12. Foot-path should be well maintained. It should not be slippery and hazardous and free from obstructions. It should be covered with planks or muck and should be kept dry.

13. For safety and efficient operation all the debris and the refuse should be kept clean, both on the surface and underground.

14. Loading of the muck and their hauling should be done with great care.

15. Only required material should be taken inside the tunnel, as extra material will cause obstruction.

16. The cars carrying rails, steel, sleepers, posts etc. should not be overloaded, as extra projections in the side may strike with the workers working inside the tunnel or with the posts and struts.

17. All trolleys or trains should be properly controlled with signals for safety during their movements.

18. All the hoists, cables and other shaft equipments should be in sound working conditions, with sufficient factor of safety.

19. All the shaft openings should be properly protected to prevent the workers and the materials from falling inside the shafts.

20. Shafts should be provided with the indicators to indicate the position of the cage.

21. All the hoists should be provided with automatic braking arrangements, which should be in operation during power failure.

22. Proper signals should be provided to prevent accidents of the cage moving in the shaft, with the trolleys or train moving in the tunnel.

23. All the shafts must be provided with safety ladder for use during emergency for exit and access.

24. All the persons working in the tunnel should have metal hats, to prevent the heads against accident from the falling stone pieces.

25. Fire-fightings equipments should be provided at all key points for safety purpose.

26. All the workers should be medically fit for working inside the tunnel. Workers should also be medically examined from time to time.

27. After the blasting inside the tunnel, the inside poisonous gases should be removed before allowing the workers to enter in the tunnel.

28. Telephone facilities should be provided inside the tunnel at various places, for sending the information of the tunnel position outside from time to time.

29. In case of compressed air tunnels, extreme care should be taken to prevent fire. Smoking should not be allowed in any case, as fire catching gases may cause accidents inside the tunnel.

30. First aid facilities with equipments and doctors should be available all the times at the site, for immediate help in case of accident.

31. No unauthorised person should be allowed to enter the tunnel.

32. Authorised visitors should be equipped with safety hats and other necessary equipments. They should be allowed to visit the site only with the authorised guide meant for this purpose.

33. Safety sign boards should be provided at various places.

34. Emergency lights should be provided inside the tunnel at all the working places, which should give light in case of power failure.

35. Double power supply system should be provided, so that during failure of one supply, the power can be resumed by the standby unit.

36. In risky places special care should be taken to prevent the accident.

PROBLEMS

9.1. Explain the importance of hill road drainage and state why it requires special consideration.

9.2. State in brief the method of planning of hill roads.

9.3. How hill road geometries are different from that of roads in plains. Discuss briefly the salient features.

9.4. Describe briefly the problems encountered in the construction of hill roads.

9.5. Discuss the importance of hill road drainage with the help of neat sketches, show the drainage system of hill roads.

9.6. Explain with the help of neat sketches the salient features of hill road construction. What type of special precautions are taken to ensure economy and safety in hill road construction.

9.7. Write short notes on :

(i) Bridle path

(ii) Scupper

(iii) Breast wall

(iv) Catch water drain

(v) Kerb and channel drain

9.8. State the advantages and disadvantages of tunnels.

9.9. State the method of transferring centre-line and grade of a proposed tunnel from the ground surface to the inner surface.

9.10. Write short notes on :

(i) Tunnel grade

(ii) Spade and plug

(ii) Curvilinear tunnel

(iv) Shafts

(v) Half tunnel

9.11. Describe with the help of a sketch the construction of a shaft in a tunnel.

9.12. State in brief the method of tunnelling in hard rocks. How various problems are encountered.

9.13. State the difference in the method of tunnelling in hard rocks and soft rocks.

9.14. Write short notes on :

(i) Full face

(ii) Heading and benching

(iii) Drift

(iv) Mucking

(v) Rotary excavator

9.15. State in brief the compressed air method of tunnelling in soft rocks.

9.16. State briefly the method and importance of mucking and hauling in tunnels.

9.17. Write short notes on :

(i) Car changers

(ii) Muck cars

(iii) Loading machines

9.18. Elucidate the method of tunnel lining. When tunnel lining becomes essential ?

9.19. State the objects of ventilation of tunnels. What is mechanical ventilation?

BIBLIOGRAPHY

1. Recommendations about the alignment, survey and Geometric designs of hill Roads IRC : 52–1973.
2. Location of roads in hilly and mountainous terrian – B Chandhry—Journal Indian Road Congress Vol. XXIII – 3 and 4.
3. Landslide investigation and treatment G.C. Sharma –I RC journal Vol. XXXIII 3 and 4.
4. Road Drainage Practice – I.R.C. special Publication – 78.
5. Gupta and Gahlowet – Hill road construction.
6. Ahuja and Birdie – Tunnel Engineering
7. Geology in the design and construction of hill roads – V.A. Khair I.R.C. XXXI, 2 and 4.
8. Geometric Design of hill roads – L.L. Bhired I.R.C. journal XXXI – 2 and 4.
9. Highways – O' Flaherty' 82.

10

Highway Machinery

In this Chapter you will study,

• Road Machinery • Tractor • Dozers • Rooter or Ripper • Scraper • Grader • Shovels • Basic Shovel • Attachments to Shovel • Dipper Shovel • Drag Shovel or Hoe or Pull Shovel • Dragline • Clamshell • Comparison of various Machines • Road Rollers • Rolling • Out-put or Road Roller (Approx.) • Dumpers • Bitumen Mixing Machinery • Concrete Mixing Machinery

GENERAL

Today is the age of machines. With the help of machines, many complicated and laborious works have been simplified. For the construction of roads also, an engineer has to depend on machines. However, in India, highway machinery is only used for big projects as the financial conditions of this country do not afford the use of such machines. But a highway engineer must be familiar with the use of modern road machinery. It will be beyond the scope of this little volume to describe each of this machinery in detail. However the particular use of important road machinery and plants have been outlined in brief, so that the reader must have a fair idea about the use of all such machinery and plants used in road construction.

10.1 ROAD MACHINERY

Road making plants and machinery is used in various jobs. The various works included in the road construction are :

(a) Clearing trees, shrubs, jungles etc.

(b) Forming the soil, to level by cutting ridges or filling hollows.

(c) Spreading the metal and rolling.

(d) Surfacing or applying wearing course.

I. The following tools and plants are generally used for clearing the jungles, shrubs etc.

(1) Bull dozer

(2) Rooter

(3) Tractor

(4) Towed scraper

(5) Shovel units with trucks

II. For cutting, filling and preparing the formation or subgrade for a road, the following machinery is generally used :

(1) Tractor

(2) Dozer

(3) Grader

(4) Shovel

(5) Clamshell

(6) Trucks

(7) Trench cutter

(8) Plough

(9) Rollers

III. For spreading the road metal or aggregates on the prepared sub-grade and rolling the same, the following tools and plants are used :

(1) Crusher

(2) Trucks

(3) Aggregate distributor and

(4) Rollers

IV. The surfacing generally consists of (a) Bituminous surfacing, and (b) concrete surfacing.

(a) For Bituminous pavements following tools and plants are used :

(i) Bitumen Boiler

(ii) Bitumen Sprayer

(iii) Aggregate spreader

(iv) Bitumen mix spreading machine

(v) Grouting machine

(vi) Rollers

(b) For cement concrete pavements the following tools and plants are used:

(i) Central batching and mixing plant

(ii) Concrete mixers

(iii) Concrete pavers

(iv) Concrete spreader and finisher

(v) Concrete vibrators

In the following articles some of the important plants and machinery will be briefly described.

10.2 TRACTOR

It is an important machinery which is used for different type of works. The tractor works with diesel engine having horse power varying from 20 to 200. There are two types of tractors:

(i) Track type tractor

(ii) Crawler type tractor

The track or wheel type tractors move on pneumatic tyres. This type of tractor is used for even and smooth country. The crawler type tractor moves on

Figure 10.1 *Crawler-type tractor*

an endless chain and is used for uneven and rough country. Tractors are intended primarily to exert a powerful tractive force for towing other heavy equipments. Tractors are generally provided with various attachments such as a dozer, scraper, harrow, plough etc. Tractors with various attachments

Figure 10.2 *Track-type tractor*

are used for clearing the shrubs, roots etc. and for earth work in formations. The tractor is primarily an earth moving machinery.

10.3 DOZERS

Dozers are attachments of a tractor which are used for various works. In particular, dozers are important machines which are used for various operations in road construction. There are three types of dozers viz., bull dozers, angle dozer and tree dozer.

Figure 10.3 *Bull dozer*

A *bull dozer* consists of a blade, attached to the front side of the tractor. The bull dozer is used for excavating the material and pushing the material in a forward direction. It is mainly used for levelling a heap of excavated material.

Angle dozer consists of a blade attached to the front side of the tractor, which can be set obliquely at any angle to the direction of motion of the tractor. The maximum possible angle of inclination of the blade to the front face of the tractor is 30°. Angle dozer is used for pushing the material to the side, right or left.

A *tree dozer,* which is also sometimes called *a stumper,* consists of a slightly curve blade, with its concavity in the forward direction of motion. The blade is specially designed for uprooting the trees, shrubs etc.

10.4 ROOTER OR RIPPER

This is a very useful instrument used to remove stiff clay, soft rock or other hard soils. It is mounted on wheels and is towed to a tractor. It has one or more tyres or teeth which are driven through the ground.

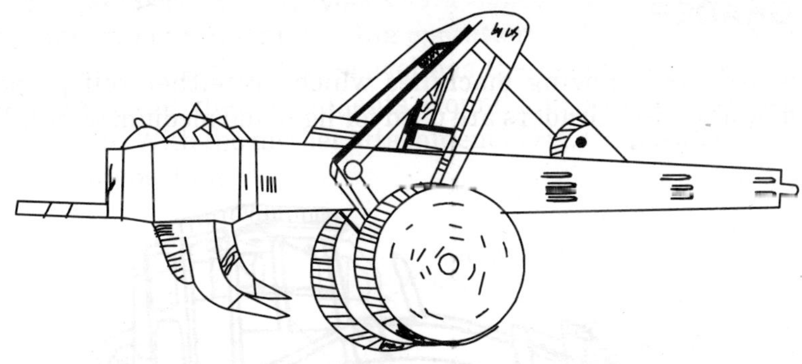

Figure 10.4 *Rooler or ripper*

10.5 SCRAPER

A scraper is an earth moving equipment generally drawn by a tractor. It is used for earth work of a light nature and where the load is also very small. It provides one of the most economical machines used during earth operations of a road. It is a self-sufficient machine, as it can do all the operations necessary for light soil viz., it can dig, load, carry, dump, and spread the earth or soil. It consists of :

superstructure. Down through the deck passes a strong vertical shaft which carries the driving power to a horizontal axle having a clutch and brake system. The track or travel system and the revolving deck containing actuating mechanism together constitute the basic shovel.

The truck mounted shovels can move from one place to another quickly whereas the crawler mounted shovels cannot move quickly and at long distances have to be transported on trailers. However maneouverability is much less in wheel mounted and are liable to get struck up on bad ground. Crawler mounted shovels are very popular on account of their great maneouverability, ability to travel on bad ground, marshy land, and hilly terrains.

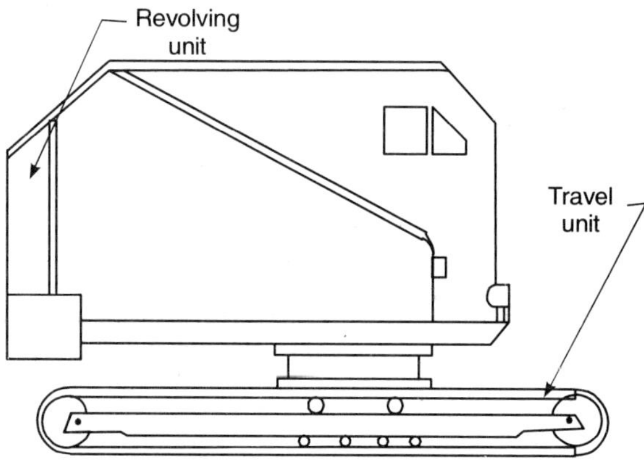

Figure 10.8 *Basic shovel (truck type)*

10.9 ATTACHMENTS TO SHOVEL

The attachment or front end or rig consists of the boom and a bucket attached to its far end. Booms are of two types viz., shovel boom and crane boom. To the boom various type of buckets, hooks and clamps are attached. With the various attachments different type of shovels are illustrated below.

1. Shovel boom and dipper stick for a dipper shovel.
2. Shovel boom, jack boom and a stick for a drag shovel or hoe.

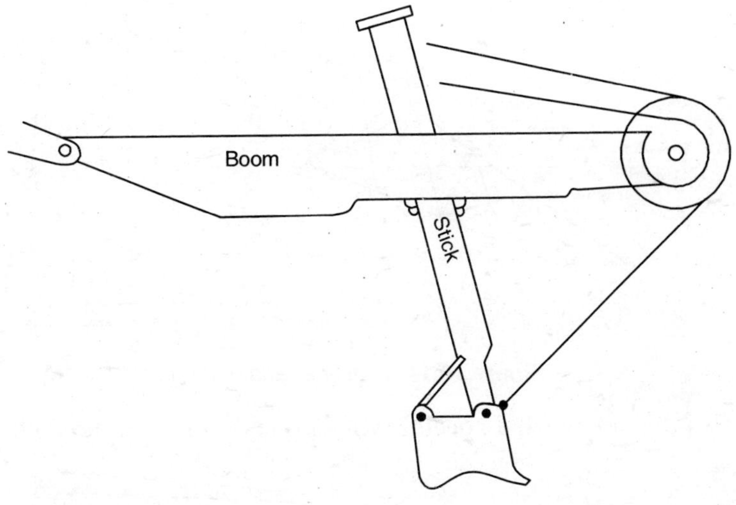

Figure 10.9 *Shovel attachment*

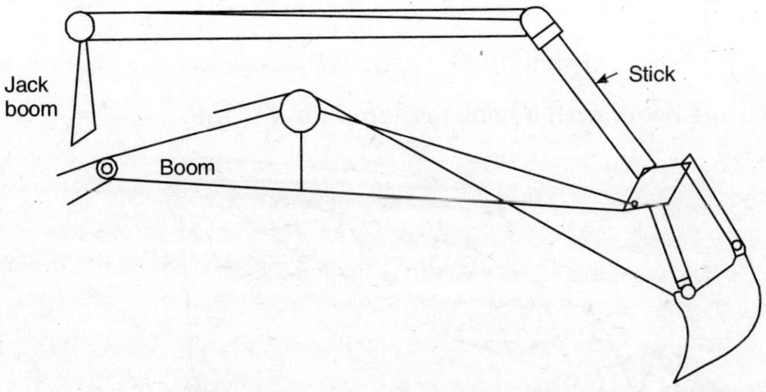

Figure 10.10 *Drag shovel and hoe*

3. Crane boom with a fairlead and a loosely attached bucket for dragline.

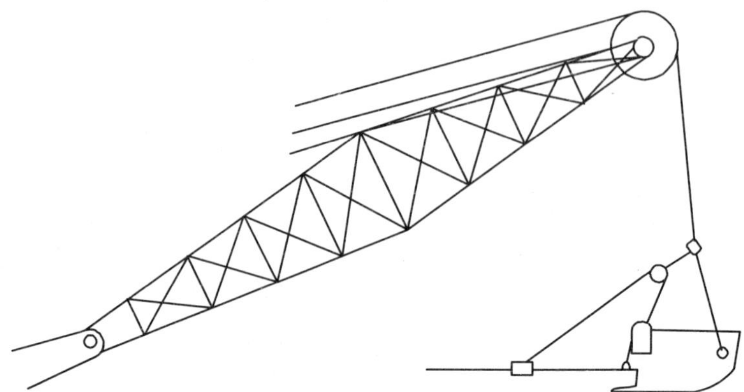

Figure 10.11 *Dragline attachment*

4. Crane boom with a special type of bucket for a clampshell.

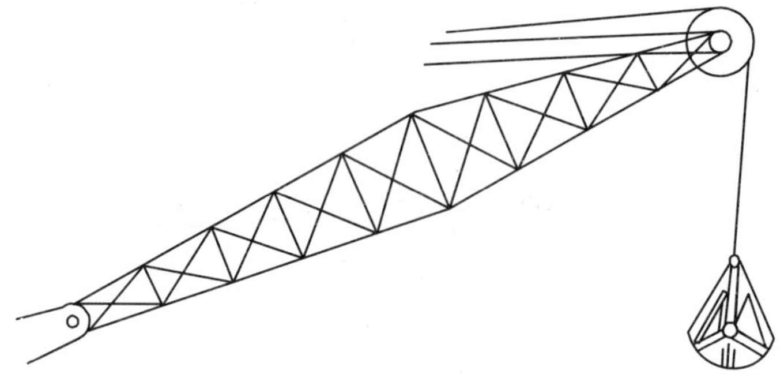

Figure 10.12 *Clampshell attachment*

5. Crane boom with a hook or clamp for a crane.

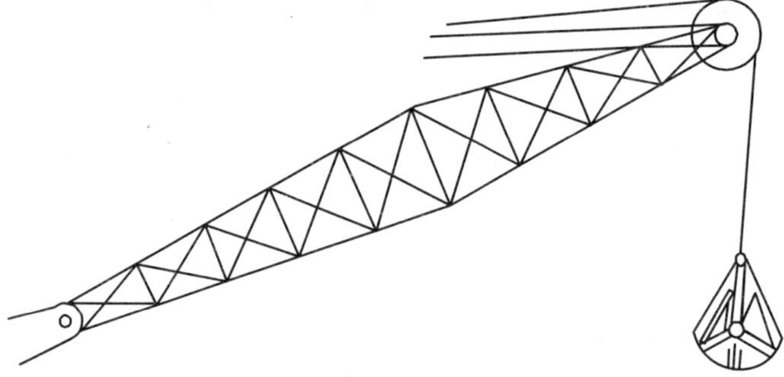

Figure 10.13 *Crane attachment*

10.10 DIPPER SHOVEL

Dipper or power shovel is the most commonly used form of shovel. It consists of a stout boom a dipper stick and a bucket. There are two type of arrangements for the boom and stick. In one case, the stick passes through a slot in the boom and in the other it is made in two longitudinal halves jointed together at the ends and mounted on the sides of the boom. The bottom end of the boom is hinged to the support brackets of the revolving deck and the upper end extended diagonally out to support the bucket.

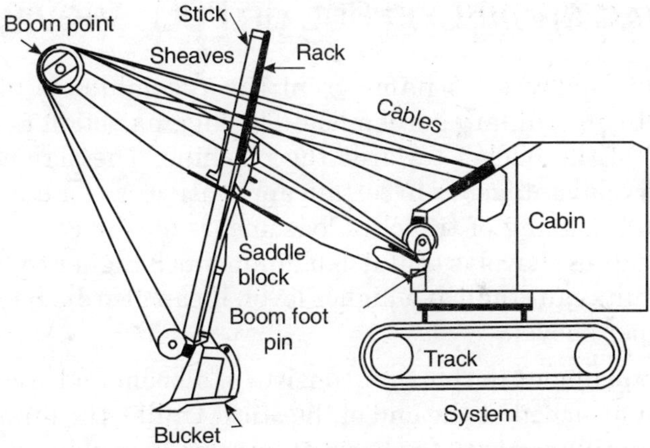

Figure 10.14 *Dipper shovel*

The bucket which is the actual digging tool is pinned to the end of the dipper stick and is further supported by braces. The bucket is a stoutly built steel

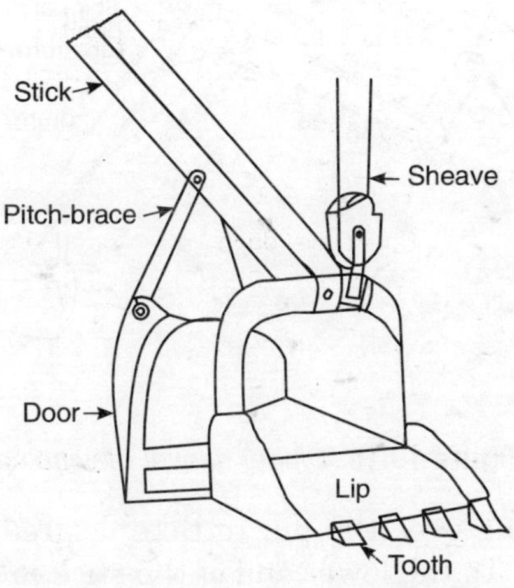

Figure 10.15 *Dipper shovel bucket*

box open at the top and provided with hinged door at the bottom. A sheave is attached to the bucket through a pin close to the point of attachment of the stick. The bucket hoist line is carried on this sheave and passes on to the boom point sheave and then to the hoist drum on the deck.

The dipper shovel is an ideal machine for close range work. The dumping can be precisely controlled and so loading of carrier units is much easier. Shovels are capable of digging very hard rocks and can easily remove heavy size boulders.

10.11 DRAG SHOVEL OR HOE OR PULL SHOVEL

This machine derives its name from the resemblance of the digging mechanism to an ordinary garden hoe. The digging action results from the drag or pull of the bucket towards the machine. The action of the hoe is therefore very advantageous in certain applications as for digging below the machine level, digging of trenches, basements etc., is concerned. Another advantage of drag shovel is that it can dump much higher than its deck and the carrier units can stand at a higher level. It can also dig harder materials just like dipper shovel.

The rig attachment to this unit consists of a boom and stick, a jack hook and a bucket attached at the end of the stick. Unlike the dipper shovel, the hoe stick is pin-hinged with the boom at intermediate point of its length and

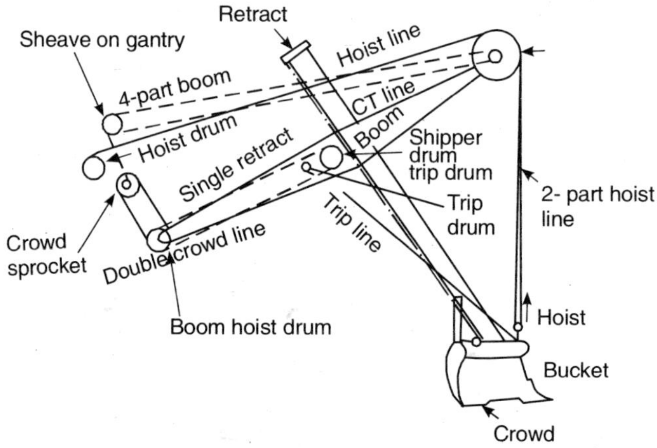

Figure 10.16 *Dipper shovel reeving cable*

is capabale of turning over the pin to take desired direction suitable for digging or dumping. To the lower end of the stick is attached to the bucket while the other end carries a sheave bearing the hoist cable which is supported by the jack boom at the other end and passes on to the hoist drum. The boom as a whole including the stick is therefore capable of movement up and down.

The boom can, therefore, take in any position in the vertical plane being conveniently lowered below the machine level and raised for dumping at the desired height. It is not a very efficient machine and is very slow in working particularly in dumping. The dumping operation is not efficient as much material falls down in the operation of hoisting. The best operation results when digging is close to the track level and high dumping is not required. The bucket can dress and trim the trenches also which saves manual operation.

10.12 DRAGLINE

The machine is capable of dragging the bucket against the material to be dug or hauled. Unlike shovel it has a long light crane boom and a digging bucket. The bucket is loosely attached to the boom through cables. It can dig and dump over long distances. It is not necessary to place the machine close to the digging site on account of longer boom. It can therefore handle digging of slushy material or under water excavations while the machine is away from the site on a stable soil.

The boom length is 15 m to 24 m. The bucket capacity is 8 cu m to 10 cu m. The rated capacity of the machine is 400 cu m per hour, with 1500 H.P.

Operation

The machine is to be operated in the following manner:

(i) Swing the empty bucket to the digging position.

(ii) Pull the bucket towards the machine for excavation.

(iii) Hoist, swing and dump the loaded bucket.

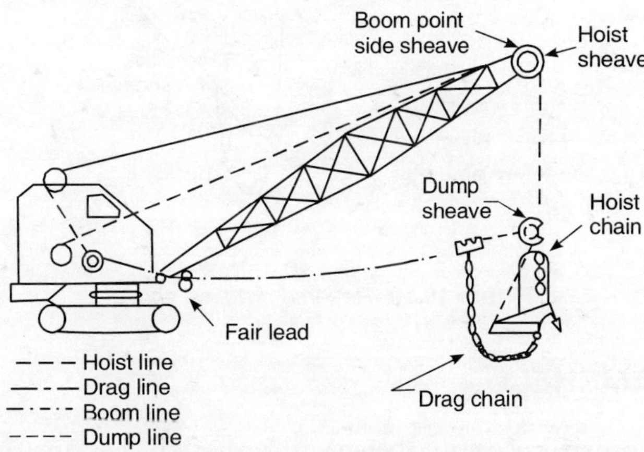

Figure 10.17 *Drag line*

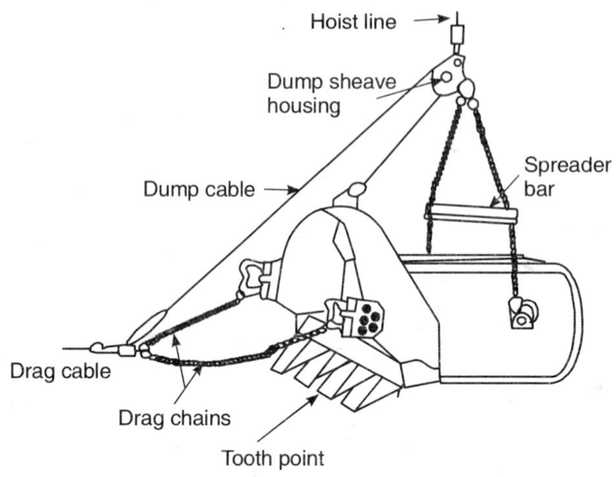

Figure 10.18 *Drag line bucket*

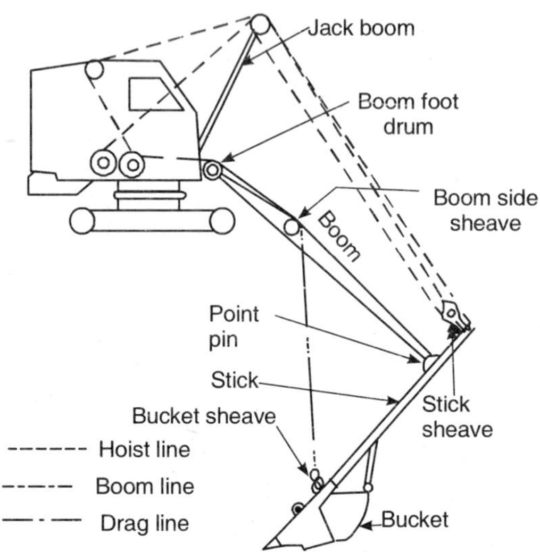

Figure 10.19 *Reverse dragline shovel*

10.13 CLAMSHELL

It essentially consists of a crane boom with a specially designed bucket loosely attached to the boom with cables. The bucket consists of two halves which are hinged at the top and provided with sharp teeth at the bottom. The digging or filling operation consists of lowering and swinging the bucket close to the material till a good contact is made. The bucket is then closed through the closing line. Due to heavy weight of the bucket it is capable to dig the material and thereby filling it. It is then hoisted and swinged for dumping. It can be used for handling or digging of loose material and medium hard material also. It is very useful when operation is required in a vertical range and placement needs precision such as charging of bins or loading a carrier, building stock piles, digging pits, foundation trenches, well and caisson foundations etc. It can dig at above or below the level of machine. It can lift large sized rock or boulders. A clamshell can be used for almost all jobs, where other machines of the shovel family can be used.

The selection of bucket depends largely on the requirement of the job and the machine performance. A heavy bucket with sharp hardened steel teeth is used for excavating hard or rocky soils. A clamshell bucket is particularly useful for lifting heavy and large size rocks. Its capacity is 3–5 cu m and takes 10–15 min for loading and discharging.

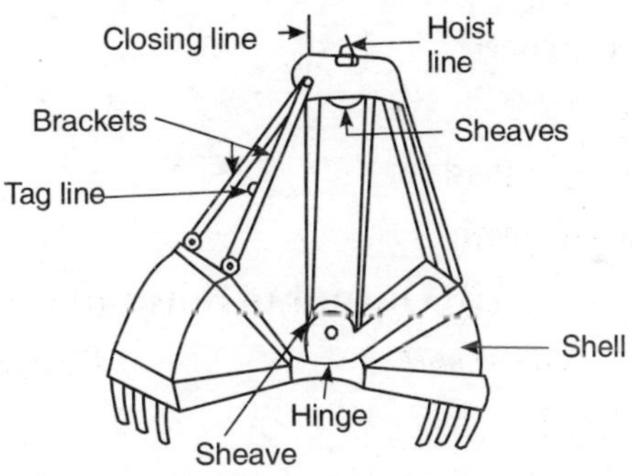

Figure 10.20 *Clamshell bucket*

10.14 COMPARISON OF VARIOUS MACHINES

S. No.	*Operation*	*Dipper shovel*	*Dragline*	*Hoe or pull shovel*	*Clamshell*
1.	Operation with hard soil or rock	Good	Very moderate	Good	Very moderate
2.	Operation in wet soil or mud	Poor Poor	Good Good	Poor	
3.	Range of work	Small	Long	Small	Long
4.	Efficiency in loading carriers	Good	Moderate	Good	V. Good

5.	Footing required	Close to work	Fairly away away	Close to work	Fairly away
6.	Digging level	At or above base level	Below base level	Below base level	Below base level
7.	Cycle time	Less	More	More	More

10.15 TABLES FOR CAPACITY OF ROAD MAKING MACHINERY

(A) Output of a three wheel roller

	Job	*Output per day*
1.	Water Bound Macadam Road surface average uncompacted thickness of 15 cm	240 sq. m
2.	Surface dressing–1st coat	650 sq. m
	– 2nd coat	920 sq .m
3.	Premixed carpet	350 sq. m
	Seal coat	900 – 1000 sq. m
4.	Bituminous Macadam	350 sq. m
5.	Asphaltic concrete	300 sq. m

(B) Output of Scrappers

	Type of soil	*Capacity for hour (Average)*
1.	Sand and soil	45 cu m
2.	Common earth	45 cu m
3.	Hard clay	40 cu m
4.	Mooram	40 cu m
5.	Wet and sticky clay	40 cu m

(C) Output of Tractor Dozer

	Output per hour	
Type and H.P.	15 m lead	30 m lead
(a) Granular type		
65 H.P.	45 cu m	30 cu m
80 H.P.	75 cu m	50 cu m
130 H.P.	85cu.m	55 cu m

(b) Wheeled Type

165 H.P.	90 cu m	60 cu m

(D) Working Life

1.	Road roller	18,000 hrs
2.	Tractor	12,000 hrs
3.	Scapper	12,000 hrs
4.	Truck	20,000 – 30,000 hrs
5.	Hot mixer	10,000 hrs
6.	Paner	15,000 hrs
7.	Grader	15,000 hrs
8.	Bitumen boiler	15,000 hrs
9.	Tractor	12,000 hrs
10.	Grader dozer	15,000 hrs

15 – 25% working life can be increased if the equipments are properly used by trained and experienced persons and properly maintained. The steam rollers have generally three wheels, a steering wheel, and two rear compaction wheels. The oil rollers have generally 2 wheels. A two-wheel roller is also called *Tandem Roller*. The rollers are built in sizes varying from 3 to 20 tonnes. It is mainly used for compacting or consolidating thick layers of macadam or road metal.

10.16 ROAD ROLLERS

It is one of the most essential road making machinery. It was the only highway equipment used in India some few years back. Rollers are generally self-propelled but some of the rollers are towed by tractors etc. The following are the common types of rollers which are generally used:

1. Cylindrical roller.
2. Flat-wheeled rollers or smooth wheeled rollers.
3. Pneumatic tyred rollers.
4. Sheep foot rollers.

As has been said earlier, a roller is an essential tool for the construction of a road. Its main function is compacting the surface such as sub-grade, base or soling ; and wearing surface etc.

1. Cylindrical Rollor. It is a light roller of iron, concrete or stone, which is drawn by hand or bullocks. These are available in various sizes and weights. Usual size is 100 cm in diameter and 150 cm long. This roller exerts about 7.0 to 10.0 kg/cm^2 pressure on the ground. These rollers are used for very light work—for rolling village roads, parks, game fields etc.

2. Flat-wheeled Roller. This is a very common type of roller and is commonly used. It can be driven by steam or diesel oil.

Figure 10. 21 *Flat-wheeled roller*

3. Pneumatic Roller. This roller consists of a small truck with two or more axles with a number of pneumatic-tyred wheels. The truck may be loaded so that the required intensity of pressure may be

Figure 10.22 *Pneumatic roller or wheeled rubber tyred roller*

attained. This type of roller is capable of consolidating the surface slowly and uniformally, without causing damage to the aggregate.

These rollers are used for compacting cold laid bituminous pavements, soft base course and layers of loose soil. These are also suitable for compacting closely graded sands, and fine-grained cohesive soils at nearly plastic limits. These are useful for finishing the embankments compacted by sheep foot rollers or on loose sandy soil.

The pneumatic rollers have got the following specialities :

(i) They are able to knead the aggregate in the binder without breaking or crushing the aggregate pieces, which is very important.

(ii) When the aggregate and the binders are spread on the road surface, the aggregate binder joint is very weak for the time being and the pneumatic wheel rollers do not have the tendency to dislodge the aggregate.

(iii) On steep gradients and sharp curves, they work very efficiently and have been found very useful in the construction of ghat roads.

4. Sheep Foot Roller. It consists of heavy metal cylinders usually of 1 to 1.5 m diameter and 1.2 to 1.5 m long. Each of these drums have 16 to 18 cm projections on their surface and each drum weighs about 3 to $4\frac{1}{2}$ tonnes.

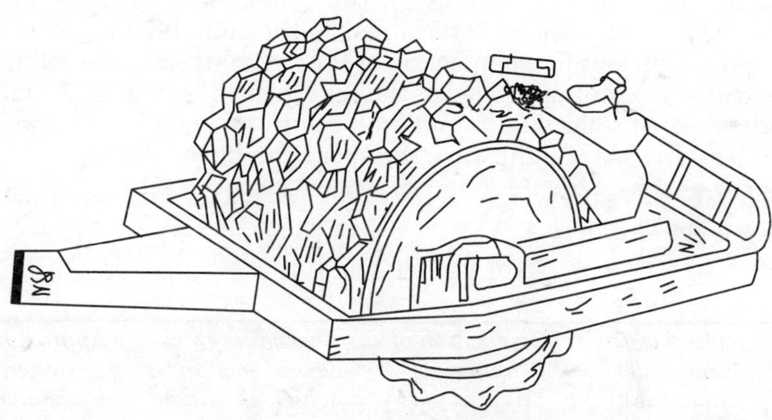

Figure 10.23 *Sheep foot roller*

The cylindrical drums are hollow from inside and they can be filled with water or sand to increase their weight. Two rollers form one set to provide flexibility of movement and control of larger area. The cylinders are generally towed to a tractor.

With this roller the surface is compacted from bottom upwards. The feet of the roller penetrate into the surface and knead it and thereby ensuring the proper compaction of soil from the bottom itself. These rollers are used for compacting filled up earth in banking.

10.17 ROLLING

The weight of the road roller to be used for compaction of road depends on the size and hardness of the stone aggregate and the thickness of the layers of the materials to be compacted. For ordinary soft stone aggregate 6 to 8 tonne roller is sufficient. For brick aggregate, 6 tonne road roller is used. For roads and highways near towns, 10 to 12 tonne road rollers are used. The actual depth of compaction of any particular road roller should be determined by the actual field tests. Usually the thickness of the loose soil for compaction should be between 15–23 cm and that of stone aggregate upto 10 cm. If more depth is to be rolled, it must be done in layers.

The operation of the rolling should be started from the road edges and it should progress gradually towards the centre from both the sides. The rolling is done parallel to the centre line of the road. In case of curves where super-elevation is to be provided, the rolling should be started from the inner edge to the outer edge of the curve. While rolling care should be taken that each pass of the roller should uniformally overlap not less than one-third of the track rolled in the preceding pass.

The actual number of passes of the roller for obtaining the desired compaction, should also be determined by the actual field test. Usually 50 passes are sufficient for compaction of the dry material. The rolling should be done at the minimum speed of the roller which is usually 3 km/hour for smooth wheeled roller and 5 to 6 km/hr with pneumatic-tyred and sheep-foot rollers. In case of light work the speed can be increased.

Table 10.1 gives the weight of the roller and maximum layer thickness for various types of rollers.

TABLE 10.1

S. No.	*Weight of road rooler in tonnes*	*Description of material time in cm*	*Maximum thickness of each layer which should be compacted*	*Approx. moisture content in the material in per cent*
1	6 – 8	Clayey sand	20 – 23	10 – 18
2	6 – 8	Lean mix concrete	20 – 23	–
3	8	Ashes, clinker etc.	13 – 15	1 – 20
4	8	Clay gravel (stable)	2 – 23	6 – 10
5	10 or more	Crushed rock or hard brick-bars	10 – 12	–

10.18 OUTPUT OF ROAD ROLLER (APPROX.)

Table 10.2 gives the output of a 8–10 tonne road roller with various type of materials for 8 hour/day work.

TABLE 10.2

S.No.	Material/work	Output
1.	Rolling of 6 mm chips with bitumen/ asphalt spread at about 1 cu. m per 100 sq. m.	2000 – 2200 sq. m per day
2.	Rolling of 12 mm chips spread at about 1.5 cu. m per 100 sq. m.	1600 – 1800 sq. m per day
3.	Consolidation of the premixed carpet	500 – 600 sq. m per day
4.	Rolling of two coats of surface dressing done together	1000 sq. m per day
5.	Rolling of 3.8 m wide road soling	1000 m per day
6.	Rolling of premixed metal (base-course)	1100 – 1400 sq. m per day
7.	Rolling of hard road metal to full compaction	20 – 25 cu. m per day
8.	Rolling of soft metal to full compaction	50 – 60 cu. m per day

10.19 DUMPERS

The dumpers are used for conveying the materials to the spot and lay them in their proper places. The dumpers are trollies fitted on trucks which can be

Figure 10.24 *Dumper*

tilted in one or the other direction. The materials are loaded, conveyed and dumped in places by dumpers.

10.20 BITUMEN MIXING MACHINERY

The machinery, which is used for laying and mixing bitumen and aggregate and applying bituminous surfacing on the prepared surface, consists of the following :

1. Bitumen boiler. *Bitumen boiler* is used for hot process, where the bitumen is required to be heated. The bitumen or tar boiler, as the name suggests, is used to heat the bitumen to the required temperature. The boiler consists of a continuous system, capable of supplying hot bitumen at the required temperature.

2. Bitumen sprayer. *Bitumen sprayer* is a bitumen boiler, fitted with a pumping and spraying equipment. In this case the boiler is fitted with a double acting pump and is so designed that one man can easily obtain and maintain the necessary pressure at the delivery point. The bitumen passing out of the pump flow through a long flexible tube at the end of which a straight metal pipe is fitted with a nozzel. Through this nozzel, the bitumen is sprayed uniformly on the surface.

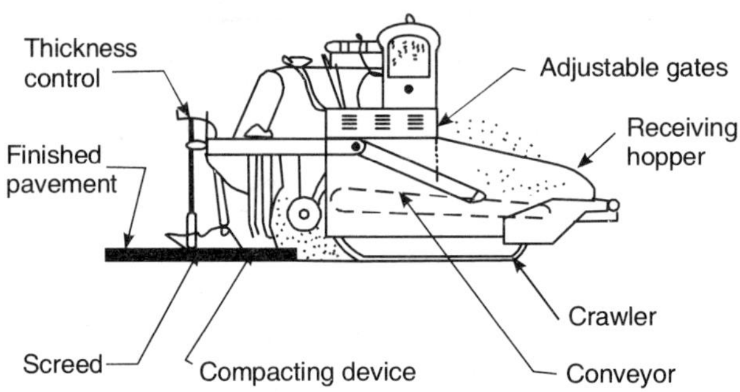

Figure 10.25 *Bitumen finisher*

3. Bitumen Mixer and Spreader. *Bitumen mixer and spreader* is a road making machinery in which the road metal or grit are mixed with the required quantity of bitumen at the desired temperature and spread on the surface. This machine is used in the case of premixed carpet and bituminous concrete roads.

4. Gritting Machine. *Gritting machine* is used for broadcasting grit of the required size uniformly over the surface. The machine is fitted with a sieve of the required size through which the grit passes and is spread on the prepared surface in uniform thickness. These machines are seldom used in India.

5. Aggregate Heater or Drier. *Aggregate heater* is an ordinary sheet metal pan in which the aggregate is heated to drive off moisture. The aggregates can also be heated in revolving drum.

10.21 CONCRETE MIXING MACHINERY

For the construction of cement concrete roads, the following machinery is generally used:

(1) Concrete mixers.

(2) Concrete pavers.

(3) Concrete finishers.

(4) Concrete vibrators.

1. Concrete Mixers. The ingredients of concrete such as fine and coarse aggregates and cement are mixed in concrete mixers along with the requisite quantity of water. The mixers can be stationary or portable, tilting or non-tilting type. The tilting type portable mixers are generally used. The tilting mixers usually have a conical or bowl-shaped drum,which is rotated either electrically or manually. They are charged by means of a feed hopper. Mixers are of various sizes which is indicated by a number equal to its rated capacity in cubic metre of mixed concrete.

2. Concrete Pavers. Concrete pavers are used when the mixing is done on the site. The concrete pavers do the mixing and laying of concrete on the sub-grade. They consists of a bucket and boom arrangement which carry the concrete and discharge it on the required position. Concrete pavers are rarely used in India.

3. Concrete Finishers. After spreading the concrete in a uniform layer, the finisher does the job of finishing. The finishers have two screeds, which work in transverse direction like a saw. The machine moves on the slides of form work on its wheels. The front screed acts as a strike off bar and brings the surface to its final level. The rear screed gives the exact shape contour in the finished pavement.

4. Concrete Vibrators. Vibrators are used to compact and consolidate the concrete. There are three types of vibrators which are commonly used for different concrete works.

(a) Surface Vibrators.

(b) Internal Vibrators.

(c) Poker or Needle Vibrators.

PROBLEMS

10.1. Name the various types of road machinery for the construction of a W.B.M. road.

10.2. How many types of dozers are there ? What are their main functions ? Describe in brief.

10.3. You are the incharge of a big road project. List the various types of road machinery you will require for the completion of the various types of works involved in the construction of the road.

10.4. What type of road machinery is generally used for the construction of a cement concrete road ? Describe in brief.

10.5. Describe the various types of road rollers. Enumerate the special features of each of these rollers.

10.6. Write short notes on :

(i) Ripper

(ii) Bull dozer

(iii) Scraper

(iv) Concrete mixer

(v) Bitumen sprayer

10.7. What do you understand by rolling ? How it is done ? What are the main purposes of rolling ?

10.8. Write short note on the output of the road rollers.

10.9. Compare the working efficiency of a dragline and power shovel.

10.10. Name and state the uses of various attachments of a shovel.

10.11. State the working operation of a Dragline and clamshell.

10.12. Name the various excavating equipments generally used for highway construction and compare their efficiencies and working capacities.

BIBLIOGRAPHY

1. Varma Mahesh Construction Equipment Metropolitan Book Company, Delhi. 1964
2. Kadiyali L.R. Road Making Machinery (High way Engineering)–1984
3. Recommendations for the size for road making machinery I.R.C. : 22, 1980
4. Specifications for heaters for tar and bitumen I.S. 2004 –1974
5. Specification for Hot-mix plants I.S.: 3006–1965

6. Harbert L. Nichols – Moving the Eartho –1964
7. Audel. Truck and Tractor guide, 1960.
8. Techmial Information and Hardy Rickner on construction equipments, Voltas Bombay.
9. Hand Book on Earth Moving Machinery, Ministry of Irrigation and Power, Central Water and Power Commission, New Delhi –1962.
10. Kellag,Construction Methods and Machinery Practice Hall, Incorporation, New York.
11. Sharma and Sharma – Highway Engineering Asia Book Company, Bombay – 1962.

11

Road Arboriculture

In this Chapter you will study,

- Objects of Road Arboriculture • Selection of Trees • Planting Operations
- Location of Trees

GENERAL

The highway engineer with his inherent ingenuity and patience can achieve wonderful results with only small cost to beautify the road. Road arboriculture is one of the architectural effects, which adds to the general or overall appearance of the road. *Arboriculture* means tree culture, that is, care and planting of trees. In this small chapter, some thought will be given to the planting and care of trees along the road-side.

11.1 OBJECTS OF ROAD ARBORICULTURE

The following are some of the objects of planting trees along the road-side :

(1) Trees merge with the natural landscape of the surrounding area and hence provide an attractive appearance to the road user.

(2) They provide shade to the traffic.

(3) By planting useful timber trees and fruit bearing trees, it becomes a source of supply of fruit and timber.

(4) They break the monotony of the road, specially in country roads or in un-built-up areas.

(5) Trees also act as wind breakers and reduce the temperature variations on the surface of the road.

11.2 SELECTION OF TREES

The selection of trees mainly depends upon the nature of the soil and the climatic conditions of the locality. However the following should be the characteristics of road-side trees :

(1) Trees should have long life.

(2) They should yield either fruit or timber.

(3) They should not be very clumsy and should grow centrally so that they should not cover the roadway.

(4) They should be dense and should yield shade all the year long.

TABLE 11.1

S.No.	*Type of Soil*	*Trees which can be planted*
1.	Loamy soil	Shisham, mango, jamun, kathal, imli etc.
2.	Clayey soil	Gold mohar, jamun, mango, malina etc.
3.	Sandy soil	Shisham, safedsin, arroo, kanjee etc.
4.	Usar	Neem, imli, babool, dhak etc.

11.3 PLANTING OPERATIONS

The tree planting along the road-side involves the following operations :

1. Excavation of pits.
2. Preparation of seedlings.
3. Transplanting.
4. Watering and care of trees.

1. Excavation of Pits. Pits are excavated of 1.2 m × 1.2 m × 1.2 m size or less and the excavated earth should be dumped along the edges of the pit in the shape of a bund. Good soil mixed with some manure is filled into the pit upto a depth of 15 cm below the ground level and the soil is allowed to settle for some time.

2. Preparation of Seedlings. Seedlings are prepared in a nursery, located in a nearby place. (A nursery is a place where plants are grown in a most favourable condition.) A nursery should be as near the road line as possible so that during transplantation the plants may not be injured.

3. Transplantation. When the seedlings are grown to a sufficient height, they are first transferred to pots and kept there till they are nearly

1.15 m high and then transferred to pits. Before the plants are received in pits, the soil in the pits is made loose and mixed with some measure. The best season for transplantation is when the rainy season has commenced.

4. **Watering and Care of Trees.** After the transplantation, the plants require regular attention and care including watering for about 3 to 5 years. The plants should be protected by tree guards made of steel, wood, earth, or masonry. They are also to be protected against the decaying agents such as fungus, white ants etc. by spraying D.D.T. or tobacco water. The plants should also be watered at regular intervals.

11.4 LOCATION OF TREES

Planting of trees along the road-side can be either on one side or on both sides. The tree should be planted in a position where it does not interfere with the traffic or damage the road surface. Generally trees are planted on both sides of the road. The minimum distance from the edge of the road and line of trees should be 2 to 3 m. The distance between two trees depends on the

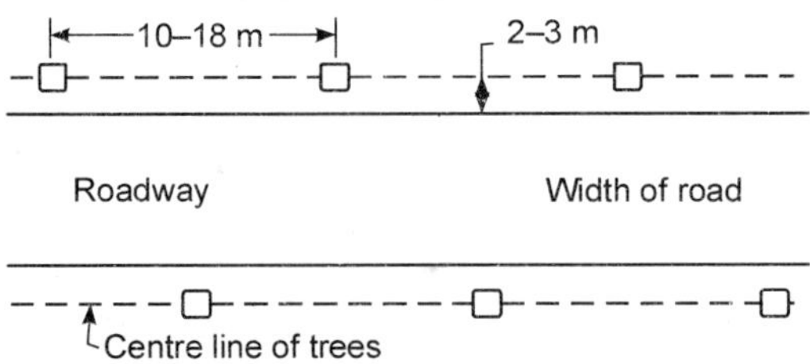

Figure 11.1 *Location of trees (plan)*

nature of the trees and it varies from 10 m to 18 m. The trees should be planted in a staggered way, that is, a tree in one row should be in between the two trees in the other row, on the other side of the road as shown in Fig. 11.1.

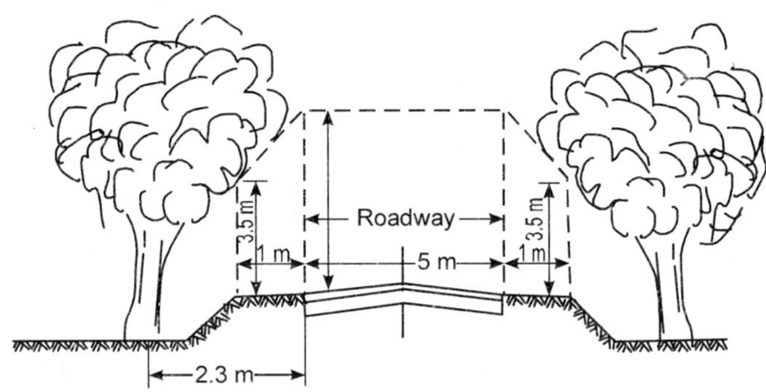

Figure 11.2 *Location of trees (cross section)*

When the trees have grown up, care should be taken to see that the branches of the trees should not interfere with traffic otherwise all branches of the trees within 4.5 m height from the road surface should be trimmed off. It should also be seen that branches of the trees do not interfere with the telegraph and telephone wires.

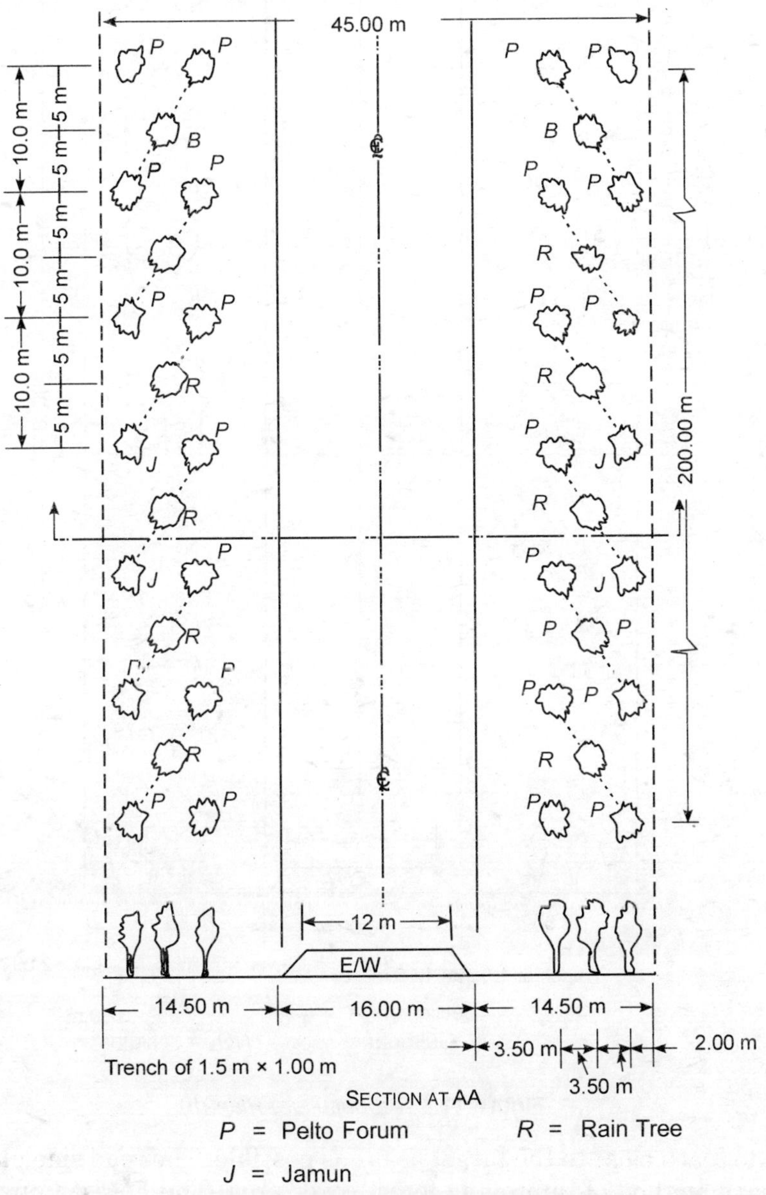

Figure 11.3 *Arboriculture Plan (I)*

The Govt. of India has laid down a forest policy for having 33% of forest cover of the total land. The present day condition is that we have only 8–10% forest. For maintaining ecological balance it is important to increase the forest

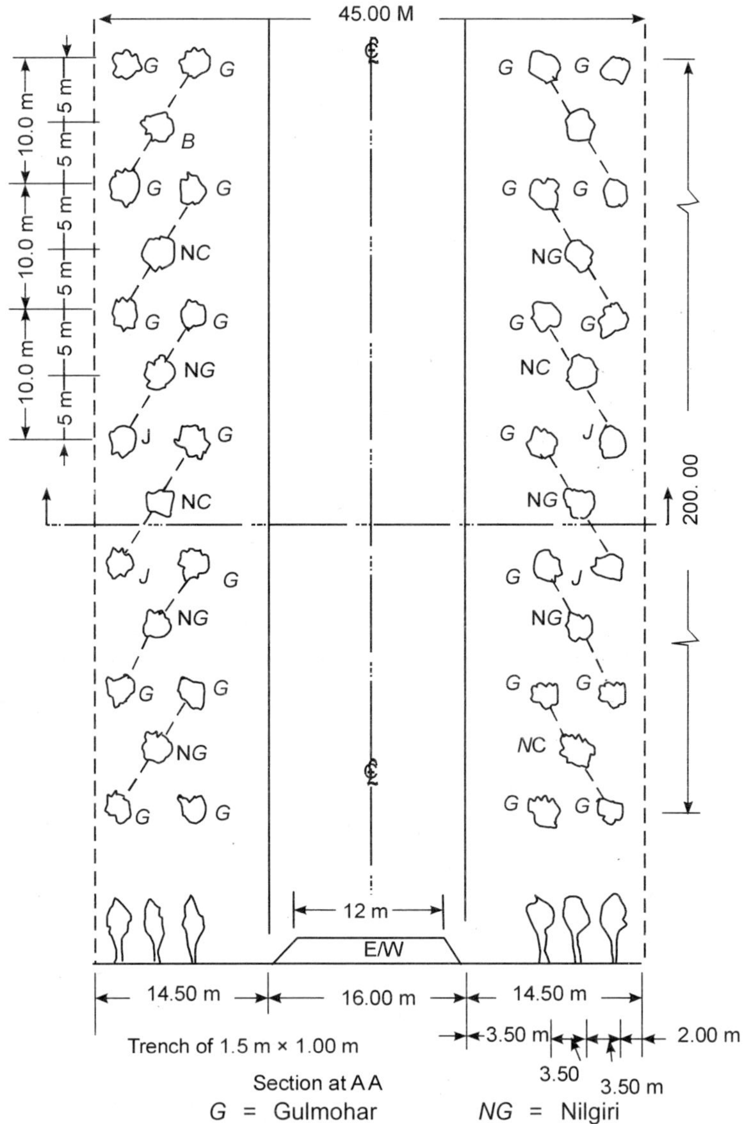

Figure 11.4 *Arboriculture Plan (II)*

cover to reach close to the target as far as possible. The road side plantation is a easy method of improving forest cover area. The Forest Conservation Act of 1980 is an awakening in the direction of improving ecology and minimising hazards of floods, droughts, silting of rivers etc. The highway engineers have come forward to contribute their bit by planning road arboriculture.

Single row plantation on both sides of the road will not serve the purpose for achieving the long term objective of improving the forest cover of the land. It is now proposed to have 3 rows of trees on either side of the road.

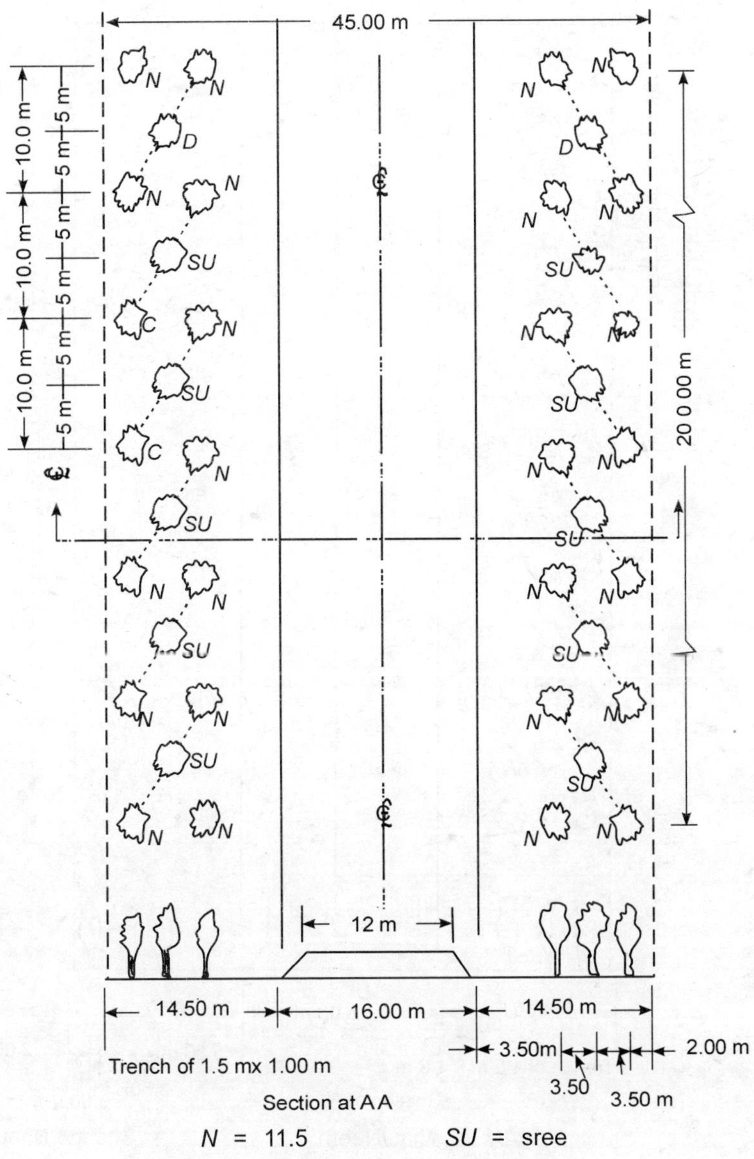

Figure11.5 *Arboriculture plan (III)*

This method has been tried by Maharashtra State P.W.D. for Amrauate National Highway Bye-pass in the year 1986–87. The distance of first row from the edge of the road has been kept at 5.5 m and the last row is 14.5 m. The middle row in staggered in between the first and the second row the middle row trees are of different variety than the other two rows. The variety of trees to be selected for the first and the last row should be such so as

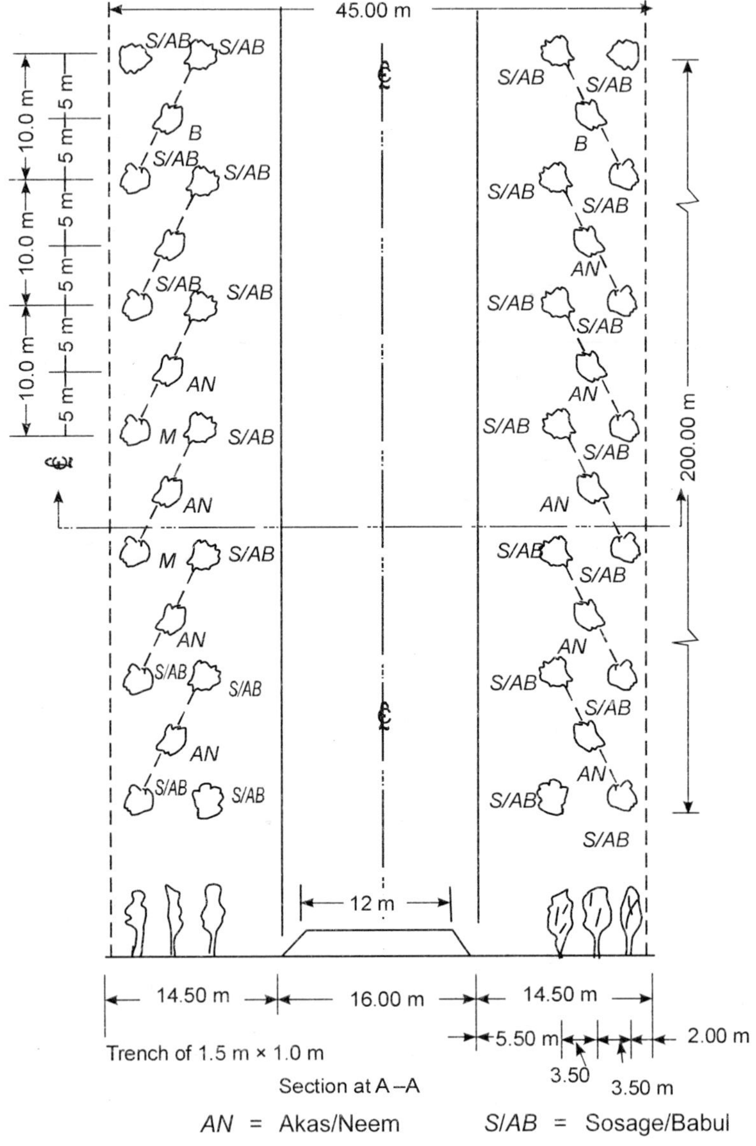

Figure 11.6 *Arboricukture plan (IV)*

to have more height and for the middle row trees should be of lesser height. This will help in increasing the shaddy area and less interference with the movement of traffic. Mango trees are generally preferred for the middle row and comparatively early growth.

11.5 PROTECTION OF TREES

It is generally felt that the protection of plants by conventional method of empty drum is not suitable around the small plants since it retains the heat inside the drum and retards the growth and branching etc. and reduce survival rate. The maintenance operation like weeding, watering prunning, tending the soil, replacing casualties etc. becomes difficult. It is suggested to have barbed wire fencing around the trees parallel to the road with some protecture mesh at the bottom. The entire area is divided into rectangular blocks with proper doors for the entry of maintenance staff. This arrangement facilitates proper and speedy growth of plants, maintenance protection, supervision and replacement of casualties etc. The key to success of arboriculture is proper protection.

Weeds should be removed as soon as they are seen. The frequency and growth is more during rainy season than in dry weather. The area around the plant 1.25 m diametric should be free from the weeds, gross etc.

The watering sequence should be two to three watering per week during summer and during winter one watering within 8–10 days is necessary for proper growth. The watering should be sumptuous and the circular bunds should be completely filled with water.

Agrochemical and nutrients should be sprayed during early days of plantation to achieve fast growth and reduce casualty. Healthy and well grown plants become self-sufficient very early. The frequency of spray should be thrice a year for the first three years. Pesticides should be applied as and when required so as to protect the plants from any disease. The objective of pesticides is to protect the plants from disease and pests if any for their healthier growth and longer life.

PROBLEMS

11.1. What do you understand by the term road arboriculture ? Why are trees planted along the roadside and what are their advantages ?

11.2. What type of trees would you recommend in loamy soil and clayey soil ? Draw a typical cross-section and a plan of road showing position of trees.

11.3. Discuss the factors that influence the selection of roadside trees.

[Hint. In India, rainfall, temperature, topography, soil etc. are different from one place to another. Trees which grow well in one place may not even grow in other type of soil. Hence according to the nature of the soil and the climatic conditions of a particular locality, the trees are selected. (*See* Article 11.3)

BIBLIOGRAPHY

1. I.R.C. – 1979, Landscaping of Roads.
2. I.R.C. – 1979, Environmental Consideration in Planning and Design of Highways in India.
3. Indian Highway – Vol. 17, No. 8 – 1989.

12

Traffic Engineering

In this Chapter you will study,

• Scope of Traffic Engineering • Traffic Studies • Traffic Volume and Characteristics • Methods of Measuring Traffic Volume • Scheduling of Volume Counts • Highway Inventory • Speed Studies • Parking Studies • Accident Studies • Road Junctions • Accident Records and Data • Accident Investigation • Speed and Skid Distance • Speed, Travel Time, Volume and Density • Traffic Regulations • Conflicts • Travel Time Studies • Visibility at Corners, Bends and Junctions • Traffic Islands • Shapes and Traffic Islands • Refuge Islands • Design of Traffic Islands • Pedestrian Crossings • Channelization of Traffic • Acceleration and Deceleration Lanes • Road Markings • Intersection Design • Basic Characteristics of a Crossing • Traffic Rotary • Advantages of Traffic Rotary • Intersection at two Grades • Interchanging Ramps • Terminology • Traffic Signals • Basic Features of Traffic Signals • Classification of Road Signals • Signal System • Timings of Signals • Conditions Warranting Installation of Traffic Signals • I.R.C. Signal Design Method • Road Signs • Highway and Street Lighting • Factors Effecting Visibilty on Roads • Design of Higway Street Lighting • Highway Capacity • Level of Service • Passanger Car Unit • Parking Places and Lay-Bys

GENERAL

Traffic Engineering is a very important branch of engineering which deals with highway planning, geometric, design, traffic control, traffic movement, forecasting, safety precautions, accidents prevention etc. It is a full-fledged branch of engineering which has recently attracted many highway engineers to put their heads together to solve multiple problems of safety, planning and smooth movement of traffic. The traffic engineering deals with the following :

(i) Effective design of roads and road facilities.

(ii) Comprehensive traffic and transport studies in all major cities to effectively forecast the anticipated future traffic and to design the required traffic facilities.

(iii) Provision and design of crossings, parking places and development of adjoining land.

(v) Standardize traffic signals and speed breakers.

(v) Geometric standards for urban roads safety.

(vi) Planning of street lights.

(vii) Publicity of traffic safety aspects.

(viii) Traffic flow regulations and control.

(ix) Collection, compilation and analysis of accident data.

(x) Speed limit regulations.

(xi) Standards for grade separators.

(xii) Road user cost studies.

(xiii) Planning of channelizing traffic, decongestion from control business places.

(xiv) Locating areas prone to skiding and accidents.

(xv) Control of air and noise pollution.

In short the responsibilities of a traffic engineer is to ensure safe, convenient and early flow of men and material on the road by devising ways and means of fulfilling the above objects of traffic engineering. Traffic engineer also deals with the fitness of vehicles, their up keep, over loading and preventing mis-use of the common facilities on the roads by regulations and mandatory provisions.

12.1 SCOPE OF TRAFFIC ENGINEERING

The traffic engineering covers the following aspects:

(i) Traffic characteristics

(ii) Traffic operations

(iii) Traffic planning and

(iv) Traffic administration.

Traffic Characteristics. Traffic characteristics comprises of

(i) Average weight of vehicles

(ii) Average speed of vehicles

(c) Limitations of drivers, their skill, hearing and vision ability, fatigue, impatience, ability to follow rules etc.

Traffic Operations. It consists of the following:

(i) Traffic regulations

(b) Traffic control devices

(c) Traffic segregation and channelization

(d) Traffic islands or refuge islands.

Traffic Planning. It consists of the following :

(a) Construction planning or programme

(b) Planning of main streets, freeways, terminals etc.

(c) Planning of parking spaces.

Traffic Administration. The traffic administration consists of

(a) engineering

(b) enforcement and

(c) education.

12.2 TRAFFIC STUDIES

To form the basis of establishment for planning of highway a number of traffic studies have to be conducted. For conducting traffic studies the following points should be kept in mind :

1. Type and volume of traffic
2. Geometric design
3. Structural design
4. Highway safety
5. Future development.

For achieving the above objectives, following studies are conducted:

1. Traffic volume or characteristics
2. Speed
3. Speed and delay

4. Parking
5. Origin and destination.

12.3 TRAFFIC VOLUME AND CHARACTERISTICS

Traffic volumes studies are conducted by

(a) *A.D.T.* Average Daily Traffic, it is the average of 364 days, 24 hours.

(b) *Classified volume* involving the composition of traffic i.e., heavy vehicles such as buses, trucks; light vehicles such as cars, scooters three wheelers, motor cycles; slow moving traffic such as cycles, rikshaw; iron wheeled traffic such as tanga, bullock carts etc.

(c) *Hourly volume studies.* It is generally conducted in cities only for determining number of lanes, geometric design, parking demand etc.

(d) *Directional distribution studies* measures the direction of vehicle in a particular direction or destination during a particular time period and reverse direction.

(e) Cordon count volumes determines the accumulation of vehicles at a particular time for deciding parking places or zones.

(f) Pedestrian volume decides the planning of cross walks, footpaths and signals.

12.4 METHODS OF MEASURING TRAFFIC VOLUME

Traffic volume is counted either manually or mechanically.

Manual Counting. In this method, a team of experts is posted at strategic points to count the various category of vehicles such as heavy vehicles, light vehicles, iron wheeled vehicles etc. The method is very tedious and unpracticable during all the 365 days and round the clock. Although the method is very accurate and dependable. To cut short the process statistical sampling technique is adopted. The fluctuation of traffic volume and the intensity of traffic during peak hours is recorded with the help of statistical analysis, peak hourly traffic volume as well as average daily traffic is calculated. This method is generally used due to its specific advantages.

Mechanical Counting. Mechanical counting is done by automatic recorders. The recorders can be permanent type or portable type. The permanent records may be

(i) Magnetic detectors

(ii) Pressure sensitive recorders

(iii) Electronic detectors

Permanent recorders are fixed on the road junction or at strategic points for continuous recording. Portable recorders are smaller in size and are actuated by air switches. These are moving car counters also which are fixed in a moving car. This method consists of counting the number of vehicles crossing, area overtaking in one direction and in the opposite direction, time taken by the vehicle (observer) to cover a specified distance.

If t = erage time taken to travel a specified distance then

$$t = t_w - \frac{y}{g}$$

where

t_w = time taken by the observer

y = number of vehicles overtaking minus number of vehicles over taken by the observer

q = traffic flow in one direction (number of vehicles per minute)

$$q = \frac{x+y}{t_a + t_w}$$

where

x = number of vehicles crossing from opposite direction

t_a = time taken by the observer while travelling in the opposite direction.

Mechanical counting though easy and quick but is not reliable as the impulses caused by light vehicles may not be recorded. Also it is not possible to record pedastrians.

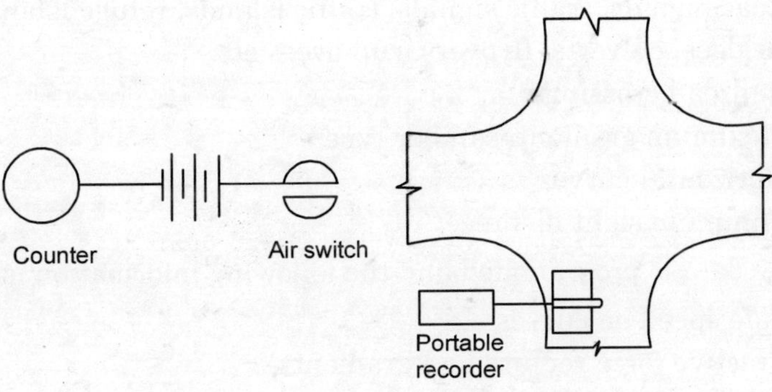

Figure 12.1. *Scheduling of volume counts*

12.5 SCHEDULING OF VOLUME COUNTS

The following schedule should be maintained for urban roads:

1. For population of a town varying between 8 to 10 million, 1 to 3 permanent count stations should be maintained whereas for population 10 to 50 million. 4 to 6 stations and with population above 50 million atleast 10 permanent count station should be maintained.
2. Cordon and screen counts are after made during peak hours only.
3. Evening peak hour counts are made at the intersection of two major streets.
4. Quality counts on representative street or lanes is made for providing appropriate repressions plans.
5. Coverage count to provide current date.

12.6 HIGHWAY INVENTORY

The purpose of road inventory is to collect information of highway planning surveys. In this study a physical check is made with provisions for revision, improvement, re-construction and maintenance. Data is collected and incorporated in a map showing all public roads inside and outside the city, location of important structures on the road side, principal highway connections etc.

Items to be recorded in highway inventory are :

(i) Type of road
(ii) Real width, number of lanes, type of surface and its condition
(iii) Sidewalks, footpaths their width
(iv) Movement along side the road
(v) Road signals, traffic signals, traffic islands, refuge islands etc.
(vi) Bridges, culverts, flyovers, run overs etc.
(vii) Railroad crossings
(viii) Maximum gradient, camber type
(ix) Horizontal curves
(x) Minimum sight distance

The above data is programmed and the following information is computed :

(i) Safe speed on curves
(ii) Tractive force required on gradients
(iii) Length of transition curves
(iv) Correctness of super elevation
(v) Widening of roads on curves

(vi) Future separation plans

(vii) Danger or accident prone zones

(viii) Maximum speed limits

(ix) Desirable width of lanes.

12.7 SPEED STUDIES

The speed studies mainly deals with the variation of speed between the slow moving traffic and the fast moving traffic between the heavy and light vehicles. Following are the objectives of this study :

(i) Planning traffic control, planning speed zones, planning traffic signals, and traffic signs.

(ii) Computing the speed trends

(iii) Controlling and minimising accidents

(iv) Analysis of high accident locations

(v) Suggestions for the attraction of traffic design.

The speed study should be conducted on the following locations and time periods:

(i) All major roads

(ii) All accident prone zones

(iii) All major crossings

(iv) All installation of road signs traffic signals etc.

(v) All representative strategic location

Time

Between 0600 to 0900 hours

Between 0900 to 1200 hours

Between 1200 to 1500 hours

Between 1500 to 1800 hours

Between 1800 to 2400 hours

Methods of Speed Studies. Following are the methods for conducting speed studies.

(i) **Manual Method.** This method used to be adopted by Central Road Institute, Delhi upto late sixties. The method consists of using stop watches noting down the time taken by the vehicle covering a fixed distance and then calculating the average speed.

(ii) **Electro-mechanical Method.** In this method, a condenser is used for recording the impulses and the time period. The electric charge begins discharging with the first impulse and ceases to do so after the second impulse. The speed is recorded by direct caliberation.

(iii) **Radar Method.** It is operated on Doppler's principles. An electromagnetic wave is deflected by the vehicle. The speed of the vehicle changes the wave length and returns to the receiving unit of the metre. This change in wave length is directly caliberated to its speed on the metre. A graphic recorder is connected for directly recording the speed.

(iv) **Enoscopic Method.** It is a simple L-shaped box with mirrors fitted at 45°. Both ends of the box are open. The reflection of the vehicle is produced on the mirrors and the time between the successive reflections is noted with a stop watch.

(v) **Photographic Method.** Pictures of the moving vehicles are taken by a movie camera. From the line film speed of each and every vehicle can be measured.

12.8 PARKING STUDIES

In congested localities, parking of vehicles is a very big problem. Parking studies aims at determining the parking demands of the area and its location. The parking studies can be conducted by parking inventory. It consists of collecting information regarding the number and type of vehicles during peak hours on a particular location, kerb spaces, area of foot paths, shoulders and side walks etc., parking spaces available but not used. From this date the following informations is computed :

1. Total parking demand of the area
2. Deficiency of parking space
3. Origin and destination of vehicles demanding parking
4. Turn over of parking spaces
5. Distance to be walked by the occupants of the vehicles

Methods of Parking. Vehicles can be parked in one of the following ways:

1. Parallel Parking. This type of parking is preferred on wide and less busy lanes in non-commercial areas. The vehicles are parked parallel to the centre line of the road on either side of it.

A minimum distance of 1.2 m is left between the two vehicles. Parallel parking have been found satisfactory from safety point of view, but vehicles occupy more space putting vehicles in the unoccupied spaces between the vehicles, is sometimes difficult job.

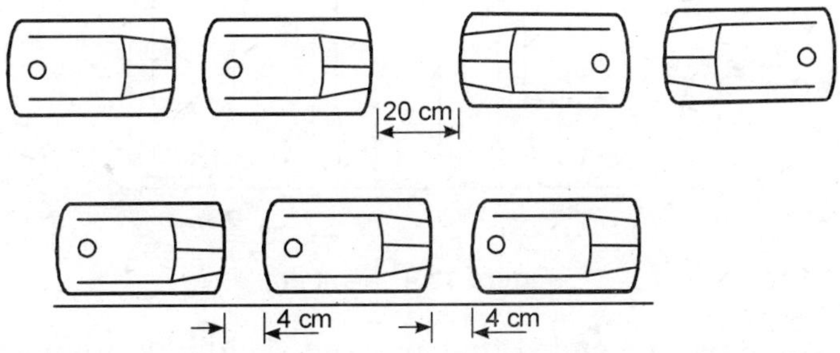

Figure 12.2 *Parallel parking*

2. Angle Parking. Angle parking provides more space than the parallel parking. It is easier and quicker to park but difficult to impark. It occupies more space on the lane, hence it is not recommended on both sides of the lane. This type of parking is recommend for spaces only marked for parking.

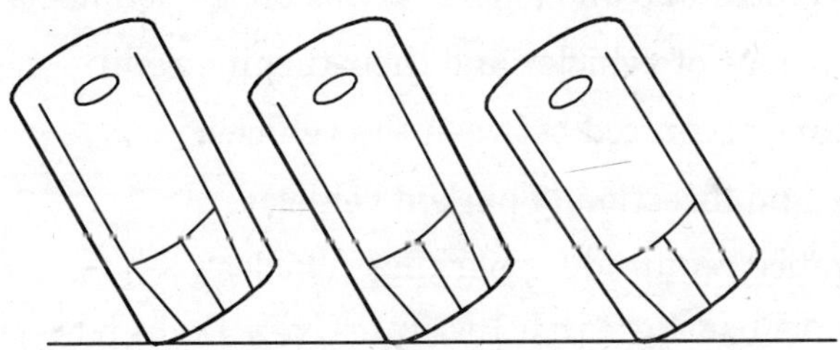

Figure 12.3 *Angle parking*

3. Right Angle Parking. Right angle parking is used on steep slopes, but should not be used on through lanes. This type of parking is recommend in residential localities.

Parking should be prohibited on the following places :

1. Side walks or shoulders and cross walks.
2. Drive-way entrance
3. Within 5 metres of fire hydrants
4. On refuge islands
5. Bridge or culvert entrance

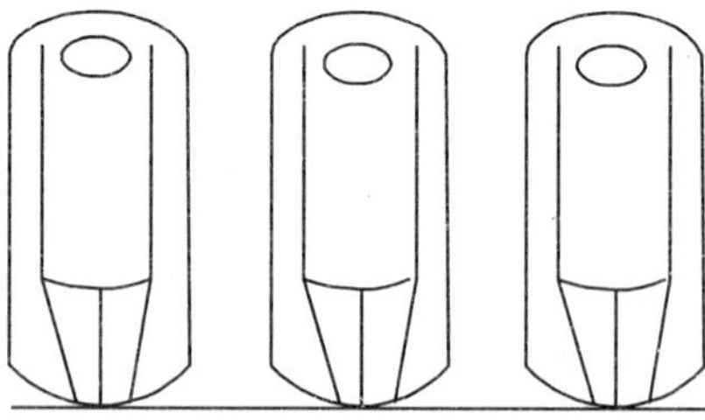

Figure 12.4 *Right angle*

It is further recommended that parking should be prohibited within 6 metres of cross walks and 10 metres of stop signs or other warning signs. No parking on roads having a carriage way of less 8 metres should be allowed. Parking should be allowed on one side of the road only if the lane width is between 8 m and 10 m. During peak volume hours parking should be prohibited on these roads also.

Stall Size and Aisle

Design of parking stalls and aisle depends on the following factors :

(i) Dimensions of vehicles and their turning radii

(ii) Clearance required between the vehicles

(iii) Angle and direction of parked vehicles

(iv) Clearance required for parking vehicles

The Central Road Research Institute, New Delhi has recommended the following dimensions for different vehicles plying on Indian roads :

Types of vehicle	*Overall length*	*Overall width*
Auto rickshaw	2.44 m	1.37 m
Cycle rickshaw	2.59 m	1.19 m
Tanga	4.11 m	1.55 m
Bullock cart	5.87 m	1.73 m
Cycle	1.91 m	0.53 m
Car	3.6 m	2.1 m
Truck	6.0 m	3.0 m
Scooter	1.63 m	0.75 m
Motor cycle	2.0 m	0.87 m
Motor cycle rickshaw	2.82 m	1.42 m

Aisle Width. Aisles are spaces ear marked or cordoned for the parking of vehicles and their circulation within the stall. For angle parking best arrangement is a series of one way aisles alternate in direction. Normally a minimum width of 4 m for one way and 7.5 m width for two way movement is desirable for aisle.

While planning parking places, the following should be kept for consideration :

(i) *Surfacing*: The parking spaces should be properly marked and well surfaced without undulations and steep slopes. Concrete and bituminous surfaces are always preferred.

(ii) *Lighting*: Parking area should be properly lighted.

(iii) *Pedastrian Safety*: Pedastrian safety becomes a problem at the entrance and exits. Suitable walkways, shadowing of pedastrians by islands and control of pedastrian by signals may be necessary for pedastrian safety.

(iv) *Drainage*: If the parking is to be situated on a low lying area, proper drainage should be planned or the low lying area should be suitably raised.

Off-Street Parking

When on congested and busy lanes, kerb parking is not feasible a separate parking space away from the lane is provided. This parking space is known as off-street parking. The main advantage of this type of parking is that the main road remains free from the congestion but the owners have to walk down quite a distance after parking their vehicles. To find a separate parking place nearby a congested and busy market place is also a big problem. The off-street parking should be as close to the lane as possible so that the owners of vehicles should use this facility willingly. Two types of off-street parking are in use.

(i) *Parking stalls or lots*: This is a surface off-street parking and is recommended where sufficient parking space is available near the lane, at a reasonable price. The parking of vehicle is done by the owners or drivers themselves or by the attendants of the parking stalls. The *attendant parking* system is convenient and time saving than the *self or driver parking system*.

(ii) *Multi-storeyed parking*: This type of parking is provided in places where sufficient parking space on the surface is not available. Multi-storey garages are constructed at a nearby place to accommodate large number of vehicles at a time. Ramps and elevators are provided for reaching the different floors. Elevators are capable of moving horizontally and vertically to park the vehicles in their specified places. The space requirement in the case of ramps is much more than for elevators or mechanized parking. But during power failure or mechanical break down the whole process of parking and de-parking will come to a grinding halt.

The parking operation consists of the following :

1. Entrance
2. Acceptance
3. Storage
4. Delivery
5. Exit

The space required for vehicles during entrance, acceptance and exit is called *Reservoir space.*

12.9 ACCIDENT STUDIES

Accidents are the results of collision of one highway user with the other highway user. The accidents cause death, serious injuries and damage to property. A traffic engineer is not concerned with the legal cause of accidents but to prevent and avoid such accidents, he will have to give some thought to the geometric design of pavements in order to educate and provide such facilities to the traffic by suitable signals and road regulations, by means of which it may be possible to reduce or minimize the chances of fatal accidents.

Causes of Accidents. The main causes of accidents on the road are :

1. Want of segregation between fast and slow moving traffic.
2. Lack of sufficient sight distance at the corners and junctions.
3. Road junctions not controlled by proper road signals.
4. Constant speed variations of vehicle due to congestion and the traffic jam.
5. Insufficient *weaving length,* between the intersection of roads.

Methods of Prevention of Accidents. To prevent or minimise accidents, the road engineer, the road user and the legal authorities will have to go hand in hand. The road engineer will design the pavement in such a way so as to minimise all possible chances of accidents by providing the following facilities to pedestrians and vehicular traffic :

1. Providing traffic signal for pedestrians.
2. Installation of pedestrian islands.
3. Improving street lights.
4. Construction of under and over bridges where essential.
5. Proper design and marking of cross walks.
6. Efficient layout of crossings, junctions and corners.

7. Providing proper cross-section of road.
8. Providing suitable super-elevation and transition curves at all curves.
9. Provision of non-skid and uniform surface.
10. Providing traffic segregation where possible.

The road users should realize their responsibilities towards each other. A road user must realize the large difference in speed between the pedestrians and vehicular traffic especially pneumatic traffic. A driver and a pedestrian must realize the distance required by the driver to react and apply brakes to stop his vehicle, especially under unfavourable conditions such as slippery road surface.

The legal authorities should get the laws of road enforced. It is the duty of the traffic police to guide and force the traffic to obey the rules of the road, keeping in view the facilities of the road users.

12.10 ROAD JUNCTIONS

Road junctions are places where two or more roads meet or cross each other at different angles. Safety of vehicular traffic and pedestrians is very essential at such places. They must be well-planned and properly signalled. Road junctions are the places where accidents generally occur, unless proper precautions are taken in the design or lay-out. As the traffic has to take different routes, it has to cross each other, and so proper segregation and control of the traffic is essential.

Tho following principles should be kept in mind while designing and planning the road junctions :

1. The intersections must, as far as possible, be at right angles and acute-angled crossing should be avoided.

2. The main road traffic should be independent of the traffic from branch roads and for this purpose, if possible, cither an over-bridge or under-bridge should be provided.

3. Sufficient space should be provided for proper visibility as also for any future extension and improvement.

4. Proper provision should be made for pedestrians crossing the road.

5. As far as possible, change of gradients at the junctions should be avoided.

6. The camber usually allowed is avoided and as far as possible the whole area of the road crossings should be in one level.

7. *Round-abouts* should be provided especially when more than three roads cross.

8. While designing the crossings consideration of the minimum turning radii at junctions as given in Table 12.1 should also be kept in view.

3. T-junctions. These are junctions where two roads meet but do not cross at right angles. They are of the following types :

(a) All paved type.
(b) Triangular island type.
(c) Elliptical island type.

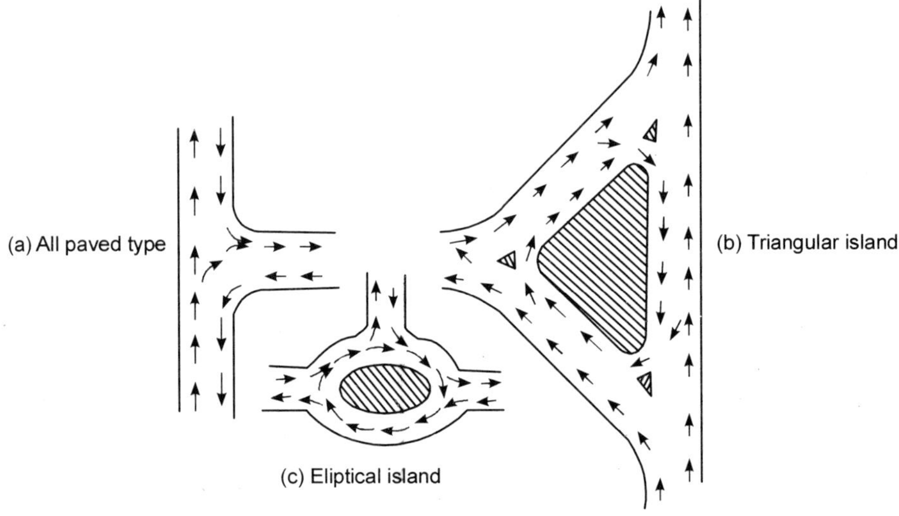

Figure 12.8 *T-junctions*

4. Y-junctions. When two roads meet, but do not cross, at an angle other than a right angle, a Y- shaped junction is formed. Y- junctions can be of the following types :

(a) All paved type.
(b) Triangular island type.
(c) Circular island type.
(d) Channelised Y.

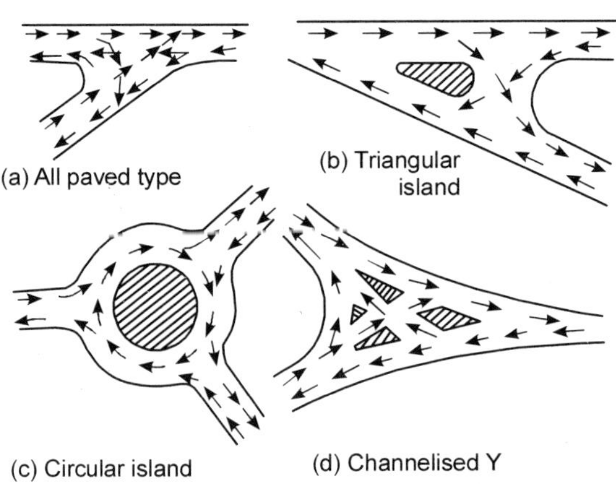

Figure 12.9 *Y-junctions*

5. Staggered Junctions. In this type of junctions the branch roads meet the main road at a good distance apart and at right angles.

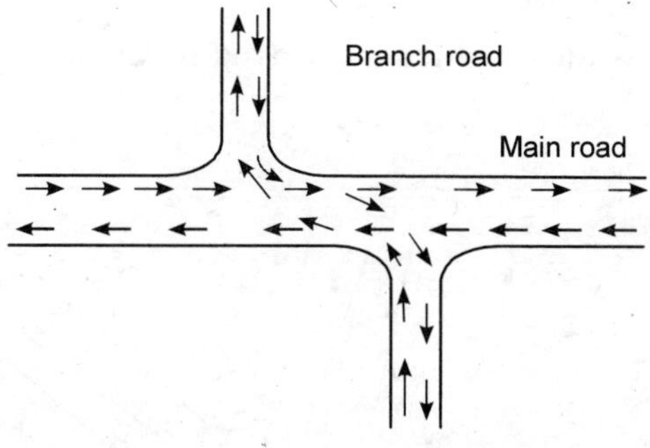

Figure 12.10 *Staggered junctions*

6. Multiple Junctions. This is a junction where more than two roads meet. This type of junctions are highly dangerous and special precautions should be taken in their planning and layout.

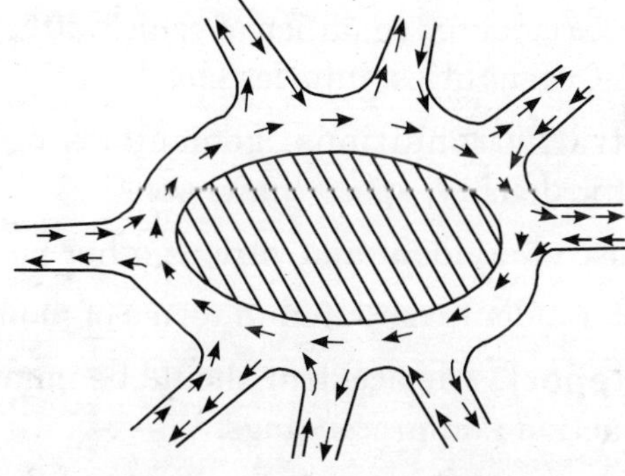

Figure 12.11 *Multiple junction*

12.11 ACCIDENT RECORDS AND DATA

Various steps involved in traffic accident studies are collection of data, preparation of reports, location file and diagrams application of these records for preventive measures etc.

(i) **Collection of accident data.** The following data should be collected:

(a) Location of accident

(b) Details of vehicle/vehicles involved in the accident

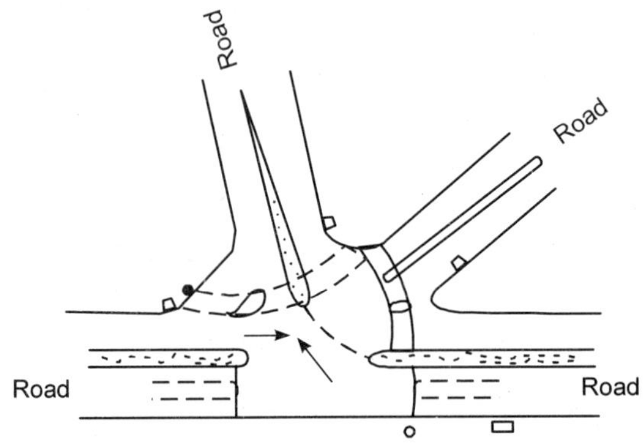

Figure 12.12 *Accidents spot diagram*

(c) Details of the accidented condition of vehicle after the accident, details of collision, damages, casualities, injuries etc.

(d) Road and traffic conditions, geometrics of the road, surface characteristics, traffic density, speed limits etc.

(c) Primary cause of accident and various other possibilities.

(f) Total cost of accident computed in terms of money.

(ii) **Accident Report.** The accident should be immediately reported to the police for legal action and proceedings.

(iii) **Accident Records.** The accident records are maintained for giving full details and particulars. The records should be in the form of location maps, spot maps, collision diagrams and condition diagrams.

Location map indicates the exact location of the accident and to identity the high density accident spots. Spot maps indicate accidents by spots or symbols. The common legends used for spot maps are shown in Fig. 12.13 (c).

A collision diagram showing all the important physical features of the accident location is prepared. This should be a scaled drawing. The drawing should indicate import and features such as curves, kerb lane, trees, electric

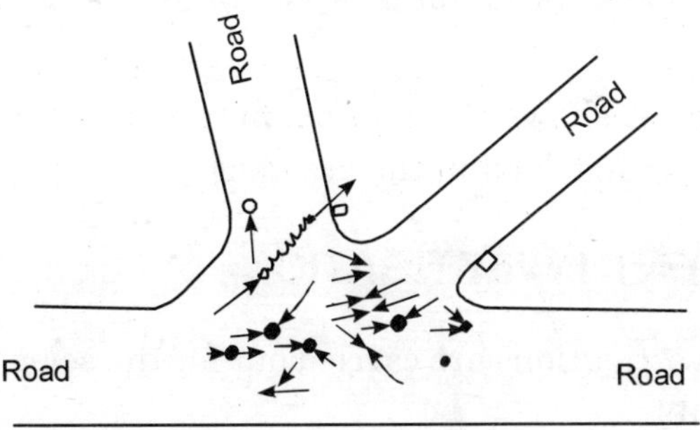

(a) Collision diagram

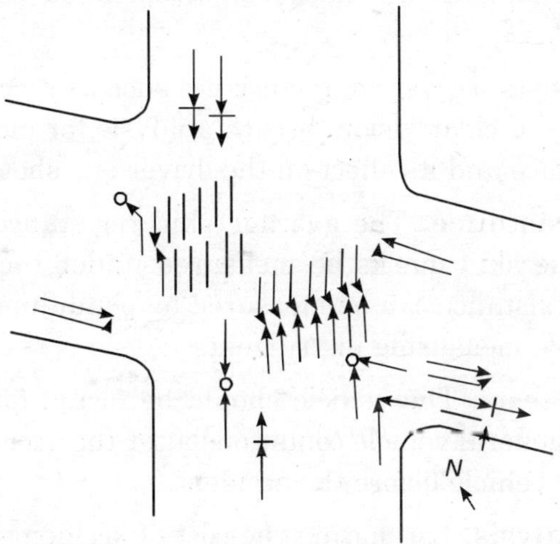

(b) Collision diagram

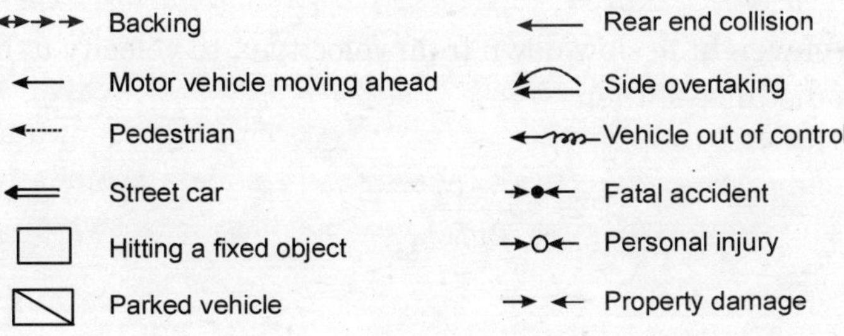

(c)

Figure 12.13

poles, road conditions, speed limits, speed breakers, building line, position of road signs etc.

Collision diagram shows the approximate path of vehicles/vehicle and pedestrian or object involved in the accident.

12.12 ACCIDENT INVESTIGATION

The following investigations are carried out for the scientific analysis of the causes of accidents :

1. General observations. It consists of measurement of skid length, relative position of vehicles and objects involved in the accident, details of accident, injuries, damages, condition of pavement surface, shoulders, light conditions etc.

2. Driver tests. Driver characteristics such as reaction time, distance judgement, angle of clear vision, breath analysis for alcohol consumption, glare on the surface and its effect on the driver etc. should be tested.

3. Skid resistance. The average skid resistance of the pavement surface along the skid marks be measured under the conditions of the accident. Skid resistance can be measured by pendulum type skid resistor and other suitable measuring instruments.

4. Vehicle tests. The vehicle should be tested for brakes, steering, tyre conditions, general vehicle conditions after the accident and presumed conditions of the vehicle before the accident.

5. **Cost analysis.** Estimating the cost of accident in terms of rupees for injuries, damage to vehicle, property damage, traffic delays etc.

12.13 SPEED AND SKID DISTANCE

If a vehicle of weight w slow down from velocity v_1 to velocity v_2 by applying brakes in a distance s, then

$$s = \frac{v_1^2 - v_2^2}{2gf}$$

where

gf = skid resistance of the road surface.

If the skid distance s is measured from the skid marks, the initial speed of the vehicle can be calculated by the equation

$$s = \sqrt{v_2^2 + 2gfs} \text{ or } v\sqrt{v^2 + 254fs} \text{ km/hour}$$

Suppose a vehicle A moving with a speed v_1 m/sec skids through a distance s_1 after applying brakes, collides with a parked vehicle B and the two vehicles skid through a distance s_2 before coming to a stop, then

$$v_1^2 = v_2^2 + 2gfs_1$$

After the collision, both the vehicles move with a velocity v_3 m/sec a distance s_2, then equating the momentum

$$v_2 = \frac{W_A + W_B}{W_A} \times V_3$$

where

W_A and W_B are respectively the weights of two vehicles.

Now $$v_1^2 = \frac{W_A + W_B}{W_A}^2 \times V_3^2 + 2gfs_1$$

Substituting-the value of v_3

$$v_1^2 = \frac{W_A + W_B}{W_A}^2 \times (2\,gfs_2)^2 + 2\,gfs_1$$

12.14 SPEED, TRAVEL TIME, VOLUME AND DENSITY

Travel time is defined as the time taken by a vehicle to cover a unit length of a highway. Travel time is inversally proportional to the speed of a vehicle. If T minutes/km is the travel time and V km/h is the speed, then

$$T = \frac{60}{V} \text{ or } T\,(\text{sec/km}) = \frac{3600}{V}$$

Traffic volume is defined as the number of vehicles crossing a section of the road per unit time at any selected period. Traffic volume is used as a quantity measure of flow and the unit is vehicles per day or vehicles per minute.

When the speed of the traffic decreases and becomes zero the density becomes maximum, volume becomes zero. With higher speeds there is no appreciable saving in travel time. An almost straight line relationship

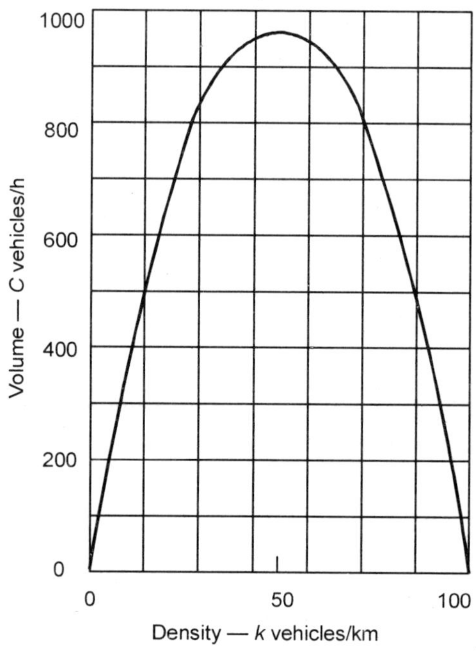

Figure 12.17 *Volume and density*

exists between speed and density for a good range of speeds particularly when the speed is not very high. It is difficult to measure density in practice hence for measuring density, relation between speed volume and density is used.

The maximum speed value is called *free mean speed* V_m and the maximum density at zero speed is called *jam density k*. The maximum flow q_{max} or the capacity flow q_c occurs when the speed is $\frac{V_m}{2}$ and density is $\frac{k}{2}$ and therefore

$$q_{max} = \frac{V_m \cdot k}{u}$$

12.15 TRAFFIC REGULATIONS

For the effective control of traffic all aspects of the vehicle control and the driver control is necessary. The regulations should be rational and in the interest of the road user. Traffic regulations give legal coverage for strict and effective control of traffic. The traffic regulation should cover the following.

1. Vehicle control. This includes registration of all vehicles, periodic checking by competent authority for their road worthiness. Vehicle control also deals with the requirement of vehicles, equipment and accessories, maximum dimensions and weight of the vehicles. Engine power is also included in the regulations.

2. Driver control. The vehicle driver must have a valid licence to drive a vehicle. Different type of licenses for heavy and light vehicles should be obtained. The driver control also includes fitness of driver to drive a particular vehicle, his vision, colour blindness, education to understand traffic signs and signals and follow the rules.

3. Flow control. Regulation of traffic flow control consists of direction, turning and overtaking, speed limits, prohibitory signs, pedestrian control etc.

4. General control. In case of accidents the rivers are required to report the matter to the nearest police station, removal of damaged property or vehicle, removal of injured persons to the hospital, cleaning of road for traffic are some of the general control regulations.

The Motor Vehicle Act, 1939 and subsequent modification (in 1989) and several ordinances appending the Act have covered several traffic regulations.

12.16 CONFLICTS

Conflicts points are those points where vehicles cross each either from the opposite direction or from sideways. The various type of conflicts of an intersection are

(i) Crossing conflicts (ii) Merging conflicts (iii) Diverging conflicts

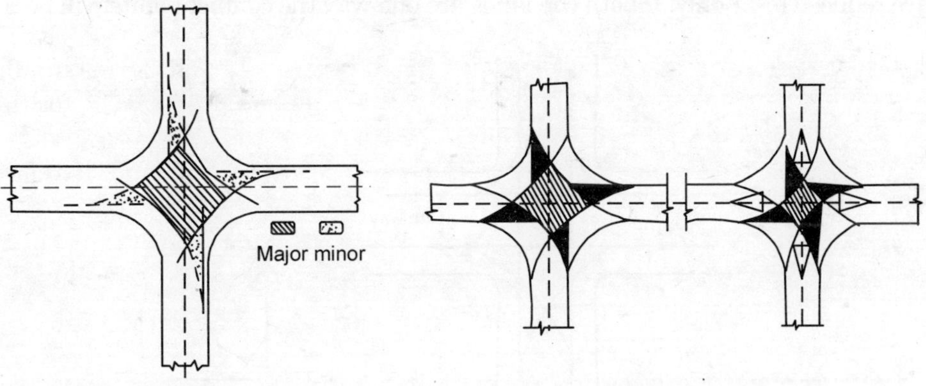

(a) Conflict zone (b) Areas of conflict

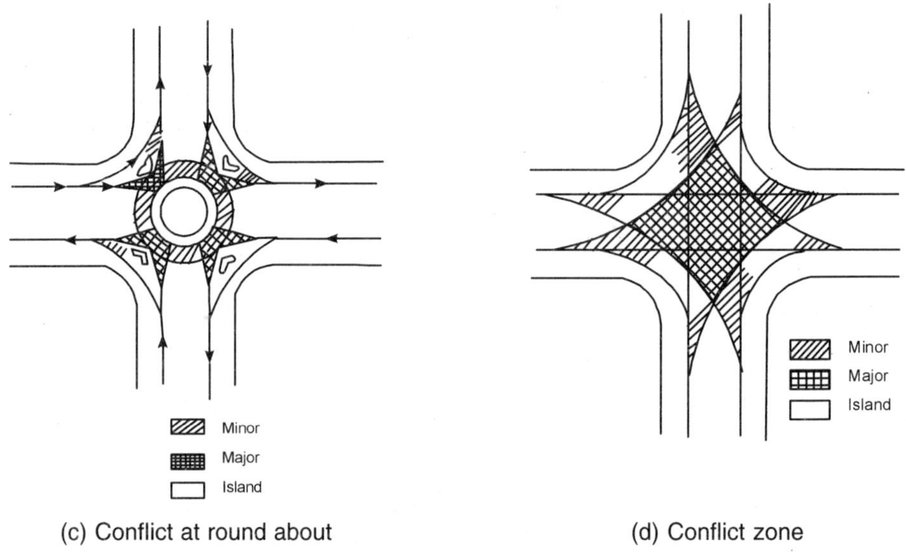

(c) Conflict at round about

(d) Conflict zone

Figure 12.16

In congested areas one of the methods of reducing conflict points and ensure smooth flow of traffic is by regulating traffic along *one-way* streets. The traffic is allowed to move in one direction only. Such a type of regulation is possible only when there is a network of roads connecting two bigger roads so that additional distance to the transversed by some vehicles is not much, one-way traffic ensures greater capacity increased average speed, improved pedestrian movement and reduction in accident rate.

On a right angle intersection of two-way traffic lanes the total number of conflict points are 24, where if one of the lane is one way, the conflict points are reduced to 11 only. If both the lanes are one-way the conflict points will be 6.

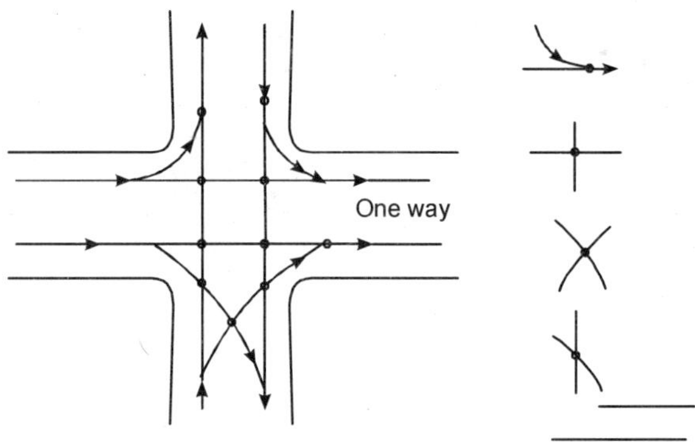

(a) Two way and one way lanes

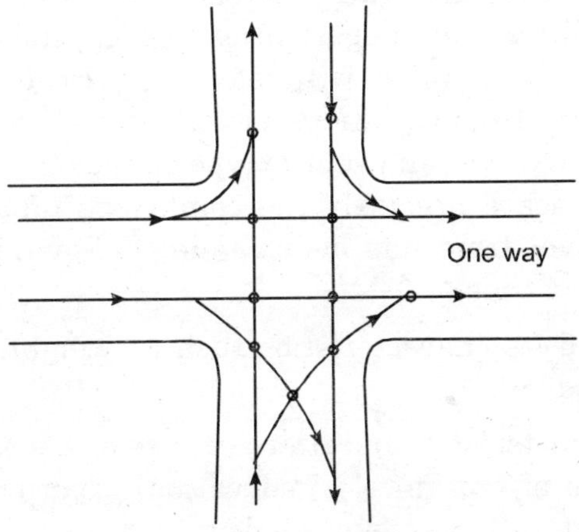

(b) Both lanes one way

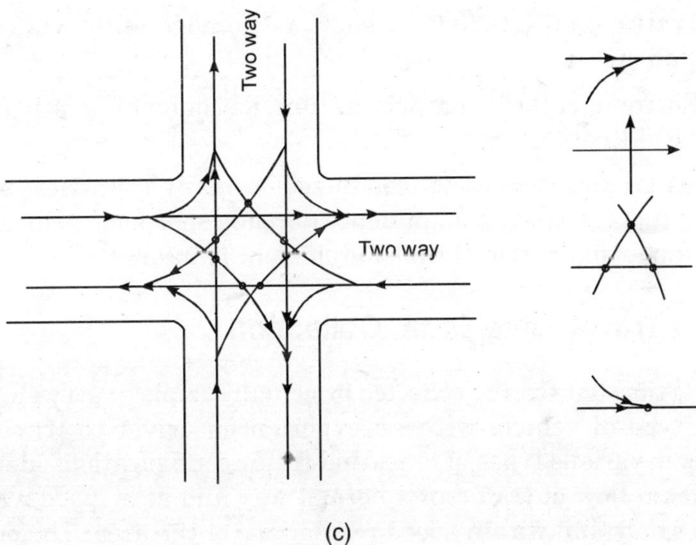

(c)

Figure 12.19

12.17 TRAVEL TIME STUDIES

Data on various aspect of travel time constitutes one of the basic prerequisites for various traffic and transportation problems. The traffic engineer and the transportation planners require travel time to analyse travel demand to evaluate the efficiency of traffic flow, recommend improvement facilities and suggest priorities in implementation. It is a varying phenomenon as it differs with reference to roadway as well as traffic characteristics, apart from variations associated with hour, day and season on the one hand and driver and vehicle type on the other hand. Travel time should clearly indicate precisely the conditions under which such an estimation of travel time has been made. Following are the various parameters of travel time:

(i) Vehicle and its characteristics such as weight, engine condition, kilometre done, body etc.

(ii) Driver characteristics and traits such as age, sex, vision, occupants of vehicle, purpose of trip, level of education, experience, reaction time, distance judgement, influence of alcohol etc.

(iii) Road characteristics such as pavement condition type, curvatures, sight distance, width and shoulder conditions, type and number of intersections, gradient and the number of lanes.

(iv) Traffic characteristics, such as homogeneity, volume, speed, composition etc.

(v) Environmental factors such as time, day or night, weather conditions, geographical locations etc.

Various studies have been conducted in many countries of the world including India to study the influence of road conditions, vehicle condition, traffic volume on the travel time for efficient planning.

12.17.1 Travel Time Data Collection

The travel time data can be collected by actually employing a well conditioned average type of vehicle with an experienced driver to travel specified distances in various types of lanes and traffic configurations along with the main stream flow of traffic in a normal way and at a speed which can be attained safely and within speed regulations of the area. The other way of collecting travel data is by noting down the registration number and exact time of arrival of different vehicles at two different test stations. Time taken by the vehicles to pass the specified road length is obtained by the difference between the observed time reading at the entry and exit points. For the purpose of traffic time and composition of traffic, the traffic is divided into three groups viz.,

(i) fast category consisting of two wheelers, three wheelers, cars and jeeps etc.

(ii) bicycles

(iii) slow moving traffic such as bullock cart, cycle rikshaw, hand driven trollies, tongas etc.

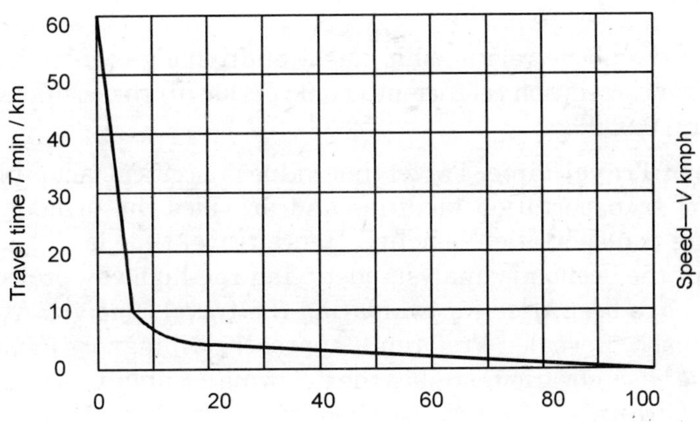

Figure 12.20 *Speed and volume*

It has been observed that travel time and volume of traffic has a linear relationship to a great extent as shown in the graph. Following relationships have been established by performing actual field studies :

(i) Travel time = $A_1 + B_1$ Fast + C_1 Bic + D_2 slow

Here the value of $A_1 = 66.2212$

$B_1 = 0.0063$

$C_1 = 0.2156$

$D_1 = 2.7381$

(ii) Travel time = $A_2 + C_2$ Bic + D_2 slow + E_2/Fast

Here the value of $A_2 = 60.4862$

$C_2 = 0.2252$

$D_2 = 3.1418$

$E_2 = 301.0723$

(iii) Travel time = $A_3 + B_3$ Fast + C_3 Bic + D_3 slow + E_3/Fast

The value of $A_3 = 58.6086$

$B_3 = 0.0132$

$C_3 = 0.2199$

$D_3 = 3.0680$

$E_3 = 382.65$

and

Fast = Fast moving traffic

Slow = Slow moving traffic

Bic = Bicycle traffic

Each of the above relationship has a multiplying coefficient. For (i) the multiplying correlation coefficient is 0.9377, for (ii) the coefficient is 0.9394 and for (iii) 0.9396.

Value of Travel Time. Travel time value is a critical factor in forecasting the use of transportation facilities and provides the primary means for estimating transportation benefits. Hence travel time is the major benefit in most of the economic analysis justifying the highway projects. Various methods have been tried for evaluating the travel time viz. Average wage rate approach, Revealed Preference approach, willingness to pay approach etc. On the basis of various studies the following assumptions and conclusions have been drawn.

12.17.2 Assumptions

1. It is assumed that the value of travel time and value of comfort varies with sex, age and income group.

2. Commutors attach more importance to travel time and comfort particularly during peak hours.

3. Commutors attach more importance to the travel time when the journey is undertaken for work or educational trips.

4. Commutors pay more for comfort if the travel time is reduced.

12.17.3 Conclusions

1. Female have less value for travel time than males of the same category and during the same peak hours.

2. With increase in age, value of Travel Time decreases.

3. Regular commuters attach more importance to travel time.

4. The travel time value for waiting and in-bus travel is different for different category of commutors.

5. It is also revealed that commutors with income value travel time less than commutors with no income (students) or in other words incomeless commutors (students) are time savers and income group commutors are money savers.

6. The increase in income from Rs. 200 per month to Rs. 5,000 per month resulted in increase in the value of in bus travel time.

12.18 VISIBILITY AT CORNERS, BENDS AND JUNCTIONS

In the interest of road safety it is of primary importance that vehicle drivers should have full view of the approaching vehicles at all corners, bends and junction. Figure 12.20 shows the minimum sight distance which is required in the towns and the cities. Appropriate traffic signs are provided in the towns and the driver have sufficient warning of the presence of the junctions, corners or bends. At such places the speed of the vehicle is normally below 30 km/hour.

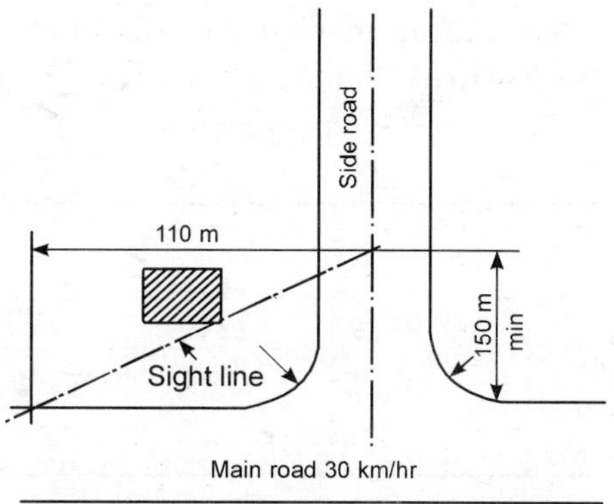

Figure 12.21 *Minimum sight distance*

Figure 12.21 shows the method of providing sight distance at corner by cutting a corner of the building by 4.6 m on each side. The figure of 4.6 m is taken by assuming that the width of the footpath on both sides of the road is

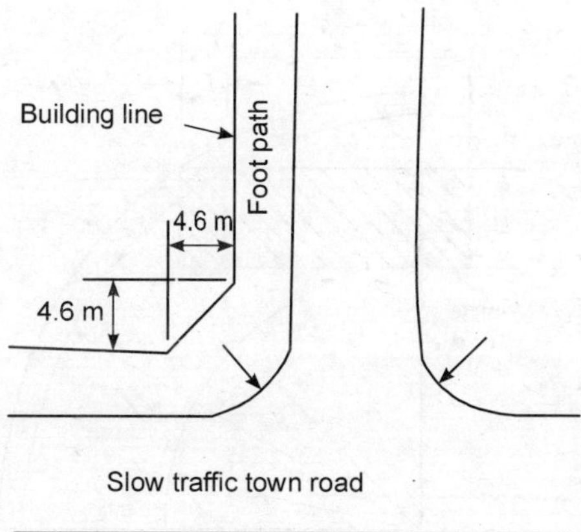

Figure 12.22 *Method of providing sight distance at corner by cutting corner of a building*

4.6 m. In the town, adjustments and changes in this 4.6 m can be made according to the need of the crossing.

Road crossings are mainly classified under two groups, depending on the importance of the road :

Group 1. Priority intersections, where the minor road meet with the major road. At such a crossing the driver moving on the major road has virtual precedence over the other driver moving on the minor road.

Group 2. Uncontrolled intersection, or the crossings where both the roads are of the same category or equal importance.

Table 12.2 gives the minimum visibility distance along major roads at priority intersection for Rural Roads as per I.R.C.

TABLE 12.2

Design speed of major road in km/hour	*50 km/hr*	*65 km/hr*	*90 km/hr*	*100 km/hr*
Minimum visibility distance along major roads at priority intersection for rural roads	110 m	145 m	190 m	220 m

Minimum visibility distance of 15.0 m along the minor road is taken. On this basis the sight triangle at priority intersections should be formed by measuring 15.0 m along the minor road, and the distance given in Table 12.1 along the major road.

The minimum turning radius R of a vehicle is the radius of the sharpest curve that can be traced by its outer front wheel as shown in Fig. 12.23. On the

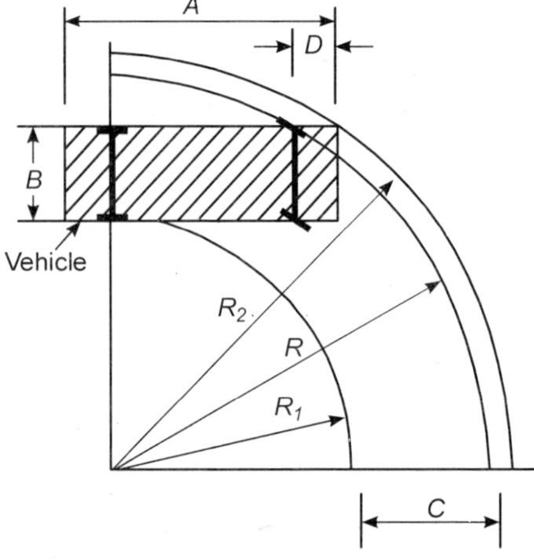

Figure 12.23 *Minimum turning radii of a vehicle*

curves when a vehicle is turning its rear wheels do not follow the tracks of the front wheel. The transverse position of the road kerb is usually determined by the radius R_1 of the inner rear wheels of the vehicle as given in Fig. 12.22.

In congested town areas which are used by light vehicles, a kerb radius of 4.6 m should be provided. The kerb radius for buses, trucks, tractors and trailers should be 17.0 m minimum. When the turning crossing is more than 90°, greater radius will be required.

Table 12.3 gives the minimum turning radii for different classes of vehicles with reference to Fig. 12.22.

TABLE 12.3 *Minimum turning radii for different classes of vehicles (Figure 12.22)*

Type of vehicle	*Vehicle dimensions*		*Width of the path C*	*Vehicle overhang D*	*Minimum desirable radius*			*Minimum right angle turn radius*		
	Length A	*Width B*			*R*	R_1	R_2	*R*	R_1	R_2
	m	*m*	*m*	*m*	*m*	*m*	*m*	*m*	*m*	*m*
Car	6.1	2.1	2.6	0.40	11.7	9.1	12.3	8.5	5.9	8.9
Bus or Truck	10.67	2.44	3.9	0.61	18.7	16.8	19.3	13.7	11.4	14.3

12.19 TRAFFIC ISLANDS

Traffic islands or rotary is an enlarged intersection where all converging vehicles are compelled to move round an island in one direction before they can weave out of the road, into their respective direction radiating from the island. They also act as shelter for pedestrians, while crossing the road. The efficiency of a rotary depends upon the diameter of the circle. The traffic circles should be free from all obstructions, as the drivers should be able to see ahead and take necessary precautions. They are marked with special colour lights, kerbs of special types and painted with broad strips of blacks and white, so that they may be conspicuous even at night.

Roundabouts are usually provided where the traffic density exceeds 500 vehicles (cars) per hour on all the intersecting roads. Some engineers recommend that rotaries should be provided at the crossings where right

2. It acts as a guidance to the drivers in negotiating on junctions.

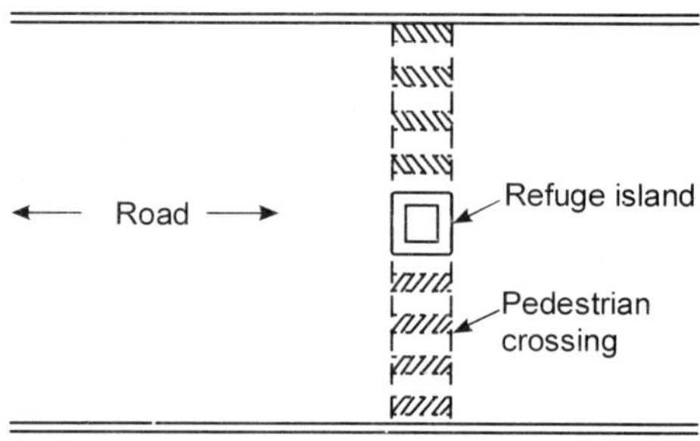

Figure 12.26 *Refuge island*

3. It increases the efficiency of traffic.
4. It segregates the traffic into proper channels.

12.22 DESIGN OF TRAFFIC ISLANDS

The common shapes of the traffic islands are circular or elliptical (elongated). The choice of shape mainly depends upon the site conditions, angle of intersections of the roads and the traffic density.

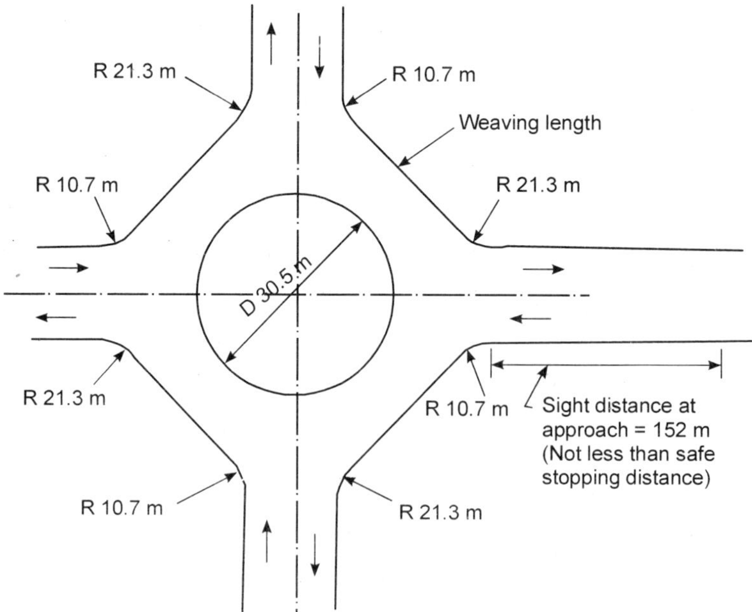

Figure 12.27 *Round about*

The circular shape is most common and economical where roads carry nearly equal volumes of traffic and the roads intersect at nearly equal angles. The elliptical or elongated shape is adopted at intersections where the cross traffic is small. The elongation of such islands should be kept in the direction of the greater flow and it should be so designed that the radius of curvature at the sharpest point is not less than 15.2 metres.

The exact shape of the island is determined by connecting the sections of the road round the central island to form a closed figure giving at least the minimum weaving length between the adjacent radial roads.

At the intersection of two or more roads the basis of the design of the rotary should be on the highest design speed of any of the highway, irrespective of the greatest volume of the traffic it carries.

Practically it has been seen that a vehicle can travel at 25 km/hr speed round an island of 30.0 m diameter and at a speed of 40 km/hr round an island of 46 m diameter. As far as possible the diameter of the central island should not be less than 30.0 m.

Indian Roads Congress has recommended a radius of 25 m to 35 m for design speed of 40 km/hr and 15 m to 25 m for design speed of 30 km/hr for radius of entry of rotary. Similarly the radius of the exit curve has been recommended to be 1.5 to 2.0 times that of the entry radius.

Indian Road Congress has recommended that the radius of the central island should be 1.33 times that of entry radius. The radius should be as large as possible and practicable.

Table 12.5 gives the width of that carriage way at the entry and the exit and radius of curve at the intersections having rotary.

TABLE 12.5 *Width of carriageway at entry and exit (Based on IRC 65–1976)*

S. No.	*Width of the approach road or carriageway*	*Curve radius at entry*	*Width of carriageway and entry and exit to rotary*	
1	7.0 m (Two Lane)	25 m – 35 m	6.5 m	Inclusive of widening required on account of curvature
2	10.5 (Three Lane)	25 m – 35 m	7.0 m	
3	7.0 m (Two Lane)	15 m – 25 m	7.0 m	
4	10.5 m (Three Lane)	15 m – 25 m	7.5 m	

Indian Road Congress has also recommended that minimum width of 5.0 m with necessary widening to account for the curvature of the road should be provided.

Width of the carriageway around the island. This should not be less than one-fourth the total width of all me radial roads and not less than one-half the width of the widest roads plus width of one lane, whichever is more.

Weaving length and width. The minimum weaving length of 4.5 m for 40 km/hr speed and 30 m for 30 km/hr designed speed should be provided. The maximum weaving length should not be more than twice the above lengths. The width of the weaving section of the rotary should be one traffic lane of 3.5 m wider than the mean entry width thereto.

Width of the carriageway around island. This should not be less than the weaving width. It should be equal to the widest single entry road into the rotary.

12.23 PEDESTRIAN CROSSINGS

The pedestrian accidents are more frequent and fatal and hence special attention is to be given to the pedestrian crossings. The pedestrian crossing may be at the same level, above or below the road level. Crossings at the road level are cheapest and are mostly used. The safety of the pedestrians depends upon the pedestrians themselves and the drivers. Strict supervision at such places is very essential and the pedestrians should be allowed to cross the road when the traffic is held up temporarily.

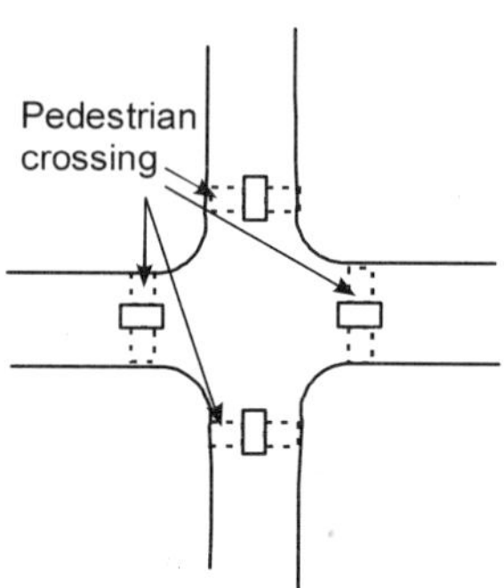

Figure 12.28 *Pedestrian crossing*

The crossing is marked on the road by means of rubber or steel studs embedded in the road. The studs should be flush with the road surface or at the most of 5 mm projecting out from the road surface and are placed at a distance of nearly 35 cm centre to centre. The pedestrian crossing should belocated just near the intersection of roads where the vehicle reduces the speed (generally). The minimum width of crossing should be 2.5 m. The crossing should be clearly visible from pedestrian crossing a distance to the drivers of the fast moving vehicles.

12.24 CHANNELIZATION OF TRAFFIC

Channelization is the process of segregation of traffic into different direction so as to avoid conflicts. This is done by providing obstacles called *islands* at the intersections or where there are chances of conflicts. These obstacles help in guiding the traffic in different channels. The objections of channelizing the traffic are :

(i) To segregate the traffic particularly pedestrian and vehicular traffic.

(ii) To reduce the conflict areas.

(iii) To increase safety.

(iv) To maintain streamline flow of traffic.

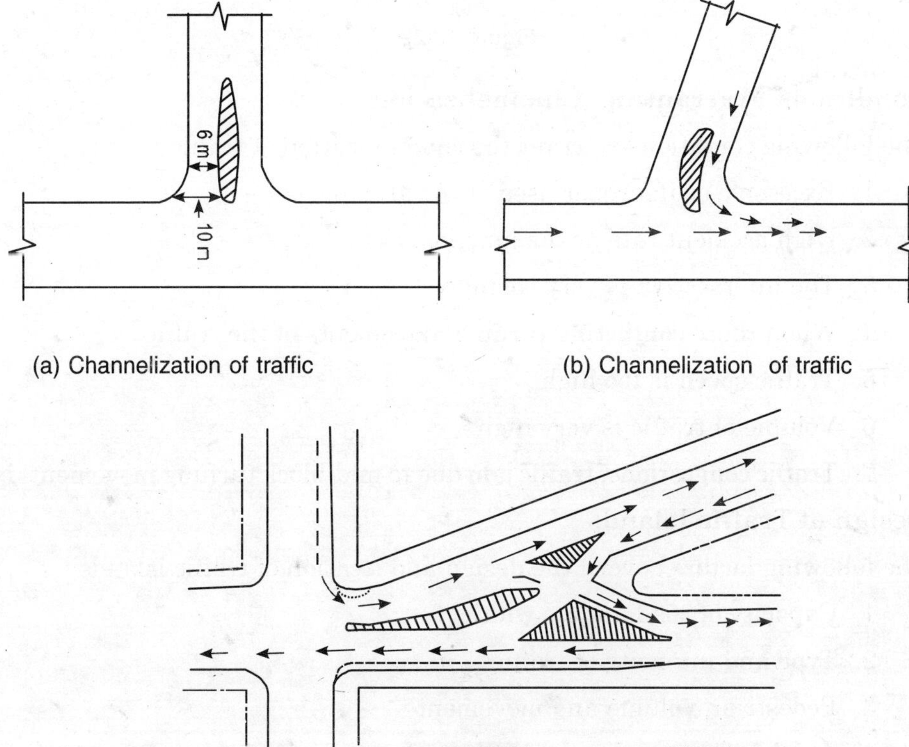

(a) Channelization of traffic

(b) Channelization of traffic

(c) Channelization of traffic

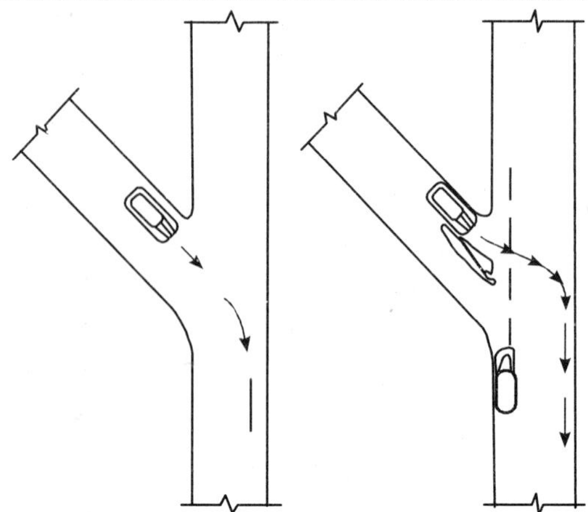

(d) Channelization of traffic

Figure 12.29

Conditions Warranting Channelization

The following conditions warrant the channelization of traffic :

1. Excessive or improper used of the area.
2. High accident rate at the junction.
3. The intersection is very complex.
4. When there conflicting turning movements of the traffic.
5. Traffic speed is too high.
6. Volume of traffic is very high.
7. Traffic congestions, traffic jam due to mid-block turning movements.

Design of Traffic Islands

The following factors govern the design and location of traffic islands :

1. Capacity of road and its width
2. Type and intensity of traffic
3. Pedestrian volume and movement
4. Accident rate
5. Conflict area

6. Gradient at the intersection
7. Sight distance
8. Type of road surface
9. Cost of improvement
10. Lighting conditions
11. Location of buildings and other structures
12. Existing traffic control devices.

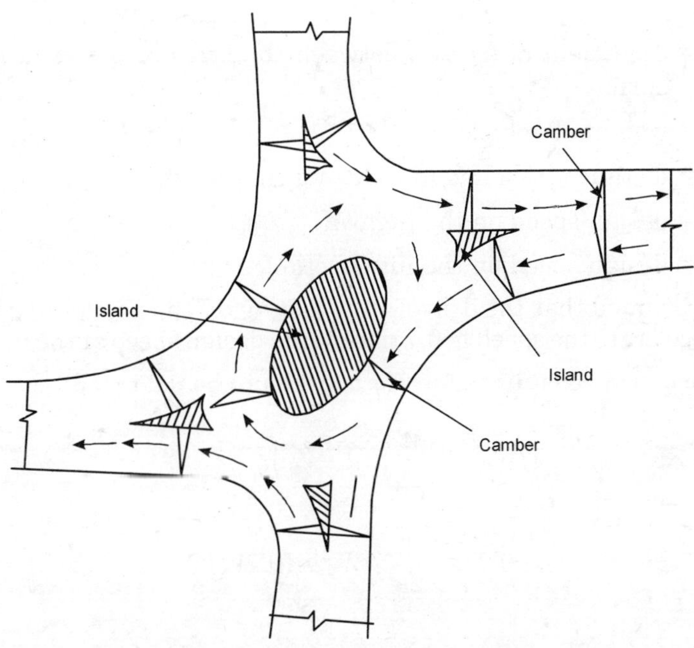

Figure 12.30 *Traffic island*

12.25 ACCELERATION AND DECELERATION LANES

On important and busy roads provision should be made for the vehicles to accelerate their speed after emerging from the intersection or decelerate the design speed before entering the intersection. The vehicle has to pull itself to the left so that it can declerate to the design speed of the intersection. It is not necessary to provide both, depending upon the movement of traffic or cross traffic. Acceleration and deceleration lanes may not be a full width lane. It is only a mean to enable the traffic to accelerate to the design speed after emerging from the intersection or to decelerate their speed to the design speed of the intersection as soon as they enter the junction.

Suppose a vehicle is moving at a design speed of V_1 km/hr. The design speed at the intersection is taken as 0.7 times the design speed of the vehicle.

Now

$$fWd = \frac{1}{2} \cdot \frac{W}{g} \left(V_1^2 - V_2^2\right)$$

where

f = coefficient of friction between the tyres of the vehicle and road surface

W = weight of vehicle

d = distance travelled in reducing the speed

V_1 = design speed on the highway

V_2 = design speed on the intersection.

It is presumed that the driver after *footing off* the accelerator will take 3 secs to decelerate the speed to 0.7 times, (the design speed at the intersection).

For normal speeds the value of f is taken to be 0.2 to 0.3.

$$fWd = \frac{W}{2g}\left(V_1^2 - V_2^2\right)$$

$$0.2 = \frac{1}{2g} \times V_1^2 - (0.7V_1)^2$$

$$= \frac{1}{2g} \times V_1^2 - 0.49V_1^2$$

$$= \frac{1}{2g} \times \left(0.51V_1^2\right)$$

$$d = \frac{0.51V_1^2}{2g \times 0.2} = \frac{51V_1^2}{4g}$$

From this equation the length of the lane for acceleration and similarly for deceleration of speed can be worked out. It has been observed that for normal traffic speed of 60 km/hr, the length of acceleration lane varies between 80 m to 100 m.

12.26 ROAD MARKINGS

For smooth, safe and efficient movement of traffic, the pavements, kerbs, vertical side of islands, curves etc. are being properly marked with lines, symbols, patterns, reflectors etc. Traffic markings are made to regulate, guide and control the traffic. The markings can be made by different methods using contrast colours of paints, reflectors etc. The markings should be visible from a comfortable distance so that the driver can react according to the situation.

12.26.1 Pavement Markings

Pavements or carriageways are generally marked with white paints. Yellow paints are used to indicate parking space. The markings are made with the help of lines 10 cm thick. Longitudinal continuous or solid lines are used as guiding or regulating lines. Transverse lines are meant for stop indication for the vehicles.

Pavement markings are of the following type.

1. Centre Line. These are meant to separate the traffic approaching from opposite direction on an undivided two-way traffic lane on the country side single broken line of 10 cm thick may be painted on the straight reaches. The length of broken lines is 4.5 m with 7.5 m gaps. This standard practice is a dopted on National Highways and State Highways. On horizontal curves the length of the segments may be reduced to 3.0 m with 6.0 m gaps. On other roads other than NH and SN, the length of the segment on straight reaches will be 3.0 m with 6.0 m gaps. On a four or more lane undivided roads two solid continuous lines of width 10 cm are painted 5 cm to 10 cm apart.

For city roads with less than four lanes, the centre line consists of 10 cm thick white broken lines of segment length 3.0 m and gaps 4.5 m. The gaps should be decreased to 3.0 m on curves and approaches to intersections.

2. No Overtaking Lines. These markings are to indicate prohibited area for overtaking of vehicles.

3. Stop Line. These markings are meant to indicate that all vehicles are to stop near the pedestrian crossings and intersections.

4. Overwalk Lines. These markings are meant to indicate that location of pedestrian crossings. The width of pedestrian crossings or zebra crossings varies from 1.5 m to 3.0 m depending on the location conditions.

5. Lane Line. Lines are drawn to designate traffic lanes so as to guide the traffic in their respective lanes. This will also help in utilizing properly the carriage width.

6. Turn Lines or Markings. Near the intersections arrows are marked for proper placement of vehicles before turning in their respective directions.

12.26.2 Kerb Markings

These marking consists of black and white lines 10 to 15 cm thick to indicate the end of the carriageway of a lane particularly on crossings and change of grades.

12.26.3 Road Delineators

These are indicators to indicate to the drivers about the alignment of the road ahead. These are actually visual treatment or devices to outline the roadway or portion of it. On curves or change of gradients or before mantatory signs.

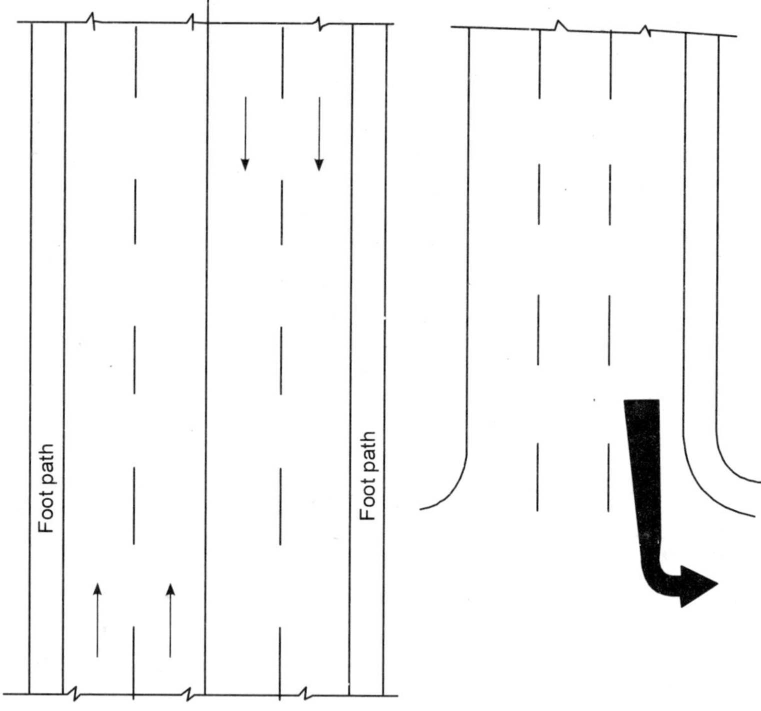

Figure 12.31 *Road makings*

Roadway delineators are in the form of 0.8 to 1.0 m high posts painted with black and white strips with or without reflectors intended to delineate the edges of the pavement particularly on the outer edges of horizontal curve so as to guide the drivers about the alignment. Hazard markings are approximately 1.2 m high plates and posts with three red reflectors or markers with yellow and black strips at 45° towards the side of the obstructions, to indicate hazards in the way like island, speed breakers, crossings, steep gradients etc. Object markers with circular red reflectors arranged on a triangular plate are also used.

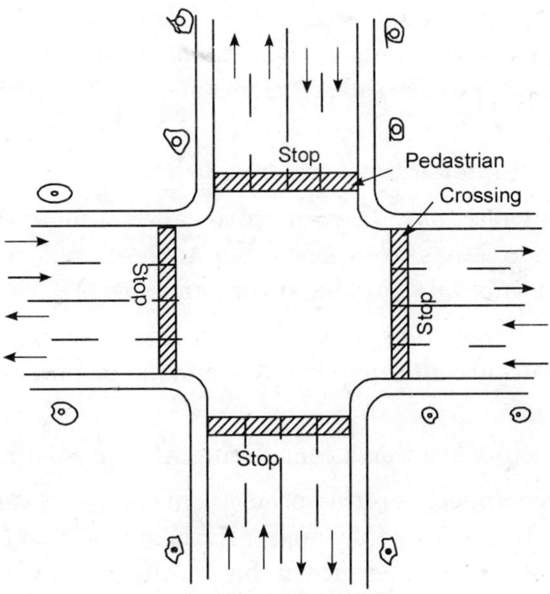

Figure 12.32

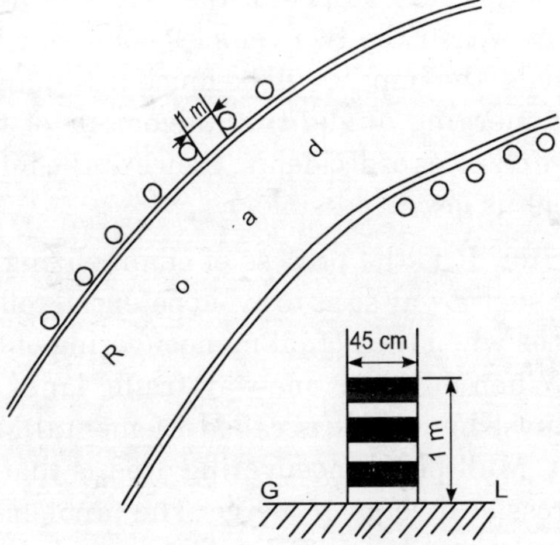

Figure 12.33 *Road delimeators*

12.27 INTERSECTION DESIGN

Intersections or cross traffic is an unavoidable problem. At the intersections there is a through traffic, turning and crossing traffic. Different type of traffic is handled in different ways depending upon the type of intersection and its design. Unless we provide sub-ways or over bridges or fly overs intersection or cross traffic problem will have to be handled. The intersection should be so designed so as to provide safe, efficients and speedy traffic at a reasonable cost.

Following type of intersections are provided:

(i) *Level intersections.* They are also called intersection at the grade. When two or more lanes cross each other at right angle or any other angle and all the lanes are more or less at the same levels.

(ii) *Grade intersections or Grade separated intersections.* When the intersecting roads are at different levels and are separated by difference in levels.

The following points are worth considering while designing intersections :

1. **Relative Speed.** It is the speed of convergence of vehicles in different or separate traffic lanes as they approach the point of potential collision. It is also defined as vector difference in the velocity of two vehicles in the same traffic flow and is the sum of the speeds of approaching vehicles from opposite direction. Relative speed is an important factor in traffic flow at grade and is dependent on the absolute speeds of the intersecting vehicles and the angle between them. When the two vehicles collide at low relative speed smaller merging angle, the impact will be much less than at higher relative speed with higher merging angle, the judgement of the driver will be inaccurate and more chances of accidents. Hence at the intersections relative speed should be kept as low as possible.

2. **Manoeuvring.** It is the process of channelizing and directing the conflicting traffic in such a way so as to avoid potential collisions. Manoeuvre areas are those areas when the actual manoeuvering of the traffic is made to avoid collision. When only two one-way traffic lanes cross or merge or diverge the manoeuvering area it is called Elemental Manoeuvering area and is the simplest. Multiple manoeuvering area is that where more than two traffic lanes cross or diverge or merge. The problems of traffic at such places are much more complex and should be avoided as far as possible. Manoeuvring areas are the actual conflict areas where potential of severe accidents are much more.

12.28 BASIC CHARACTERISTICS OF A CROSSING

Following are some of the requirements of an intersection :

1. Conflict points and conflict area should be avoided or minimised as far as possible.

2. The relative speed and the angle of approach should be as small as possible.

3. Adequate sight distance should be made available.

4. Appropriate warning and mandatory signs should be placed at appropriate places before the intersection.

5. Pedestrian crossings should be provided on each lane before the intersection.

6. Sudden changes in the flow of traffic should be avoided at the intersection.

12.29 TRAFFIC ROTARY

Following points are to be kept in mind while planning a traffic rotary :

1. Design Speed. The approaching vehicles should reduce their speed to the extend that traffic moving in the same direction may not collide with each other. The vehicles have to slow down from their designed highway speed. It has been observed and recommended by IRC also, that the maximum speed at the intersection should not exceed 40 km/h. Desirable speed at the rotary is 30 km/h.

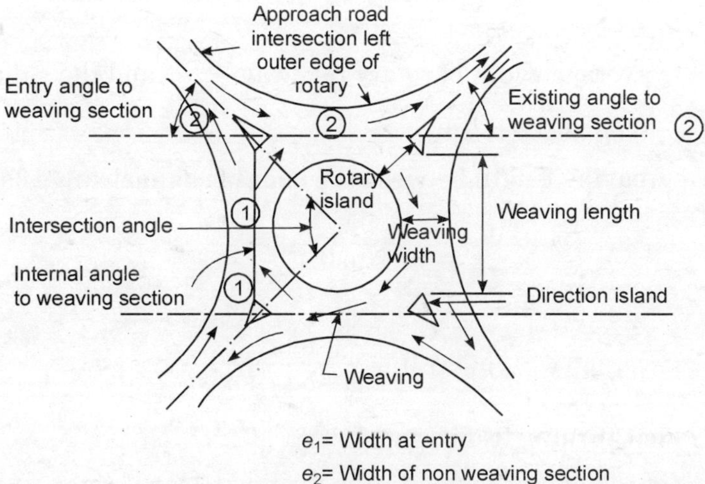

Figure 12.34 *Traffic rotary*

2. Weaving Angle and Weaving Distance. Weaving angle is defined as the angle between the incoming vehicles and the outgoing vehicles from two adjacent roads of a rotary or intersection. All vehicles before weaving out have to merge in the main flow. The weaving operation can take place between the two channelizing islands of the adjacent lanes of the rotary. The length of the rotary roadway between the adjacent lanes of converging and diverging lanes is called *weaving length.*

3. Width of Rotary Lane. All the traffic entering the rotary way or lane have to go round a short distance around the island. The actual roadway width around the island varies from section to section.

If e_1 be entry width and e_2 be the exit width of a rotary roadway, W be the width of the weaving section and should be one traffic lane wider than the mean width of the entry and the non-weaving section, then weaving width

$$W = \frac{e_1 + e_2}{2} + 3.5 \quad \text{metre}$$

4. **Capacity of the Rotary.** The capacity of a rotary can be calculated by the formula

$$Q_v = \frac{280\ W(1 + e/w)(1 - p/3)}{(1 + W/L)}$$

where

Q_v = practical capacity of the rotary

W = width of rotary roadway of the weaving section

e = average width of rotary roadway $\frac{e_1 + e_2}{2}$ and the value of e/w is taken as 0.4 to 1.0

L = weaving length between the ends of channelizing islands and

$$\frac{W}{L} = 0.12 \text{ to } 0.4$$

p = average weaving traffic $\dfrac{b + c}{a + b + c + d}$

a = left turning traffic

b = crossing or weaving traffic turning towards right

c = crossing or weaving traffic turning towards left

d = right turning traffic.

5. Cambar and Superelevation. A vehicle passing or traversing along a rotary island, traverse and reverse the curve while entering and diverging from the rotary. The camber at the entry and exit points should be just the minimum and the inward slope should serve as a superelevation for the traffic going round the rotary. But while designing a rotary it should be presumed that there is no superelevation, as super elevation is not required.

6. Sight Distance. Rotary should be free from all obstructions and maximum sight distance should be provided but in no case the sight distance should be less than the stopping sight distance. According to I.R.C. recommendations the minimum sight distance should be 45 m and 30 m for a design speed of 40 km/hr and 30 km/hr respectively. It is always preferred to locate the rotary at a level surface rather than on raised or higher level than the normal level of the intersecting lanes.

12.30 ADVANTAGES OF TRAFFIC ROTARY

Following are some of the advantages of providing rotary :

1. The traffic at the rotary can proceed in the direction without any interruption.

2. All the traffic has equal opportunity irrespective of the fact whether it moving straight, right or left.

3. At the signalled crossing the traffic has to stop for a moment and then proceed whereas at the rotary traffic can proceed without interruption thereby saving fuel cost i.e., the viable cost of operation of automobile is less at the rotary.

4. Crossing manoeuvring is converted into weaving or merging and diverging traffic and the journey becomes more consistent and comfortable.

5. Traffic rotary functions at their own and hence no signals or traffic police is required.

6. Rotary will have further advantages when the number of intersecting roads is between four and seven.

7. Of all the intersections rotary has the maximum capacity of accommodating vehicles.

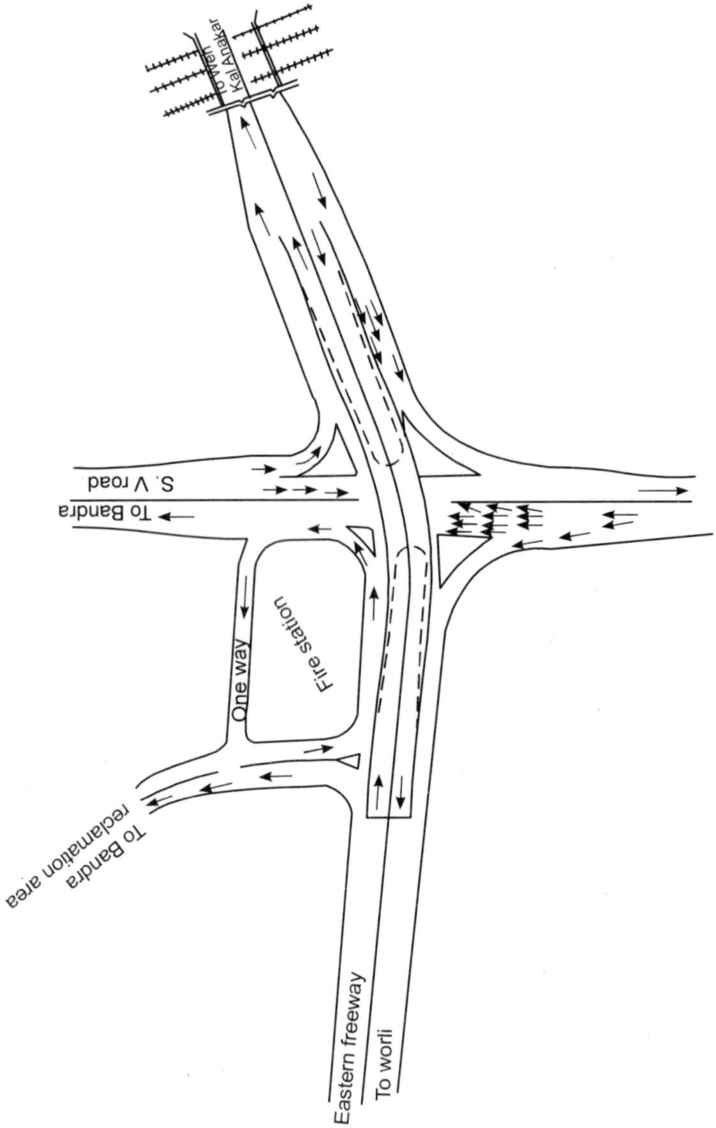

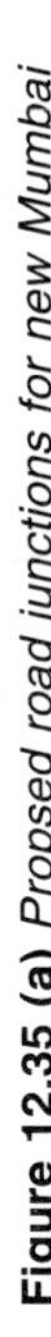

Figure 12.35 (a) *Propsed road junctions for new Mumbai*

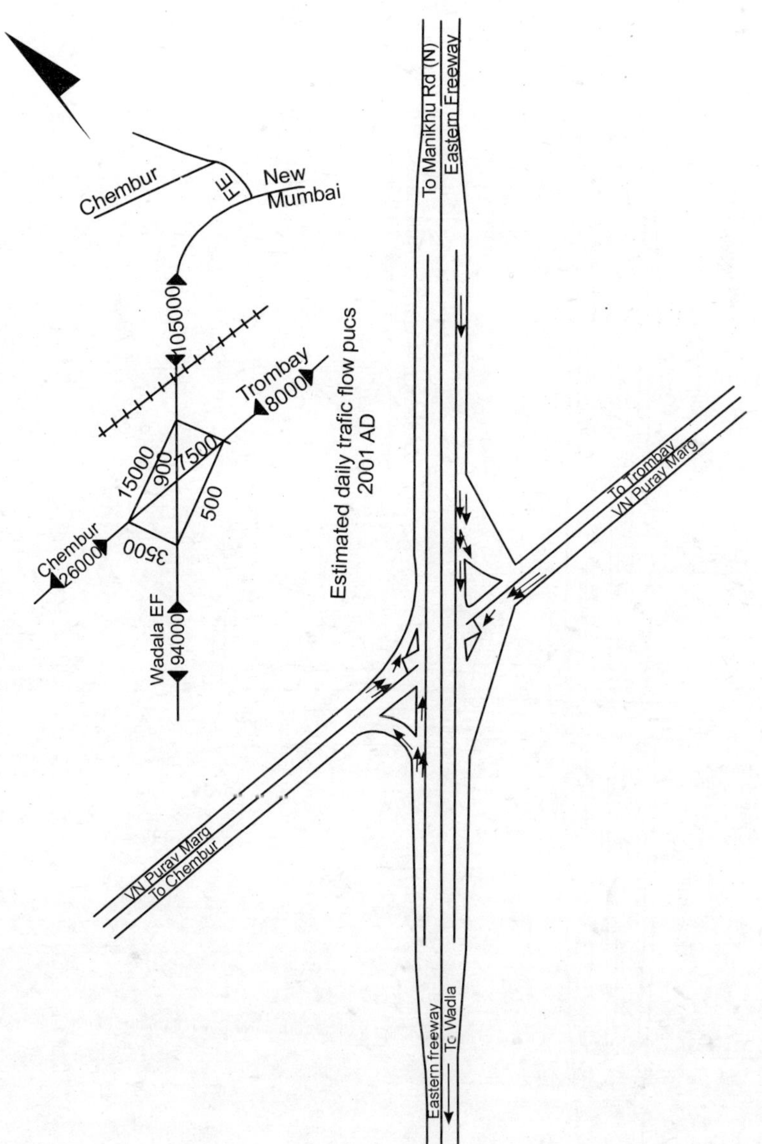

Figure 12.35 (b) *Proposed road Interchanger for New Mumbai*

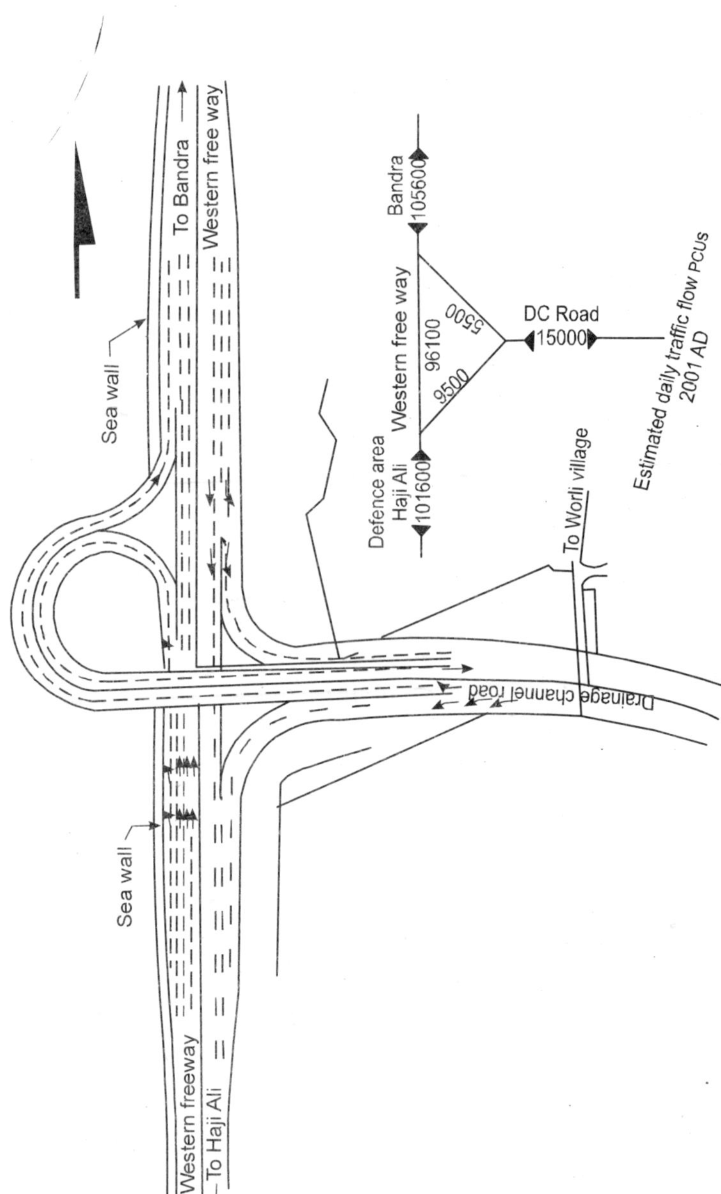

Figure 12.35 (c) *Road Interchanger planned for New Mumbai*

12.31 INTERSECTIONS AT TWO GRADES

When two highways at two different levels intersect each other or a highway grade separation is achieved by means of a vertical level difference between the intersecting roads by means of a bridge, grade separated intersection design is the best form of intersection as it eliminates all crossing conflicts thereby eliminating all hazards of accidents. Grade separation may either be an over bridge or flyover or an under bridge. When the major road is at a higher level and passes over the minor road or roads the grade separation is called over bridge or flyover. Similarly when the major road passes under the minor road it is called under bridge or under pass. Following are some of the advantages Grade separation:

1. All conflict points of crossings are eliminated.
2. Maximum comfort is provided to the crossing traffic.

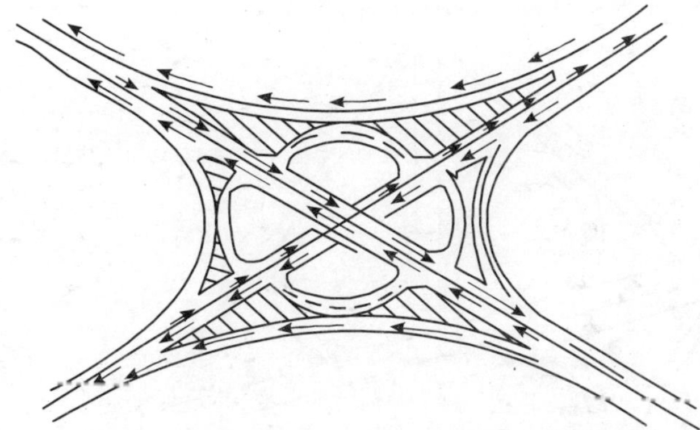

Figure 12.36 *Diamond*

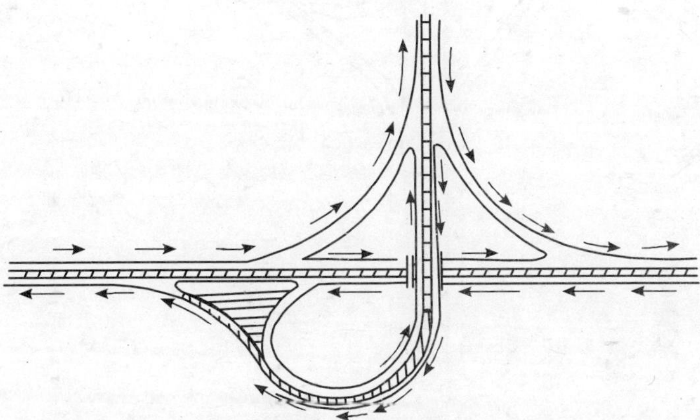

Figure 12.37 *Trumpet*

3. The turning traffic including the right hand turning traffic is provided with increased safety by constructing indirect interchange ramps.

4. There is an overall increase in comfort and convenience of motor traffic particularly by saving travel time and cost of fuel.

5. Grade separation is possible for all angles and layout of intersecting roads.

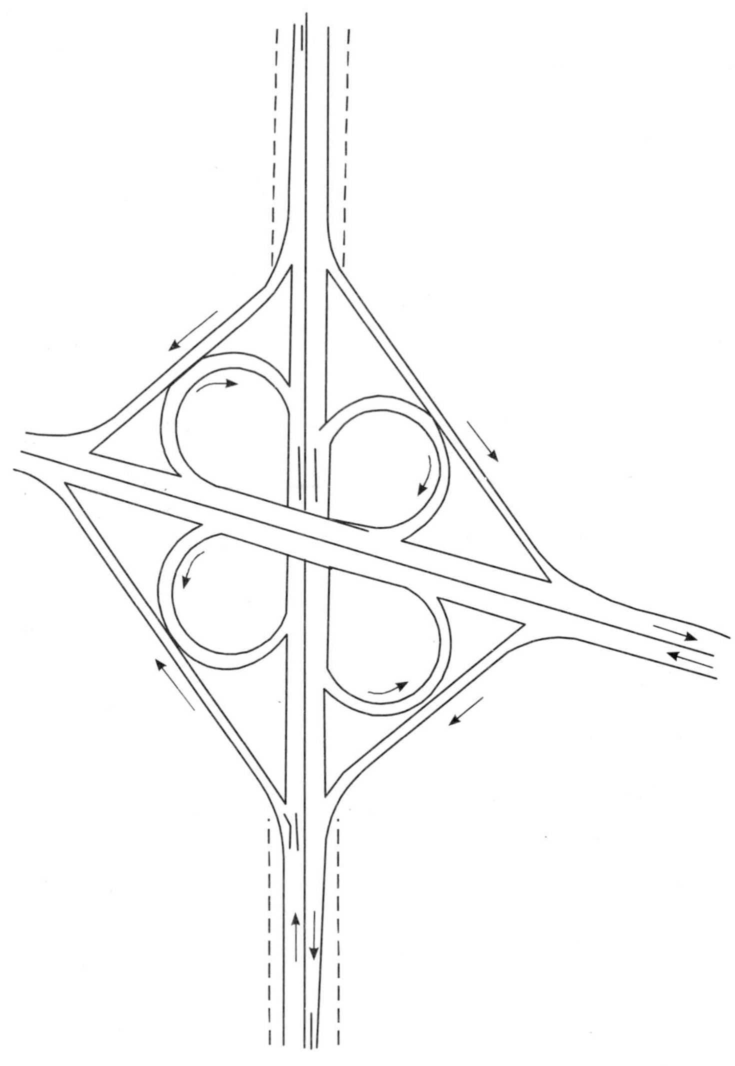

Figure 12.38 *Cloverleaf*

12.32 INTERCHANGING RAMPS

Transfer of route at the grade separation is produced by interchange facilities consisting of ramps. Interchanging ramps are of three types viz., direct, semi-direct and indirect ramps.

The *Direct Interchange Ramps* provides facility of turning to right and merging with the main stream from the right. Semi-direct interchange ramps allow diverging to the left but merging from the right side. The indirect interchanging ramp allows diverging from the left and also merging from the left. Merging and dimerging from the left is very much simpler and less hazardous but the distance to be traversed is much more than the direct or semi-direct interchanges.

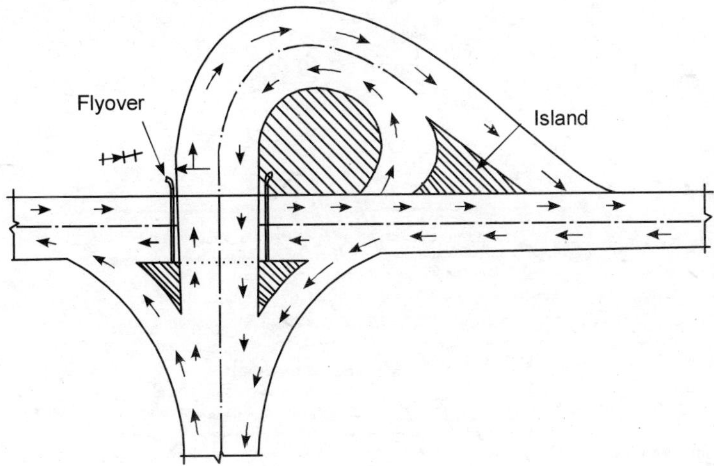

Figure 12.39 *Flyover*

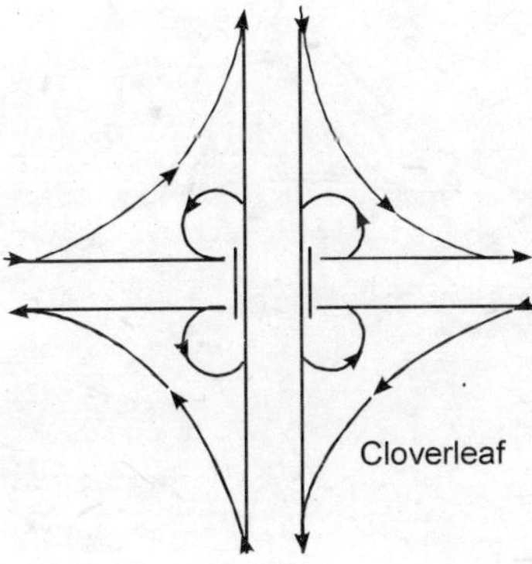

Figure 12.40 (a) *Flyover*

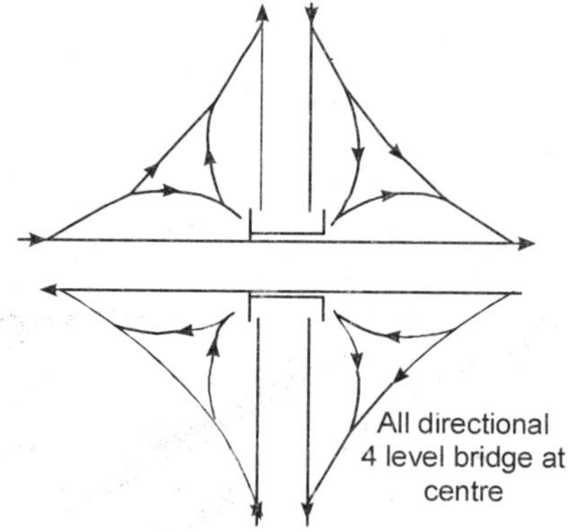

Figure 12.40 (b)

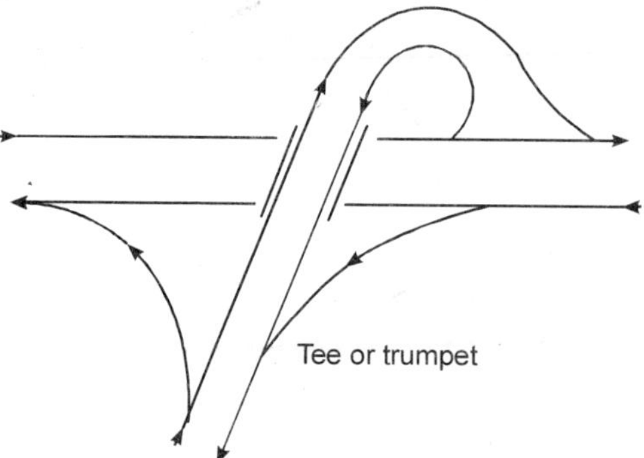

Figure 12.40 (c)

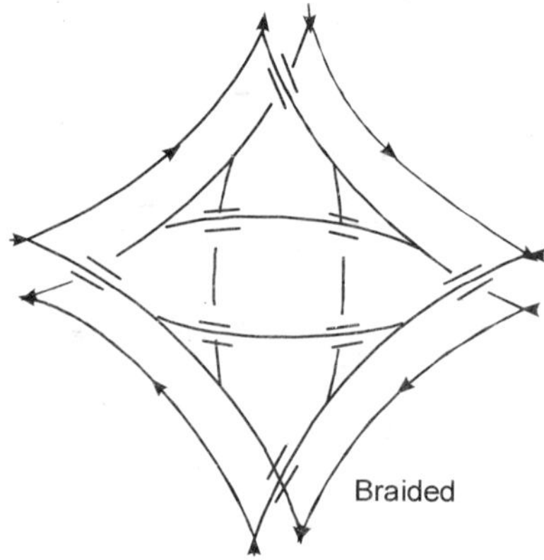

Figure 12.40 (d)

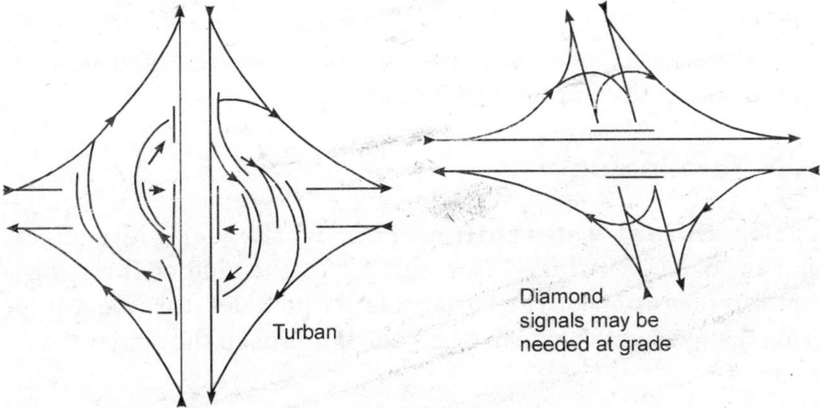

Figure 12.40 (e) **Figure. 12.40 (f)**

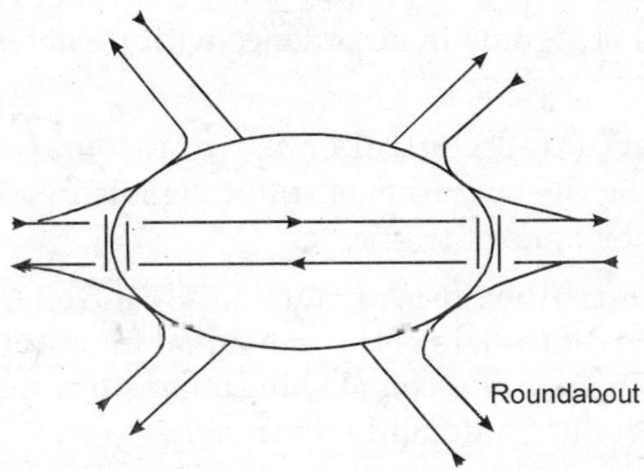

Figure 12.40 (g)

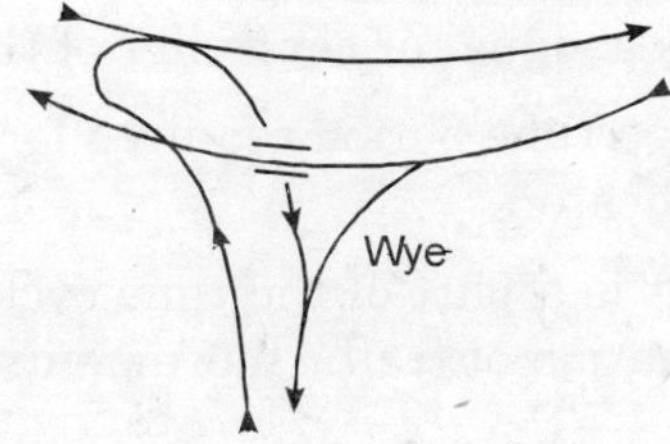

Figure 12.40 (h)

Grade separated intersection with complete interchange facilities is essential to develop a highway with full control of traffic. When there is intolerable congestion and high intensity of traffic hazards, grade interchange is a must, some of the important interchangers are given below.

12.32.1 Terminology

1. Traffic Signal Controller. This is the complete electrical mechanism, usually installed in a cabinet at the side of the road which controls the operation of traffic signals. It includes turning equipment operating the mechanism which lights the lanterns in the signal heads. The controllers are:

(i) Automatic controller is a self operating mechanism which operates automatically the traffic signals. Automatic controllers are generally fitted with a facility switch which enables the signals to be changed manually by depressing and releaving a push button.

(ii) Pretimed controller is an automatic controller for supervising the operation of signals in accordance with predetermined fixed time cycle.

(iii) Traffic actuated controller is an automatic mechanism for supervising the operation of traffic signals in accordance with the varying demands of traffic.

2. Master Controller. It is an automatic controller for supervising a system of secondary controllers of the individual intersections, maintaining definite interrelationships or accomplishing other supervisory functions. The secondary controllers are automatic controllers.

3. Local Controllers. It is a mechanism for operating traffic signals at an intersection or two or three adjacent intersections, which may be isolated or included in a signal system.

4. Traffic Detector. It is a device by vehicles or pedestrian are enabled to inform a traffic actuated signal or controller of their presence.

5. Time Cycle. It is a time period required for one complete sequence of signal operation or indicators.

6. Traffic Phase. It is a part of the time cycle allotted to any traffic movement or any combination of traffic movements recovering the right of way.

7. Intergreen Period. The time between the end of the green period of the phase and the beginning of the green period of the phase. It is also called Amber Time.

12.33 TRAFFIC SIGNALS

Traffic control by road signals has been proved to be a very effective system of controlling and guiding the traffic, especially where there is heavy vehicular traffic on the intersection of roads. The traffic signals have recently been introduced in big cities like Delhi, Mumbai, Kolkata, Poona, Allahabad, Lucknow etc. The traffic or road signals serve the following purposes :

1. Provide for an orderly movement of traffic.
2. Reduce the frequency of traffic accidents.
3. Control the speed on the main and cross roads.
4. Help in directing the traffic on different roads.
5. Control the traffic on rail-road crossings.
6. Co-ordinate the movement of vehicles in an area and allow them to flow cautiously.
7. Intercept heavy traffic in order to allow other vehicles and pedestrians to cross the road.

Apart from the above advantages the traffic signals have the following disadvantages too:

1. It may increase certain types of accidents such as re-entrant collision.
2. When improperly located, it promotes disrespect for this type of control device and improperly turned it will cause excessive delays.
3. It may encourage the drivers to prefer alternate routes thus dispersing traffic on minor streets, rather than concentraing on major streets.

12.34 BASIC FEATURES OF TRAFFIC SIGNALS

A typical signal head is composed of three lanterns arranged vertically above each other with red lens on the top and green at the bottom and amber between the two. The lenses are usually 20 cm in diameter and each is being illuminated from behind by independent light source usually a bulb.

The normal sequence of signal is red, amber and green. Sometimes red and amber glow together and green and amber glow together. The standard time of red light is 2 seconds and amber is 3 seconds. The function of amber light is to indicate that the signal is going to change so that the traffic should enter the intersection before the lights are changed so that 'ousted' or 'lost' time is minimum. The amber light also acts as a warning for the approaching traffic of a coming change is signal indirection and acts as a clearance internal for vehicles and pedestrian within the intersection as well as for those vehicles that are to close at the stop.

Signal posts are erected on the foot paths just behind the heals or suspended by means of cables over the centre of the intersection. It is essential that the location of the signal be such that approaching drivers can become aware of their resistance in sufficient time to be able to stop along the indication without difficulty. The overhead system should satisfy all the conditions of the intersection. Signal heads on post are preferred. The usual height is 2.25 m above the surface. The height can be increased where there are varying grades.

Sometimes in congested localities the signal post and lights are not clearly visible and hence more than one signal head is used. There is a serious danger that traffic signals in urban areas may fail to be seen due to being confused with the background light of shops and advertisement and hoardings. Some amelioration of this problem can be achieved by placing a dark contrasting backing board behind the signal head at such locations. Signal lights should be checked and replaced if burnt-out. Burnt-out lanterns with incorrect signals will play the role of traffic hazards.

12.35 CLASSIFICATION OF ROAD SIGNALS

Road or Traffic signals may be classified as

1. Traffic control signals
2. Pedestrian signals
3. Special traffic signals

1. Traffic control signals are again classified as
   (a) Fixed time signals
   (b) Traffic actuated signals

The traffic actuated signals are of three types viz., (i) fully actuated signals (ii) semi-actuated signals and (iii) speed control signals.

(a) Fixed Time Signals. These signals are set to repeat regularly a cycle of red – yellow – green lights at regular intervals of time. The duration of green and red light is generally the same, whereas the duration of yellow light is smaller. They have the advantage of having the minimum delay for a group of vehicles travelling along a series of intersection with interconnected signals. The disadvantage is that it may hold traffic in one direction when there is not traffic in the other direction, resulting in delay and decreasing capacity.

(b) Traffic Actuated Signals. The main feature of the traffic actuated signals is that they recognise the demand of the flow of traffic and distribute the time cycle accordingly. Time cycle is the time in seconds required to complete a cycle from green to red and vice versa. The main advantages are:

1. Reduce delays.
2. Adaptable to short term fluctuation of traffic flow.
3. Capacity of lane/lanes increases.
4. Provide continuous operation under low volume and is especially effective at multiple phase intersection and are more effective in isolated intersections.

Fully Actuated Traffic Signals. These signals have detectors located on each approach and assign the right of way to different type of traffic movement on the basis of demand.

Semi Actuated Traffic Signals. Semi actuated signals are those in which the detectors are placed only on the minor roads. These signals are used where a high density, heavy volume and high speed or express way meets a relatively very light cross traffic.

The detectors are pressure sensitive instruments which operate the signals according to the flow of the traffic. Detectors are used in the actuated signals. The detectors may be (i) pressure sensitive detectors (ii) magnetic detectors (iii) radar detectors (iv) sound detectors and (v) light sensitive detectors.

12.35.1 Disadvantages of Traffic Actuated Signals

1. They are un-economical and the initial cost is two to four times the cost of Fixed Time Signals.
2. The electronic circuits are much more complicated than ordinary signals, thereby increasing the maintenance problems.
3. Detectors to be used in actuated signals are very sensitive and very complicated in installation.

Pedestrian Signals. Pedestrians signals are meant to give right of way to pedestrian during the walk period when the vehicular traffic is stopped by red or stop signal. The traffic signal on the main lanes and pedestrian crossings or *zebra crossing* are interconnected in such a way that the GO signal of one and the STOP signal of other operates simultaneously.

Special Signals. These are also called *Flashing Beacons* and are meant for warning the traffic for stopping well before the nearest cross walks at an intersection or at a stop line. Flashing yellow signals are warning signals and caution the driver to proceed with caution.

12.36 SIGNAL SYSTEM

Following signal systems are used :

1. **Simultaneous System.** In this system all the signal are connected together on a series of intersections on the same highway and show the same indications simultaneously. In this system as all the signals are connected together, so only one controller is used for all the intersections.

2. **Alternate System.** In this system the signals fitted at alternate crossing on the same length of highway give alternate indication i.e., if one crossing gives red or stop indication, the second one will indicate green or go indication and the third in the series will also indicate as that of first one i.e., red. In this way the alternate crossings will indicate same light. This system can also be operated by a single controller.

3. **Simple Progressive System.** For continuous flow of traffic without interruption, this system is used. After crossing the first intersection, by the time the vehicle reaches the second intersection it will receive a go or green indication thereby causing no delay. In this system, signals controlling the intersection give green indications to a pre-determined schedule to permit continuous flow of vehicles along a road at planned speed.

4. **Flexible System.** In this system the signals are adjusted automatically by varying the length of cycle, time schedule and time division. This is the most effective system.

12.37 TIMINGS OF SIGNALS

The timings of signal cycles should be determined on the following basis, keeping in view that minimum cycle timings are the best.

1. Pedestrian crossing time on all crossings, assuming pedestrians walk at the rate of 1 m/sec or 1.2 m/sec.

2. Stop time or red phase of a signal is the sum of go and clearance intervals or green and amber phases for the cross flow. During this time the pedestrian crossing time should also be included.

3. Towards the end of the red phase there may be a short gap when the amber lights are put on along with the red light in order to indicate get set and go.

4. Clearance time or amber phase is provided just after the green phase before the red phase.

5. Go or green time is decided on the basis of approach volume during peak hours to enable the lined up vehicles to clear up.

The cycle timings are decided by the following two methods :

(i) *Trial cycle method.* In this method, the green, amber and red time periods are assumed as G_1, G_2; A_1, A_2 and R_1, R_2 for the two road number 1 and 2. Number of vehicles crossing lane no. 1 and 2 are n_1 and n_2 per hour, then

$$G_1 = \frac{2.5\, n_1\, c_1}{60 \times 60}$$

and
$$G_2 = \frac{2.5\, n_2\, c_2}{60 \times 60}$$

Assuming also that it takes 2.5 sec for a vehicle to cross the lane.

The amber period for a normal speed of 40 km/hr are assumed to be 3 to 5 seconds or 2 to 4 seconds, c_1 and c_2 are number of cycles for each lane.

(ii) *Approximate method.* This method is based on the following assumptions:

1. Based on the approach speed of vehicles at the intersection, amber timings are assumed as 2.3 or 4 seconds.

2. Assume walking speed of pedestrians as 1.2 m/sec, the clearance time is calculated.

3. Red time should be equal to the pedestrian clearance time plus initial start time.

4. Similarly green timings is taken as equal to pedestrian crossings and the initial start interval or it should be equal to red time of cross road minus amber time for the cross road.

5. The actual green time needed is then increased proportionately for the heaviest traffic volume per hour per lane. The cycle length so obtained is a djusted for the next higher 5 sec interval the extra time is then distributed to green timings proportionately.

12.38 CONDITIONS WARRANTING INSTALLATION OF TRAFFIC SIGNALS

On each and every crossing traffic signals cannot and should not be installed. The following points should be considered before deciding on the installation of a traffic signal:

1. It has been observed by experience and actual observation that if the total number of vehicles crossing a road on both approaches during peak eight hours on monthly average base is less than 650 motor vehicles per

hour, traffic signals are not required in our country. Further the number of vehicles approaching the crossing on minor streets is atleast 200 vehicles per hour on single lane and 250 vehicles per hour on double lane.

2. When the number of vehicle per hour during eight peak hours on the major street is 1000 to 1200 and on the minor street 100 to 150 vehicles per hour, there is unnecessary delay in the movement of traffic on the minor lane or street.

3. The minimum pedestrian volume of 150 per hour cross the major street with over 600 vehicles per hour on both approaches and 85th percentile speed of 60 km/hr.

4. If the accident rate on that intersection has a rising trend and other methods have failed, then traffic signals warrant.

Example 12.1 *Two major streets 10 m and 8 m wide are to be signalled at the intersection. The 10 m wide lane has 600 vehicles per hour and the 8 m wide lane has 450 vehicles per hour. Average approach speed on the two lanes is respectively 60 km/hr and 40 km / hr. Calculate the cycle time for the two lanes. Assume any other data not given.*

Solution Assume a cycle time of 50 seconds

$$\text{No. of cycles per hour} = \frac{60 \times 60}{50} = 72$$

Green time for 10 m wide lane

$$= \frac{60 \times 2.5}{75} = 20.8 \text{ secs}$$

(2.5 sec is the average time headway during green phase)

Green time for 8 m wide road is

$$= \frac{450 \times 2.5}{72} = 15.75 \text{ secs}$$

Assume amber time for the two lanes 4 sec and 3 sec respectively

Total cycle length = 20.8 + 15.75 + 4.0 + 3.0

= 42.55 seconds

An average cycle time of 45 seconds will be appropriate.

More than one assumptions should be made to approach a near approximation.

Example 12.2 *A 15 minute traffic count on the intersection of two roads of width 10 m and 12 m is 180 and 200 vehicles per 15 minute respectively. If*

the amber time for the two lanes is 3 sec and 4 sec and the average headway time for each vehicle is assumed to be 2.5 sec, calculate the cycle time.

Solution Assume 50 second as the cycle time

$$\text{No. of cycles per 15 minute} = \frac{15 \times 60}{50} = 18 \text{ sec}$$

Green time lor 10 m wide lane

$$= \frac{180 \times 2.5}{18} = 2.5 \text{ sec}$$

Green lime for 12 m wide road

$$= \frac{200 \times 2.5}{18} = 27.77 \text{ sec}$$

Amber time for the two lanes is 3 sec and 4 sec respectively (given)

Total cycle length = 25 + 27.77 + 3 + 4

= 59.77 seconds

Hence 60 seconds cycle time will be appropriate.

Example 12.3 *A pedestrian indication with an isolated signal is to be installed on a right angled crossing of two roads 15 m and 12 m wide. The peak volume traffic on the two roads is 300 and 250 vehicles per hour respectively. The approach speeds are 55 km/hr and 60 km/hr respectively. Design the timings of pedestrian and traffic signals.*

Solution *Design of Traffic Signals.*

Assume amber period for the two roads as 4 sec and 3 sec based on approach speed. Pedestrian walking speed is assumed to be 1.2 m/sec.

$$\text{Pedestrian clearance period for the first road} = \frac{15}{1.2} = 12.5 \text{ seconds}$$

$$\text{Pedestrian clearance time for second road} = \frac{12}{1.2} = 10 \text{ seconds}$$

Minimum red time for the first road = 10 + 3+4 = 17 sees (G_2) for the second road = 12.5 + 3 + 4 = 19.5 secs (G_1)

Minimum green period for first lane

= 10 + 7 4 = 13 second

For second road = 12.5 + 7 – 3 = 16.5 sec

Using the relation

$$\frac{G_1}{G_2} = \frac{n_1}{n_2}$$

where G_1 and G_2 are green timings and n_1 and n_2 are approach volume

$$G_1 = \frac{300}{250} \times G_2$$

$$= \frac{300}{250} \times 17 = 20.4 \text{ sec}$$

Total cycle length $= G_1 + A_1 + R_1$

$= G_1 + A_1 + G_2 + A_2$

$= 20.4 + 4 + 17 + 3$

$= 44.3$ seconds

Hence adapt cycle length of 50 seconds.

The additional period of 50 – 44.3 = 5.7 sec is proportionately divided to green periods of lane 1 and lane 2 according to the approach speed

Hence

$$G_1 = 20.4 + 3.6 = 27 \text{ sec}$$

$$G_2 = 17 + 2.1 = 19.1 \text{ sec}$$

Red time for road 1 $= G_2 + A_2$

$= 19.1 + 3 = 22.1$ sec

Red time for road 2 $= G_1 + A_2$

$= 27 + 4 = 31$ sec

Design of pedestrian signal

The red period of first lane is the WALK period for pedestrian of lane 2 and DO NOT WALK period of pedestrian signal at road 1 is the red period of road 2

$DW_1 = R_2 =$ 31 sec

$DW_2 = R_1 =$ 22.1 sec

$DW_1 =$ Do not walk for road 1

$DW_2 =$ Do not walk for road 2

Pedestrian clearance interval for the two roads respectively is 12.5 sec and 10 sec.

12.39 I.R.C SIGNAL DESIGN METHOD

The I.R.C has provided the following guidelines for the design of traffic signals:

1. The pedestrian green timings for the major and minor roads should be calculated on the basis of the walking speed of 1.2 m/sec and initial walking time of 7 seconds. These are the minimum green timings required for the vehicular traffic on the minor and major road respectively.

2. The cycle time is calculated after allowing amber time of 2.0 sec each.

3. The green time required for the vehicular traffic on the major road is increased proportionately according to the approach volume.

4. The minimum green time required for clearing vehicles during a cycle is determined for each lane assuming that the first vehicle will take 6.0 sec and the subsequent vehicls will be cleared at the rate of 2.0 sec/vehicle. The minimum green time required for the vehicular traffic to clear a lane is limited to 16 sec.

5. The optimum signal cycle is calculated using Webber's formula

$$C_0 = \frac{1.5\,L + 5}{1 - \gamma}$$

where

L = Total lost time per cycle

$= 2n + R$

n = number of phases

R = red time

$\gamma = y_1 + y_2$

y_1 and y_2 are normal flow ratio

$$G_1 = \frac{y_1}{y}(C_0 - L) \text{ and } G_2 - \frac{y^2}{y}\,(C_0 - L)$$

The saturation flow values may be assumed as 1850, 1890, 1950, 2250, 2550 and 2990 PCU per hour for approach road width of 3.0, 3.5, 4.0, 4.5, 5.0 and 5.5 metres and for widths above 5.5 metres, the saturation flow may be assumed as 525 PCU per hour per metre width. The lost time is calculated from the amber time, inter-green time and the initial delay at 4.0 sec for first vehicle on each lane.

6. The signal cycle time and the phase may be revised keeping in view the green time required for cleaning the vehicles and the optimum cycle length is determined.

Example 12.4 *The average normal flow of traffic on cross roads namely Mahatma Gandhi and Kasturba Gandhi Roads in New Delhi, during design period are 400 and 250 PCU / hour respectively, and the saturation flow values on these roads are estimated as 1250 and 1000 PCU per hour. The all red time required for pedestrian crossing is 12 sec. Design the two phase traffic signals.*

Solution

$$y_m = \frac{q_m}{S_m} = \frac{400}{1250} = 0.32$$

$$y_k = \frac{q_k}{S_k} = \frac{250}{1000} = 0.25$$

(m stands for Mahatma Gandhi Road and k stands for Kasturba Gandhi Road)

$$Y = y_m + y_k = 0.32 + 0.25 = 0.57$$

$$L = 2n + R = 2 \times 2 + 12 = 16 \text{ sec}$$

$$C_0 = \frac{1.5\,L + 5}{(1 - y)} = \frac{1.5 \times 16 + 5}{(1 - 0.57)}$$

$$= \frac{29}{0.43} = 67.4 \text{ sec.}$$

$$G_m = \frac{y_m}{y}(C_0 - L)$$

$$= \frac{0.32}{0.57}(67.4 - 16)$$

$$= 29 \text{ sec}$$

$$G_k = \frac{y_k}{y}(C_0 - L)$$

$$= \frac{0.25}{0.67}(67.4 - 16)$$

$$= 22.5 \text{ sec.}$$

All red time for pedestrian crossing = 12 sec

Providing amber time for each clearance

Total cycle time $= 29 + 22.5 + 12 + 4$

$= 67.5$ sec

12.40 ROAD SIGNS

Road signs are used to prevent accidents. Properly designed road or traffic signs placed at appropriate positions are a must for safe and efficient movements of traffic. They should be so located as to be clearly visible from a distance. They have been standardised by the Indian Road Congress and are divided into four main groups :

(1) Warning signs. (2) Prohibitory signs.

(3) Mandatory signs. (4) Informatory signs.

1. Warning Signs. Warning or cautionary singns indicate to traffic by suitable standardised symbols the approach of a place where some precaution is required for the safety of the traffic. The warning or cautionary signs are provided at a distance of 100 m on an up grade 110 m on a level road and 130 m on a down grade, from the specified danger point. These signs are meant for hair pin bends, level crossing, steep hill, round-about, school, cross road, T-junctions, narrow road bridge etc.

2. Prohibitory Signs. Prohibitory signs indicate to the traffic that the use of a particular road is prohibited by a specified class of traffic, or prohibits the use of horns in a particular area of the road, or exceed a particular speed limit as on sharp curves, or parking at a certain place, or from entering into a road.

3. Mandatory Signs. Mandatory signs or regulatory signs indicate to the traffic to comply with certain statutory regulations such as keep left, turn right etc. The mandatory signs are usually provided at round-abouts, flyovers etc.

4. Informatory Signs. Informatory signs convey certain information and guidance to the traffic such as road directions, parking places, direction to a town or city, along with distance etc.

The different road signs are shown in Figs 12.44 (a), (b) and (c).

12.40.1 Characteristics of Traffic Signs

The traffic signs should have the following characteristics :

1. The signs should be of larger size on *express ways* (high speed roads).
2. The spacing between the letters should be large and optically equal.
3. The maximum number of words should be three.
4. The signs to be read in the night must be illuminative or reflective.
5. The size, spacing, shape and colour of traffic signs must be uniform.
6. The signs should be located in accordance with the carriageway requirements.

7. Distraction or advertisement signs and other, unnecessary signs should be eliminated wherever possible.

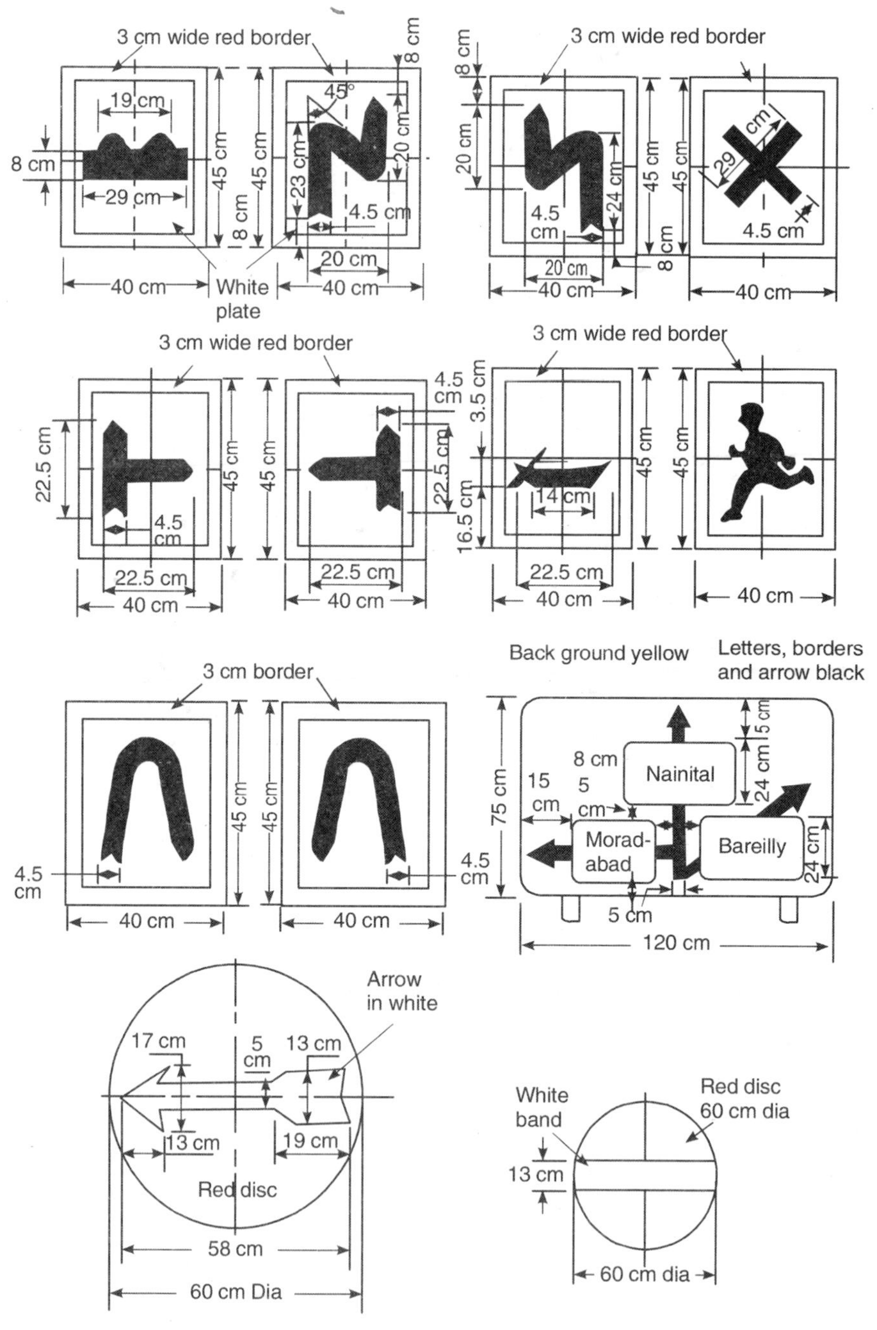

(a) Traffic signs

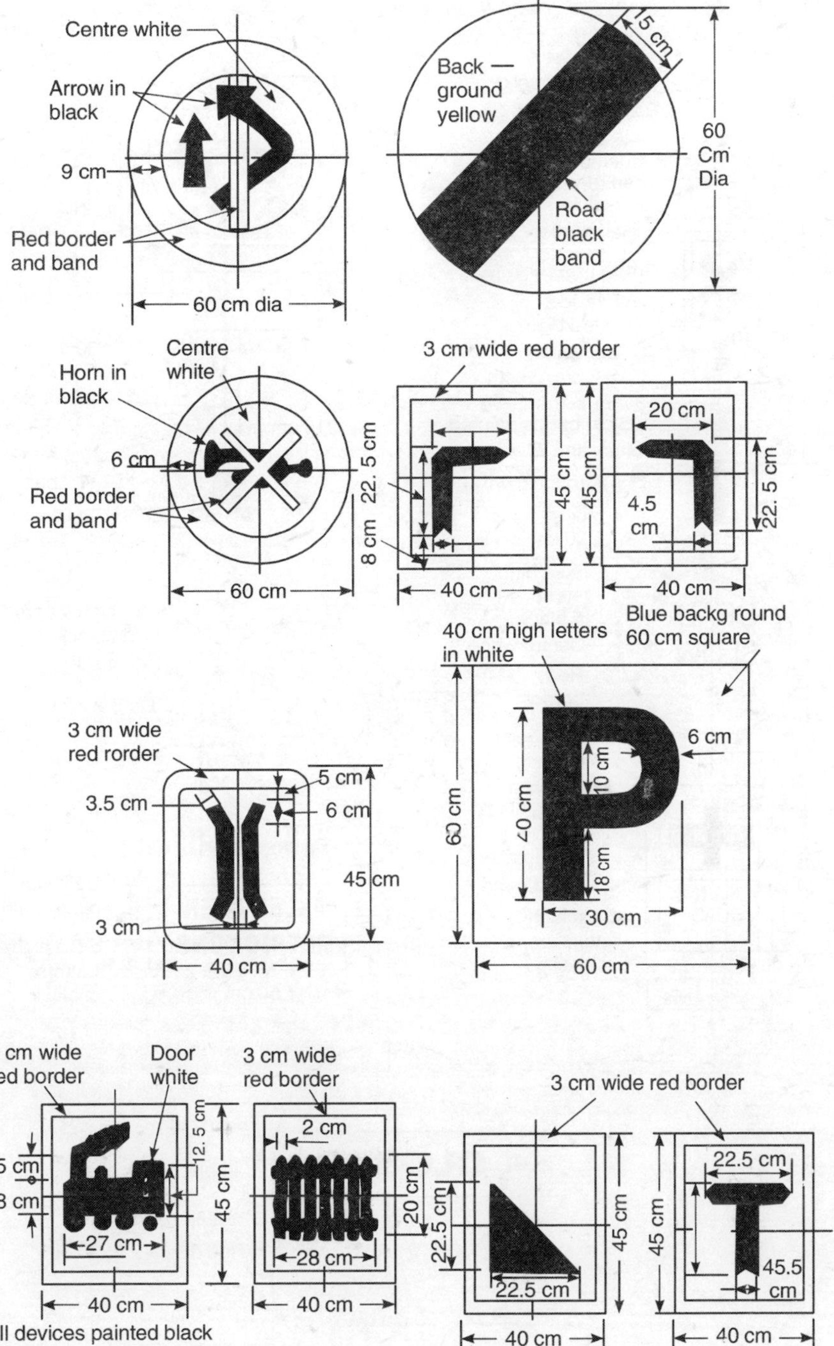

(b) Traffic signs

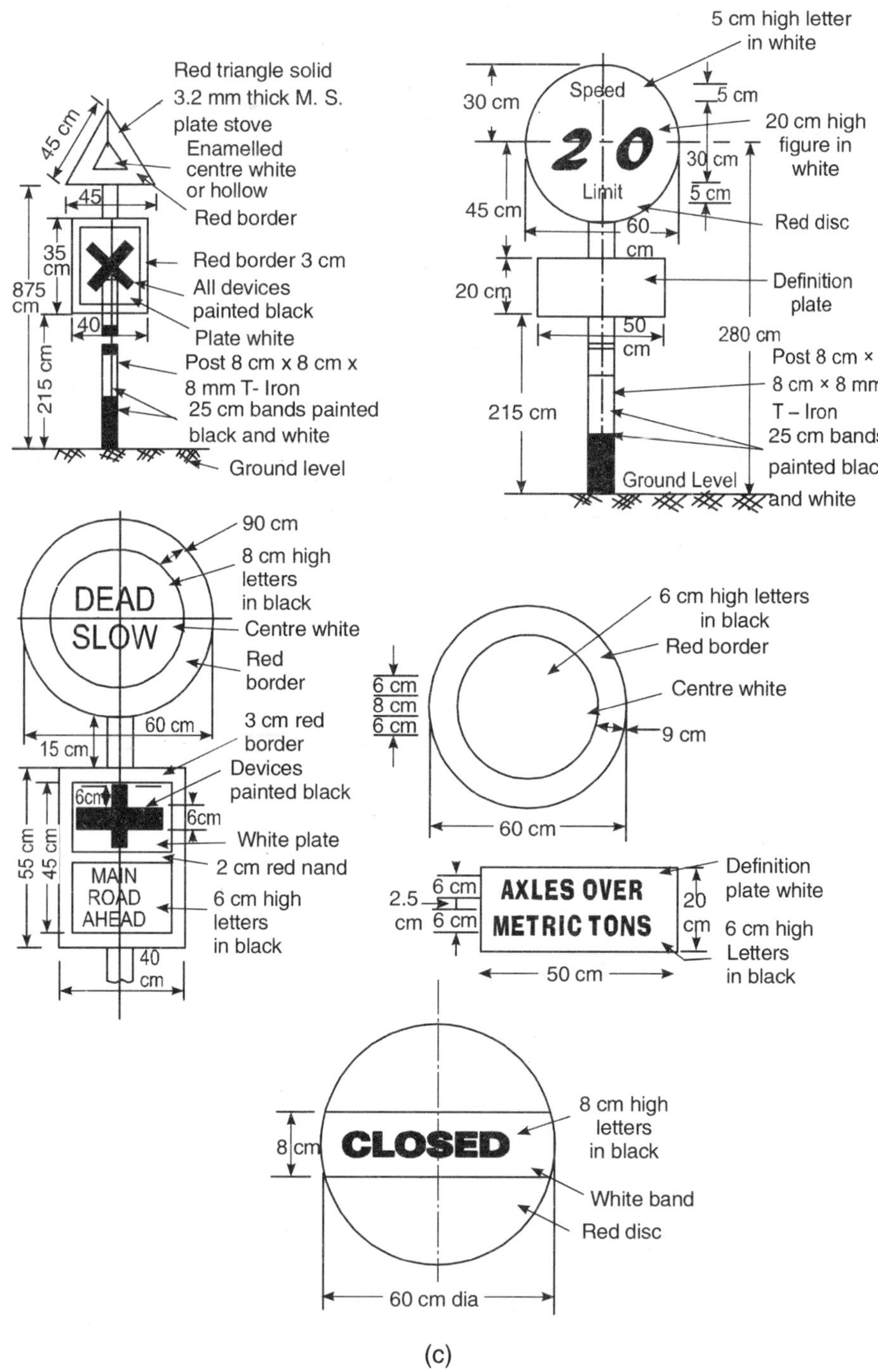

(c)

Figure 12.41 *Traffic signs*

12.41 HIGHWAY AND STREET LIGHTING

In the olden days, when there was not much of the fast moving vehicles, oil-lamps were used on the road-side. This practice is still in use in Indian villages. These oil lamps are only meant to show the path to the pedestrians and the bullock cart traffic. This was actually more a psychological than the actual benefit. With the advancement of fast moving traffic, hazards of travel for pedestrians and the automobiles have increased considerably due to dusk, darkness and impaired vision. In modern road development, efficient street lighting has become an important feature for the reduction of crime expedition of traffic.

12.41.1 Factors Affecting Street Lighting

For efficient and proper street lighting the following factors should be kept in mind:

(i) The electric posts poles should be located at proper places.

(ii) The light from the source should illuminate the entire road surface uniformly.

(iii) The posts should be spaced uniformly.

(iv) The light should not produce glare.

(v) The height of lamps should be adjusted in such a way that it should produce such a shadow which may not darken the entire line.

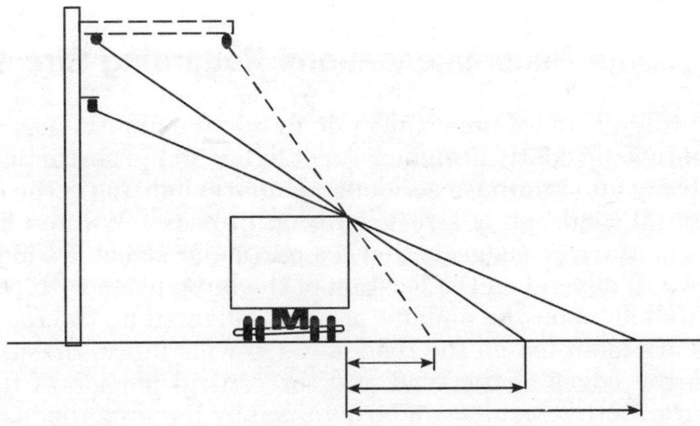

Figure 12.42

12.42 FACTORS EFFECTING VISIBILITY ON ROADS

Following factors determine the visibility on roads at night.

1. Source of Light. Source, power and type of light greatly effect the visibility on the road. Lighting on modern highways can be an ordinary filament lamp, sodium vapour lamp, mercury lamp or a fluorescent type lamp. Filament lamps are commonly used for ordinary roads for street lighting as they are cheap and easily available. Sodium vapour lamps produce distinctive yellow light are most suitable for hazardous localities. Mercury vapour lamps are being increasingly used on account of their long life and light efficiency. Fluorescent lamps are particularly used for overhead lighting.

2. Placing and Distance of Lamp Posts. Lamp posts can be located either on one side or staggered on both sides of the road or centrally overhung. Light posts staggered on both sides and centrally overhung are very efficient but costly and are used only on very important roads passing through busy and congested localities. On ordinary roads, lamp posts are placed on one side of the road only. The light posts should be spaced in such a way so as to produce uniform illumination on the road surface. It has been observed by experience that spacing between the light posts should be between 40 m and 60 m and the height of lamp should be between 3.5 m and 5.0 m.

3. Distribution of Light. It is very difficult to have a uniform brightness on the road surface during night hours, but at the same time fair distribution of light can always be achieved patchy and uneven distribution of light on the road is very dangerous. The lamp posts should be so arranged that light patches should just each other so as to avoid uneven distribution of light.

4. Glare. Glare is the main source of accidents and in the greatest cause of wastage of illumination contract lighting arrangement reduces glare.

12.42.1 Some Recommendations Regarding Street Lighting

Street lighting plays an important role in maximising the use of highways with advantage. Properly designed street lights add to aesthetic appearance of the highway and minimise accidents. Uniform lighting of the road surface is an essential condition of street lighting. Professor William E. Barrow of University of Marrine suggests that if a particular street is wide enough, its brightness will depend on the location of the lamp posts with respect to the street as well as upon the amount of light delivered on the road surface. If the lamps are mounted on the road kerbs of wide lanes, the streak of light will be on the edges of the road and the central portion of the road will become dark. Better results can be obtained by hanging the lamps through long brackets so as to produce streaks of light as near the centre of the road, as possible. Better arrangement of street lighting will be to provide lamp posts on both sides in a staggered way. Overhanging lamps on the centre of pavement or electric poles arranged on the centre line of the road, will make the surface more attractive, produce uniform brightness and improve visibility. Road surfacing is also responsible for improving, visibility, brightness and uniform illumination. A light coloured surface of seal coat is very much desirable because it reflects light to the driver. A highway with uniform horizontal illumination throughout its length presents a large variation in brightness when viewed by drivers and pedestrians. For uniform brightness, the light must be distributed to take account of the reflection of light at small, angles from the surface.

12.43 DESIGN OF HIGHWAY STREET LIGHTING

Lighting from one side of the road only usually is unsatisfactory except on bends and narrow roads. Due to economy single side lighting is usually adopted even for two lane roads. The maximum distance between two opposite rows of lights should not be more than 9.0 m. The roads upto 9.0 m width can be effectively lighted by lights mounted on the kerb-line in vertical line above the kerb.

When the width of the road is increased, brackets are used for overhanging the lights above the road. The maximum overhang of brackets over the kerb should be 1.8 m to avoid undue dark patches on the kerb and footways.

Following are the I.R.C. recommendations for horizontal clearance for lighting poles.

(i) For roads with raised kerbs (as in urban roads).	0.3 m minimum desirable 0.6 m from the edge of the raised kerb.
(ii) For roads without raise kerbs (as in rural roads).	1.5 m minimum from the edge of the carriageway, subject to a minimum of 4.0 m from the centre line of the carriage way.

The side mounted lamp posts can be arranged opposite to each other on both sides of the road or arranged staggered. The staggered arrangement is more efficient.

The usual mounting height of lamp post is 6.1 m to 7.6 m in 4.6 m (minimum) wide streets. The mounting height is between 7.6 m to 9.0 m in highways with 35 to 45 m centre to centre spacing. The ratio of spacing of poles to the mounting height is kept between 8 : 1 in congested areas and 10 : 1 in thickly populated areas. High mounting greatly reduces the blinding effect of direct glare.

The spacing of the light poles is reduced on the curves, bridges, level crossings etc. for better light requirements.

I.R.C. has specified the minimum vertical clearance required for electric power line upto 650 volts as 6.0 m. All the poles carrying overhead power and telecommunication lines except the urban areas, should be erected at least 100 m away from the nearest edge of the roadway. The distance of this line should be 5.0 m from the nearest line of avenue trees.

If the trees are interfering with the side lamps, they should be centrally suspended. On the S-curves the lamp posts should be changed from one side of the road to the other at some point in the middle of the bend.

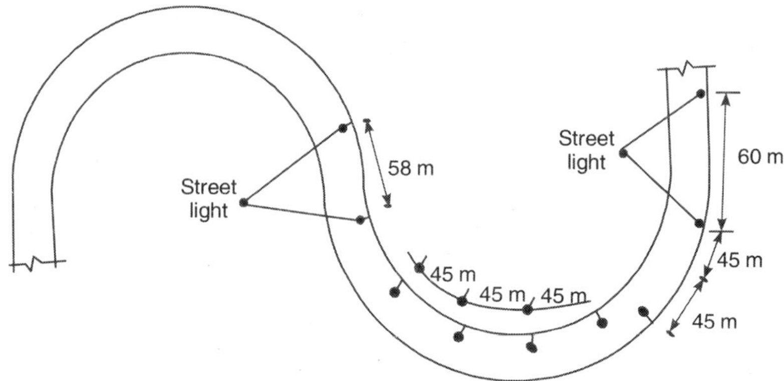

(a) Street light position on curves

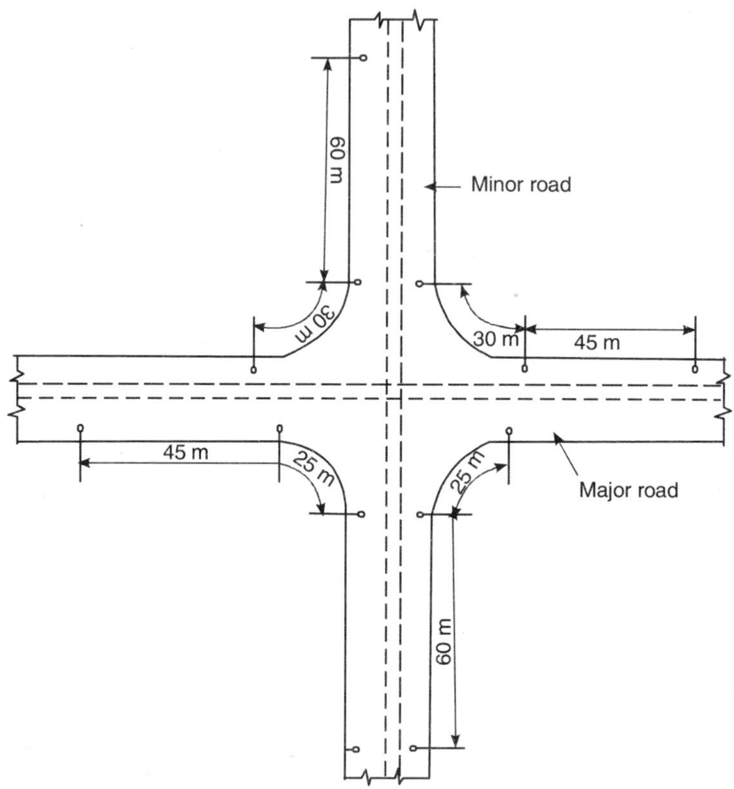

(b) Street light position on T junctions

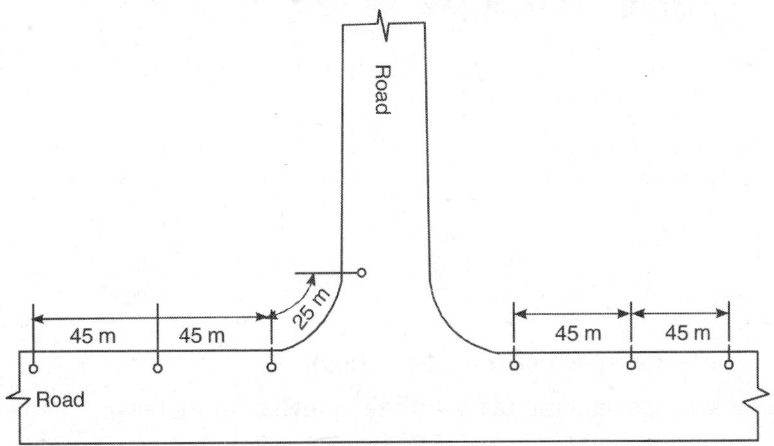

(c) Street light position on T-junctions

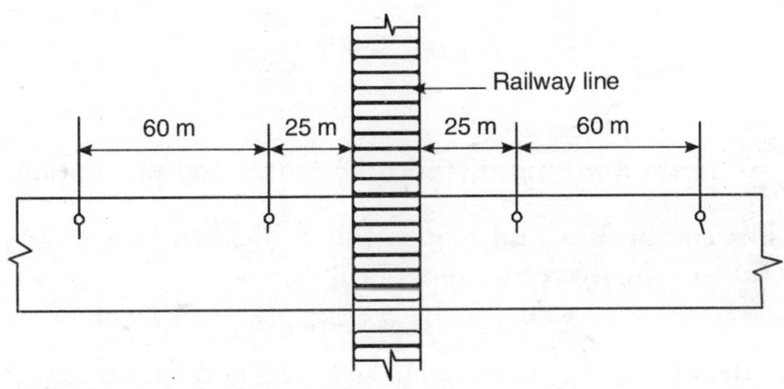

(d) Street light position on rairway crossing

Fig. 12.43

At all the road junctions and roundabouts, it is most important for safety to ensure that sources of light are so placed that a driver cannot only see from some distance away that he is approaching a junction but he can also see the route, he has to follow. At the small roundabouts upto 180 m diameter, a single light centrally mounted at a height of 9 to 10.6 m above the carriageway will be suitable. For bigger roundabouts, the light should be provided along its circumference.

At the mouth of the crossing, a bright patch must be provided. On the intersection of two roads, four light lamps at not more than 12.0 m along each road should be provided.

12.44 HIGHWAY CAPACITY

Capacity of a lane is defined as the maximum flow that can be accommodated with comfort with the increase in capacity, the speed decreases.

The maximum possible flow on a lane is attained at a particular optimum speed. The capacity of a traffic lane can be calculated by the following formula

$$C = \frac{1000\ V}{S}$$

where

V = average speed of vehicles in km/h

S = average spacing between the vehicles (in metres)

C = capacity per hour

The average distance or spacing between the successive vehicles can be calculated by the following formula:

$$S = L + 0.278\ V + \frac{V^2}{254f}$$

Here $\frac{V^2}{254f}$ is the braking distance i.e., after the perception when the driver applies the brakes and brings the vehicle to half. 0.278 V_f is the perception time of the driver to apply brakes.

L = length of vehicle

If the length of vehicle is assumed to be 5.0 m, perception time 2.5 sec and coefficient of fraction to be 0.5, then for an average speed of 50 kMp the capacity of lane for undirecuonal flow will be

$$s = 5.0 + 0.278 \times 50 \times 2.5 + \frac{50^2}{254 \times 0.5}$$

$$= 5.0 + 34.8 + 19.7$$

$$s = 5.95\ \text{m}$$

Hence capacity $= \frac{1000\ V}{s} = \frac{1000 \times 50}{5.95}$

$= 840$ vehicles per hour

Basic Capacity. It is defined as the maximum number of vehicles that can be accommodated or can pass through a lane on a given point per hour. Basic capacity is the theoretical capacity of a lane.

Possible Capacity. It is defined as the maximum number of vehicles that can pass through a lane per hour. The possible capacity is much lower than the basic or theoretical capacity. When the road conditions and traffic conditions are ideal, the possible capacity and basic capacity will be nearly the same.

Practical Capacity. It is defined as the number of vehicles that can pass a given point per hour on a lane without hazard, unreasonable delay or restrictions to the movement of vehicles. The traffic is estimated and the highways are planned on the basis of the *practical capacity*.

Factors Effecting Practical Capacity. The practical capacity of a lane will be effected by the following factors.

Lane and Shoulder Width. As the lane and the shoulder width decreases, the capacity also decreases. Narrow shoulders reduce the effective width of lane and hence reduce the capacity. The practical capacity of 3.0 m wide lane will be 76% of the capacity of a 3.5 m width lane. This shows the importance of lane width.

Clearances. If the driver of a vehicle is not in a position to see on a reasonable length of the road due to obstruction caused by parked vehicles etc. the effective width of a lane is reduced and hence its capacity. Restricted lateral distance will affect driving comfort and chances of accidents will increase. A minimum clearance of 1.85 m is desirable with reduced *lateral clearance,* the capacity of a lane reduces.

Commercial Vehicles. Large commercial vehicles occupy not only more space on the lane but influence the movement of other vehicles on the same lane as well as on the adjoining lanes, thereby reducing the capacity of a lane.

Geometries of the Road. Rough gradients, alignment and other geometric features influence the capacity of a lane. Smoother gradients, proper alignment, camber and surface will add to the capacity of a lane. Steep grades on long stretches reduce the capacity of a lane upto 65%.

Grade Intersections. Intersections do not allow free flow of traffic and hence adversely affect the capacity of a lane. At signalled crossings, the vehicles have to stop alternatively to allow the crossing traffic to pass, the capacity is further reduced. Grade separated intersections are therefore desirable on important roads to maintain optimum capacity.

12.45 LEVEL OF SERVICE

Level of service is the comfort which the driver of a vehicle enjoys while driving a vehicle on a lane. When the driver is free to drive the vehicle at his own speed or free to overtake at his convenience the level of service is fairly high. Operating speed, travel time, traffic intersections, freedom of manoeuvre, driving comfort, safety etc. are some of the factors which effect the level of service. According to Highway Capacity Manual (HCM), the following factors are considered for the level of service.

(i) Speed and travel time, this should include overall time consumed in travelling a particular distance and the operating speed.

(ii) Traffic delays, stoppages, interruptions speed changes necessary to maintain pace with the traffic.

(iii) Driving comfort including manoeuvring and freedom to maintain the desired operating speed with economy of the operating cost of vehicle.

The Highway Capacity Manual (HCM) considers the following two factors for evaluating level of service of a highway.

1. The ratio of service volume to capacity
2. Operating speed

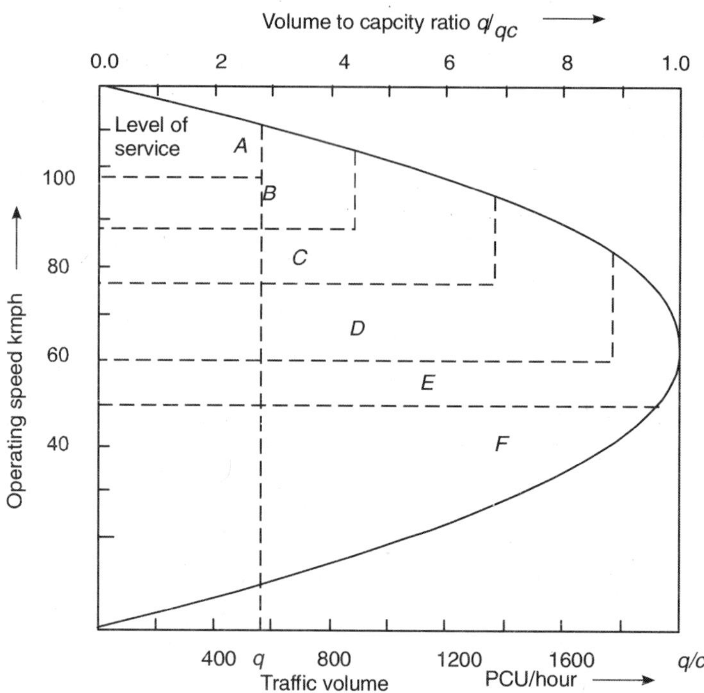

Figure 12.44 *Level of service*

From Fig. 12.44 the level of service is the indication as to how much the driver of a vehicle can service his freedom to travel at their own speed and manoeuvers to overtake, or slow down the speed as is in section A of the figure. In this section volume q to capacity q_0 ratio is very small whereas q/q_0 ratio is very large in section F and hence the level of service will be low. As is very much evident as the volume of traffic increases, opportunities to manoeuvre, will also decrease and so is the driving comfort or level of service. The stream speed of the vehicles decrease to such values which are much lower than the capacity flow condition and there will be practically no flow of vehicles and the traffic will come to nearly a grinding halt.

According HCM, traffic volume that can be served at each level of service as the service volume. For the purpose of design, a particular level of service, will automatically become the design capacity of the lane. HCM does not recommend regarding the applicability of different level of service for the design of different type of highway facility. HCM recommend level of service B for rural or country road and level of service C for urban or city roads. I.R.C recommends a level of service C for rural roads in India.

12.46 PASSANGER CAR UNIT

In our country and in most of the developing countries of the world, there is practically no segregation of different type of traffic such as fast moving traffic, slow traffic and pedestrian traffic. Cars, buses, trucks, scooters, tanga, cycle rickshaw, bullock carts and pedestrian use the same road simultaneously. This heterogeneous flow of traffic have different static and dynamic characteristics. The driver behaviour, his level of education and traffic service is also different for different category of vehicles. A bullock cart driver behaves in a very different way than a car driver particularly owner of car.

For the purpose of estimation of road capacity and density of a lane, passenger car is taken as a unit of measurement and all other type of vehicles are commented into passenger car unit (PCU) thus in drived traffic flow, the traffic volume and density are expressed in terms of PCU. Following equsatance are suggested by IRC

Vehicle type	*PCU*
Car (all types) tempo, autorikshaw, tractor (without trolley)	1.0
Bus, truck, tractor-trailar	3.0
Motor cycle, scooter, cycle	0.5
Cycle rickshaw	1.5
Tanga, Ekka	4.0
Bullock cart	6.0

It is important to note that the passenger car value for different class of vehicles do not remain constant for all times. The important factors which are taken into account for the purpose of analysing PCU value for different vehicles are

1. Average length and width of vehicle.
2. Average speed of the vehicle.
3. Average transverse and lateral gap to be allowed between the vehicles in the designed speed range.

Ministry of Transport and Shipping, Govt. of India have recommended the following PCU for different carriageway width designing :

1. Single lane width satisfactory earth shoulders	1000 PCU/day
2. Single lane road with 1.5 m wide consolidated or brick shoulders	1000 – 2500 PCU/day
3. Two lane roads under normal and favourable conditions	2500 – 10,000 PCU/day

According to HCM, a road under an ideal condition can accommodate the following number of vehicles :

Multi-lane	2000 PCU per hour per lane
Two lane, two way	2000 PCU per hour from both the directions
Three lane, two way	4000 PCU per hour from both the directions

The ideal condition for a traffic are :

1. Un-interrupted flow i.e., no interference from the cross traffic or pedestrian.
2. Only one type of traffic i.e., passenger car.
3. The width of traffic lane is 3.75 m with adequate shoulders and no lateral obstruction at least 2.0 m of the edge of pavement.
4. Clear and un-restricted sight distance.

12.47 PARKING PLACES AND LAY-BYS

For parking the cars 23 sq. m for each car is required including allowance for access roads, irregularities in manoeuvering and for opening of doors. For parking of cars parallel to the road kerb, 2.5 m wide strip is required.

For parking the cars at right angle or diagonally to the kerb, 60 m wide strip is required.

Indian Road Congress has recommended the following minimum parking spaces:

(a) For cars:

(i) 3 m × 6 m when individual parking space is required

(ii) 2.5 m × 5 m when parking is done in lots for community parking.

(b) For trucks :

3.75 m × 7.5 m parking space is required.

Lay-bys. It is the space provided on all the important roads at suitable intervals to enable the vehicles to draw-off the road for temporary parking during break-downs and repairs. In the cities lay-bys are also provided for bus-stops. 30 m × 2.5 m space is usually sufficient for lay-bys.

PROBLEMS

12.1. State the causes of road accidents. What remedies do you suggest for avoiding very frequent accidents ?

12.2. What are the main advantages of road signals ?

12.3. Why are traffic islands provided ? What are the different kinds of traffic islands, which are generally provided ?

12.4. How many types of road junctions are there ? Describe with sketches.

12.5. What are the different types of road signs ? Give example of each with appropriate sketches of each type ?

12.6. Write short note on the visibility at corners, bends and crossings.

12.7. Write short note on the design of highway and street lighting.

12.8. Write short note on the parking places and lay-bys.

12.9. State in brief the various objects of traffic engineering in the contast of planning and development of highway in the country.

12.10. State how speed and delay studies are earned out? What specific purpose do they serve?

12.11. Elucidate traffic manoeuvres to reduce conflict points and minimise accidents.

12.12. Elucidate the following terms :

capacity, weaning length, weaming angle, traffic density.

12.13. State the various type of traffic markings generally used on the highways?

12.14. State the advantages of a traffic rotary. Draw neat sketches of the various type of traffic rotarys.

12.15. Two vehicles *A* and *B* collide with each other at sight angles to each other. They skid through distance of 30 m and 20 m before collision and 20 and 35 m after collision. If the direction of skiding of the two vehicles are 45º and 130° to their original path. Calculate the speed of vehicles before collision.($f = 0.35$).

12.16. State the advantages and disadvantages of traffic signals.

12.17. State the principle of designing interchanges and grade separators.

12.18. At a right angled intersection of two roads *A* and *B*, road *A* has four lanes with a total width of 12 m and road *B* has two lanes with a total width of 6.6 m. The volume of traffic approaching the intersection during hour are 900 and 740 PCU/hour on the two approaches of roads *A* and 278 and 180 PCU/hour on the two approaches of road *B* on the basis of recommendation of IRC, design the signal timings.

12.19. State the various advantages of road signals and what should be their characteristics.

12.20. Distinguish between mandatory and obligatory road signs. Draw appropriate sketches also.

BIBLIOGRPAHY

1. I.C.R Publication on Traffic Engineering No. 39 – 1989, No. 37 – 1989, No. 35 – 1988, No. 33 – 1987, No. 26–1985, No. 28–1986, No. 25 – 1984, No. 16 – 1981, No. 14 – 1980.
2. Geometric Design Standards for Rural Roads – IRC – 1980.
3. Road User Cost Study in India-Final Report – IRC – 1982.
4. Space Standards for Roads in Urban Areas – IRC – 1977.
5. Report on Research Project R-3 scheme P.W.D (B & R), Rajasthan 1979.
6. Highway Capacity Manual – 1965 Highway Research Board Report 87, 1965.
7. Horizontal and Transition Curves I.R.C – 1947.
8. Tentative Guidelines on Capacity of Rural Roads IRC: 64 – 1976.
9. Dimensions and Weights of Vehicles I.R.C – 1973.
10. Recommended practice for the Design and Layout for Cycle Tracks I.R.C. – 1971.
11. Guidelines for Control of Access on Highway IRC: 62 – 1976.
12. Recommended Practice for Rotaries IRC: 65 – 1976.

13. Date on growth of highway traffic on NH-5 in Orissa, National Highway Division Cuttack – 1980.
14. Ribbon Development along Highway and its Prevention IRC – 15 –1974.
15. Alonso W. 'Location and Land Use' Howard University Press, 1964.
16. J.N. Bajpai—Mobility and Land use IRC: 14, 1980.
17. Khanna S.K., Sinha H.K., Arora M.G. 'A Proposed Methodology for Study on Demand for Urban Transportation' IRC – 1980.
18. Kadiyal L.R.– Geometric Design, Highway Engineering – 1984.
19. Giles C.G and F.T. Lander. The skid resisting properties of wet surface at high speeds – J. Roy or Co. U.S – 1956.
20. O' Flaherly – Highways – 1967.
21. Sona A.C., P.V Kulkarni and R.K Sorant Density Madels for work location IRC: 53 – 1982.
22. Kathi B.K, D.A. Shastri and R.H. Pathak Pedestrian Crossings Education IRC 53–1982.

13

Road Specifications

In this Chapter you will study,

• Specifications for the Consolidation of Stone Ballast • Specifications for Laying Soling • Specifications for Surface Painting • Specifications for Cement Concrete Roads

GENERAL

Engineering structures are constructed according to certain standards, which are called specifications. While preparing a project, detailed specifications or particulars for carrying out the work, are drawn, so that the work should be carried out according to these standards or specifications, as they are called. Detailed specifications are of great importance to the engineer and the contractor (or the person executing the work). Unless clear and concise specifications for a particular work are drawn up, it becomes difficult to get the things executed properly in the desired manner. The specifications should include the method of carrying out the work, precautions to be taken and the quantity of material to be used.

13.1 SPECIFICATIONS FOR THE CONSOLIDATION OF STONE BALLAST

1. Before spreading stone metals the old metalled surface shall be thoroughly cleaned of all dirt etc. The holes etc. should be roughly filled. The whole of

the old surface may be scarified if required and the surface may be brought true to template.

2. Two parallel mud walls 20 cm wide and 15 cm high, made of well primmed earth or clay shall be formed along the outer edges of the metalled surface to confine and prevent the new metal from spreading under the pressure of roller. The mud wall shall be properly aligned with flags and the clear width between them shall be the exact width of the new metalling.

3. The stone metal shall then the spread evenly over the surface true to the template. The metal shall be carefully hand-packed, the bigger size metals shall be placed below the smaller size metals in the interstices of the bigger pieces. The templates shall be placed at a distance not exceeding 12 metres in any case, and shall be fixed truly horizontal so as to ensure that both sides of the roads are at dead level. No organic material whatsoever, shall be mixed or spread over before, during or after consolidation.

4. The metal shall then be rolled with a road roller, commencing at the edges and working towards the centre. The metal shall be dry rolled until no lines of the roller are left on the surface i.e., until the metal has been thoroughly compacted together. The surface shall then be saturated with water and re-rolled until thoroughly compacted and no marks are left on the surface by the roller or any chart which may pass over it.

After this, the blinding material be spread evenly over the whole surface, and be of such thickness as well, when watered and rolled, only just fill all interstices. The blinding materials shall not contain clay or any organic matter. After the spreading of the blinding material, the surface shall be brushed slightly with *besom brushes,* backward and forward in front of the working roller and the rolling and brushing shall continue until the surface has become smooth and have been passed by the engineer incharge.

13.2 SPECIFICATIONS FOR LAYING SOLING

Soling shall always be provided except where the road is founded on a very hard natural soil such as rock etc. It shall consist of the hardest suitable material locally available or easily procurable. Stone boulders, kankar, or over-burnt bricks are suitable for the purpose.

The width of soling shall always be 30 cm more than the proposed width of carriageways. The thickness of soling should be as under :

(i) 15 cm if stone boulders are used.

(ii) 10 cm if over-burnt bricks are used.

(iii) 10 cm if kankar is used.

The size of soling material should not be less than 12 cm in any dimension if boulders are used, and 10 cm if kankar or overburnt bricks are used.

Before laying the soling coat, the sub-grade should be thoroughly levelled. Care shall be taken to see that the sub-grade is hard and well compacted and no soft patches or depressions are there while laying, stone boulders or over-burnt bricks, and these too, should be properly hand packed so that no interstices are left. The soling shall then be dry rolled with a road roller, depressions etc. being filled up with shingle, gravel or kankar bajri, sufficiently to give a hard true surface, sloped to the correct camber. Earth shall on no account be used in filling depressions.

13.3 SPECIFICATIONS FOR SURFACE PAINTING

Following is the detailed specification for surface painting :

1. Surface painting shall never be applied on a water bound kankar surface or of soft stone. Old stone or brick surface shall duly be painted when in good condition and free from pot holes and ruts.

2. When consolidation is done with the intention of painting the surface, no blinding or binding coat shall be used.

3. All new surfaces which have not been previously painted, shall be painted in two coats. The first coat may be bitumen cut-backs or bitumen and should be applied hot.

4. The grit shall be either of broken stones or coarse river bajri. The hardest material available at a reasonable price shall be used, irrespective of its shape. For example, broken stones may be of angular shape whereas coarse river bajri may be round or of any shape. For the first coat the grit shall be screened on 2 cm mesh and all the particles must be retained on 6 mm mesh. The grit for the second coat or for repainting an old painted surface, should be screened on a 12 mm mesh.

5. The period which must elapse between the completion of compaction and the commencement of painting depends on the material to be used for a time of one year and the nature of traffic. No painting shall be done in December and January. No hot painting shall be done during the rainy season. Similarly painting with cut-backs or emulsions, not to be done while it is raining or rain appears imminent.

6. In no case painting be delayed for more than three months from the date of completion of compaction or consolidation, and if it is anticipated that more than this period will elapse before the road is fit for hot paint, emulsion should be applied as soon as possible after the consolidation is complete.

First Coat with Emulsion. When the first coat is to be done with emulsion, either by choice or for the reasons stated above, this should be applied within ten days of completion of consolidation, during which time, the surface shall be protected with a blindage layer.

First Coat with Cut-back. The road must be reasonably dry but need not be bone dry right through. Two days dry weather, even during the monsoon will usually suffice to produce sufficiently dry conditions.

7. Before the application of paint, whether it is to be tar, cutback or emulsion, the road surface must be cleaned of all caked mud, crowding, with stiff brooms or wire brushes. The surface is then swept clean with ordinary brushes. Finally and immediately prior to painting, it shall be blown free of all dust by means of gunny bags or if available with mechanical blowers. The dirt and dust removed from the surface shall be collected and deposited at a distant place from the road and away from the prevailing winds. Only that much area of the road shall be cleaned and broomed which can be painted in that day. The final cleaning by blowing shall not precede the application of paint by more than half an hour.

8. A hot paint shall be heated to a temperature not higher than the specified by the manufacturer and in boilers fitted with thermometers. Heating of the paint must be gradual and only that much quantity of tar or bitumen shall be heated which may be used for nearly 30 sq. metres of the road surface. Boilers must be cleaned thoroughly every month.

9. The paint shall be applied to the road surface with specially made pouring cans with wide mouths and of known capacity. To obtain correct and even distribution of paint, the road surface shall be divided into rectangles of known area, each suitable for the contents of one pouring can and then the paint is poured longitudinally and brushed evenly over the surface with brass brooms. Brushing shall always be done from she sides towards the crown. Brooms must be cleaned at the end of the day's work. Nearly 2 kg of binder should be used per square metre of the road surface.

10. As soon as the paint has been spread, in the case of hot paint or as soon as the paint commences to break down, in the case of an emulsion or cut-back, the grit shall be broadcast on to the surface. Before doing so, the grit should be screened into two portions, one containing the larger pieces and the other smaller pieces. The larger grit shall be broadcast first, made into an even layer by hand and then the smaller grit spread over it. About 1.8 to 2 m^3 of grit will be required per 100 m^2 of road surface.

11. After the grit has been spread evenly, the surface shall be rolled with the lightest roller available. Rolling shall be continued only sufficiently to press the grit into the painting material and to fill the interstices of the metal. Excessive rolling crushes the grit and hence should be avoided. During

the rolling coolies should follow the roller, fill the lean surface with extra grit and brush all the extra grit from those places which are above the general level. Close and even packing of grit is essential for a good result. The roller must be driven at the lowest speed and care must be taken not to start or stop the roller with a jerk. The roller shall never be permitted to stop on a part of the surface which has not been finished and should never stop at the same stop twice.

12. As a general rule, the *edges* of the painted road shall be protected by bricks or water bound macadam surface, at least 30 cm wide and 10 cm thick.

13. The second coat of paint shall be applied as soon as die first coat has reached its *optimum condition.* This will be when then the paint has hardened, the surface has become smooth, exhibits a mosaic and all loose grit has been absorbed. The period of time interval, which will elapse between the first coat and second coat of paint will depend on the volume of traffic and on the material used for painting. It is essential that tar should be allowed to harden before being repainting with bitumen as otherwise it remains as a soft layer under the relatively hard bitumen and the road surface becomes uneven. In majority of cases, the time interval between the first coat and the second coat will not be less than three months.

13.4 SPECIFICATIONS FOR CEMENT CONCRETE ROADS

A cement concrete road consists of a cement slab of a specified mixture of suitable materials, laid on a suitable sub-grade. Slabs less than 8 cm in thickness are called thin slabs and those of 8 cm and more are called thick slabs. The top of the sub-grade shall not be less than 30 cm above the anticipated water level in the side drains or the highest flood level of the locality. The following are the general specifications for a Cement Concrete Road :

1. All slabs shall be anchored to the sub-grade ; the surface of the sub-grade shall be brushed so as to remove all fine and loose material such as stone, kankar or moorum, and then the concrete shall be deposited on the surface which has thus been cleaned. If the upper surface is of a water bound macadam, the surface shall be given a wash of cement and water just prior to depositing the concrete.

2. All cements to be used must comply with the Indian Standard Institution Specifications. A certificate of test shall be obtained from the manufacturers. Cement shall be ordered in quantities which can be utilized within one month and can be stored in a dry place close to the work.

3. The water to be used for mixing of concrete shall be clean, free from oils, acids and alkalies. Moreover, it should be *drinkable.* The water to be mixed must be measured properly and only measured quantity of water

shall be added, depending upon the Water-Cement Ratio specified for that particular mix. (In general, the water-cement ratio varies between 0.46 to 0.5).

4. The fine aggregates shall be of such size that they pass through a 5 mm square mesh and shall consist of sand or stone screening and shall be clean, hard, sharp and durable. Before using the sand, it must be washed and then dried. River sand need not be washed.

5. Coarse aggregate consists of crushed stone and shall be clean, hard, durable, free from thin elongated or laminated pieces. *Limestone shall not be used.* The coarse aggregate shall pass through a screen having square meshes whose linear dimensions are not greater than half the least thickness of the slab.

6. The concrete shall be mixed in the specified proportions. Hand mixing shall be resorted to only in emergency works otherwise it is preferable to postpone the work until a mechanical mixer is available.

Specified quantity of sand and cement is measured accurately and mixed together. It is always preferable to measure the quantity of sand in terms of bags of cement. All dry cement and sand shall be turned over by shovels at least three times until the mixture is of uniform colour. The specified quantity of coarse aggregate shall now be added and the whole mixture turned over again for three times. The specified quantity of water shall next be added slowly through a hose pipe attached to a watering can, while the process of turning over with shovels is being carried on. The mixing shall be constructed until the whole batch has reached an even consistency and the mortar is spread evenly through the batch.

As far as possible, machine mixing of concrete should be adopted. The various ingredients of concrete shall be dry mixed as explained above and then fed into the concrete mixture. The ingredients of concrete shall be mixed in the mixer for about 2 to 3 minutes. Measured quantity of water shall then be added. Concrete shall be for not less than $2\frac{1}{2}$ minutes after water has been added and until there is a uniform distribution of the materials and the mass is of uniform colour and consistency. The mixer shall rotate at a speed of 15 to 20 revolutions a minute.

7. Before placing the concrete, the sub-grade shall be sprinkled with as much water as it will readily absorb but there shall be no pools of water standing on it. It is preferable to have the sub-grade sprinkled with water or thoroughly wetted 12 to 36 hours in advance of placing concrete, where such a procedure seems necessary.

8. Concrete shall then be laid in its final position before setting has commenced and shall not subsequently be disturbed. Concrete shall be laid

in two layers if the thickness is more than 10 cm and in one layer if it is to be 10 cm or less. The time to elapse between the laying of two layers shall not exceed 20 minutes.

Concrete shall be laid over the entire width of the slab and between joints in one continuous operation. While being laid it shall be sliced and spaded with suitable tools and shall then be brought to correct camber by means of heavy creed or tamper of solid wood about 25 cm deep and not less than 10 cm wide. The bottom of the tamper shall be shod with sheet iron. During tampering the surface should be carefully inspected for high and low spots and any correction necessary be made by adding or removing concrete.

9. After tampering, the entire surface shall be floated longitudinally with a wooden board not less than 5 metres in length and 20 cm wide with handles at each end. The float shall be operated by a man at each end standing on a suitable platform spanning the road. The float is drawn backward and forward with strokes of about 60 cm and advancing slowly from one side of the road to the other side. When the platforms or bridges are shifted forward, they should be so placed that the next floating overlaps the first one by nearly 1 metre.

10. The surface shall then be finished by using a strong canvas belt not less than 15 cm in width and at least 60 cm more than the width of the slab. The belt is to be applied with a combined cross-wise and longitudinal motion, the strokes being not more than 30 cm but these should be reduced to 10 cm as soon as the *water sheen* has disappeared from the surface of concrete.

11. There shall be kept at the site of work, a number of strips of hessian or burlap sufficient to cover at least two bays of the road. These strips shall have bamboo or light wood strips sawn at each end and as soon as the belting and edge finishing is complete, one of these strips previously damped, shall be placed on the surface of concrete. As the work proceed more strips shall be brought forward and similarly placed so that the moment the work on any portion is finished, the concrete is protected from drying. These strips should be kept moist by light spraying. When the concrete so laid is nearly two hours' old, the strips may be replaced by damp (empty) cement bags.

On the following day, the bags should be removed and the whole surface of concrete shall be covered with earth to a depth of 15 m. The earth shall be thoroughly wetted as soon as it is laid and the whole surface shall then be divided into squares of nearly 2 m length ; which shall be kept saturated with water until 21st day of laying. The earth shall then be removed and the road opened to traffic after 28th day.

12. Immediately prior to opening the road, the joints must be filled and protected. The joints should be filled up with bitumen which should overlap about 5 cm on either side of the joint.

13. After the completion of the concrete slab, the edges shall be protected by soldering of at least 30 cm width and 10 cm thick, water-bound macadam or brick, kankar or similar material.

PROBLEMS

13.1. What do you understand by the term specifications ? What is the importance of specifications in engineering construction ? Explain briefly.

13.2. Write down specifications for laying the soling coat.

13.3. Describe briefly the constructional specifications for a cement concrete road.

14

Earth Work

In this Chapter you will study,

• Earth Work Computations • Methods of Earth Work Computations • Typical Sectional Areas in Hill Roads • Methods of Calculating Earth Work

GENERAL

For the construction of any road either in plains or hills, the irregularity of the natural ground surface must have to be removed. Earth work in its widest sense includes excavations in the soil and rock and formation of banking. Earthwork means cutting the high ground and filling the low areas, cutting of drains etc. (see Chapter 3 on Road Project).

14.1 EARTH WORK COMPUTATIONS

For the construction of a road or a railway track some amount of cutting or filling will have to be done. The amount or quantity of cutting or filling is called total earth work. Finding out the total quantity of earth work is called Earth Work Computations. Cross-secting of a road or railway track in cutting or banking is generally trapezoidal. The total quantity of earth work can be computed by taking the total length along the centre line multiplied by the average cross-sectional area for that particular length of the road or track.

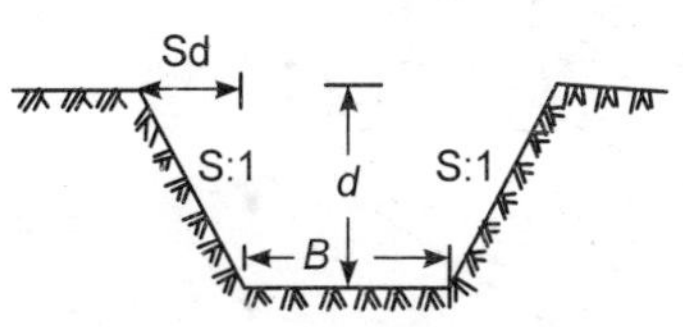

Figure 14.1 *Cutting*

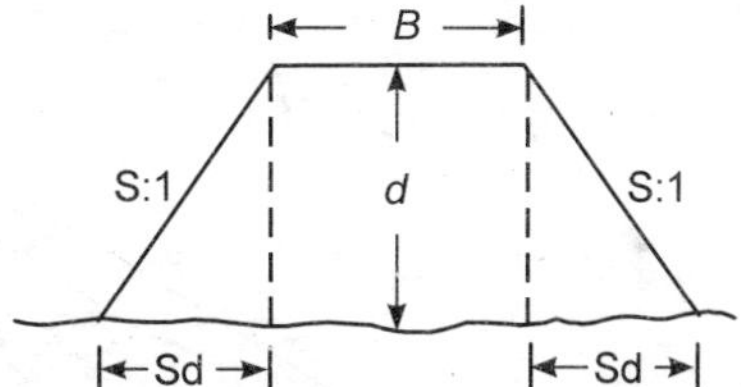

Figure 14.2 *Embankment*

Quantity or volume of Earth work

$$= \text{Average cross-sectional area} \times \text{length}$$

$$= \frac{1}{2}\,(\text{Top width} + \text{Bottom width} \times \text{Average depth} \times \text{Length})$$

or (Area of central rectangle + Area of two side triangles) × Length

$$= B \times d + 2\,(\tfrac{1}{2}Sd \times d)$$

$$= Bd + Sd^2$$

For the calculation of earth work, the longitudinal section of the length of the road on the centre line is taken, and on this longitudinal section, a number of cross-sections are taken at right angles to the centre line. Across-section should be taken at regular intervals and at such points, when the slope of ground changes sufficiently to effect the volume of earth work appreciably. On the longitudinal section, the line of formation of the road should be shown and on every cross-section, the width of formation, the side slope and the depth of cutting or filling should be shown.

14.2 METHODS OF EARTH WORK COMPUTATIONS

There are three methods of computing earth work :

(i) Mid-Sectional Area Method. Volume of earth by this method is calculated by taking the area of cross-section of the middle section by taking mean of the end sections multiplied by the length. Let d_1 and d_2 be the heights of the banks at the two ends of the embankment, b the formation width and $s : 1$ the side slope ($s : 1$ means, s horizontal and 1 vertical).

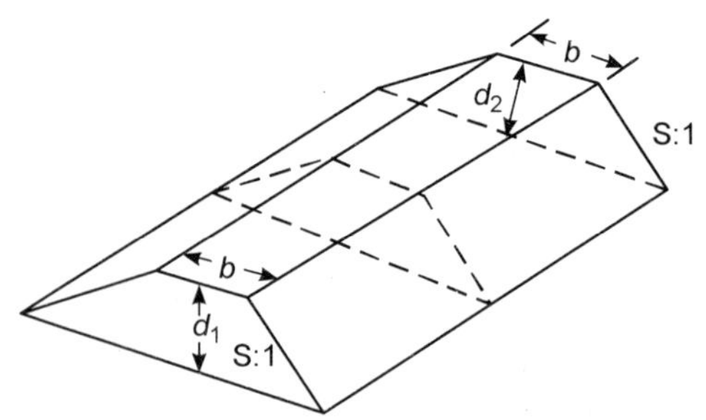

Figure 14.3

Mean height of embankment

$$= \frac{d_2 + d_2}{2} = d_m$$

Area of middle section = Area of rectangular portion + Area of the two side triangles

$$= b.d_m + \frac{1}{2} sd_m^2 + \frac{1}{2} sd_m^2$$

$$= bd_m + sd_m^2$$

Volume or quantity of earth work

$$= (b.d_m + sd_m^2) \times L$$

For calculating the earth work, the following tabular form should be used :

Station or chainage	Depth or height d	Mean depth or height d_m	Area of Central rectangle $b.d_m$	Area of side traingles sd_m^2	Total Area $bd_m + sd_m^2$	Lenght L	Quantity $(bd_m + sd_m^2) \times$ L	
							Fill-ing	Cutt-ing

(ii) Average Cross-sectional Area. According to this method, the area of cross-section of the two ends is calculated and average or mean is calculated and is multiplied by the length to find out the volume of earth.

Sectional area at one end $A_1 = bd_1 + sd_1^2$

Sectional area at the other end $A_2 = bd^2 + sd_2^2$

Mean or average sectional area

$$A = \frac{A_1 + A_1}{2}$$

Volume or quantity of earth work

$$= \frac{A_1 + A_1}{2} \times L$$

The following table is used for calculating work :

Station or chainage	Height or depth	Area of central portion	Area of side triangles	Total sectional area	Mean sectio-nal area	Length	Quantity	
							Fill-ing	Cutt-ing

(iii) Presmoidal Formula Method. In this method the area of cross-section at the two ends and mid-sectional area is obtained and mean depth of cutting or banking is also calculated, then according to the prismoidal formula, the quantity of earth work will be

$$= \frac{d_m}{6}(A_1 + A_2 + 4\,A_m)$$

where

d_m = mean depth of cutting or filling

A_1 and A_2 = Area of cross-section at the two ends

A_m = Mid-section area

This method is the most accurate method of calculating earth work.

14.3 TYPICAL SECTIONAL AREAS IN HILL ROADS

In hilly terrains, when there is a considerable longitudinal slope, the cross-sections can be wholly in cutting, wholly in banking and partly in cutting and partly in banking.

14.3.1 Wholly in Cutting

(i) *Assuming the ground to be level and the section is wholly in cutting.* Let b is the formation width, $s:1$ the side slope of the cutting and d is the depth of cutting at the centre, d_1 and d_2 be depths of cutting at the two ends, then

$$\text{Area of cross-section} = b \times d + sd.d$$

$$= d\,(b + sd) \quad \text{(i)}$$

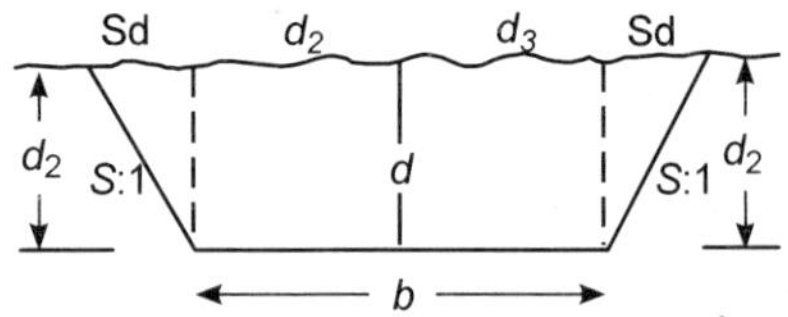

Figure 14.4

When the ground is horizontal

$$d_1 = d_2 = d$$

This formula will also hold good when the formation is in full banking and the ground is horizontal.

(ii) *Assuming the ground to be slopy and have a side long slope of m : 1 (m horizontal and 1 vertical).* From Fig. 14.5, we have

A = Area 1 + Area 2 + Area 3 + Area 4 + Area 5 + Area 6

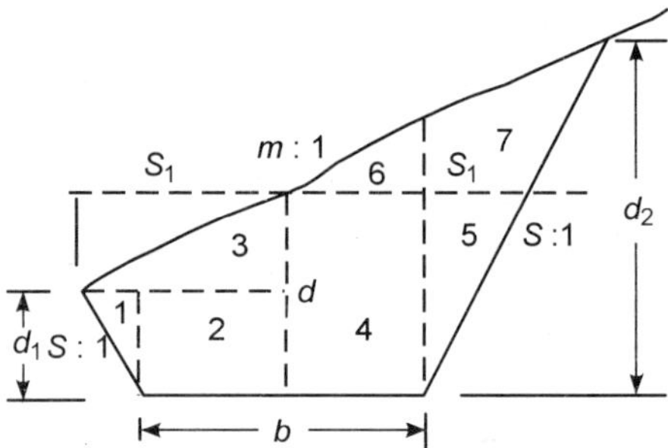

Figure 14.5

$$\text{Area 1} = \frac{1}{2} \times d_1 \times sd_1$$

$$\text{Area 2} = \frac{b}{2} \times d_1$$

$$\text{Area 3} = \frac{1}{2} \times (d - d_1) \quad \frac{b}{2} + sd_1$$

$$\text{Area 4} = d \times \frac{b}{2}$$

$$\text{Area 5} = \frac{1}{2} \quad d_2 \times sd_2$$

$$\text{Area 6} = \frac{1}{2} \times \frac{b}{2} \times (d_2 - d)$$

$$\text{Total area} = \frac{1}{2}\ sd_1^2 + \frac{1}{2}\ bd_1 + \frac{1}{2}\ (d - d_1)\ \frac{b}{2} + sd_1$$

$$+ \frac{1}{2}\ bd + \frac{1}{2}\ sd_2^2 + \frac{1}{2} \times \frac{b}{2}\ (d - d_1)$$

On simplification, we have

$$A = \frac{1}{2}\quad (s_1 + s_2)\quad d + \frac{d}{25}\quad - \frac{b^2}{25}$$

or

$$A = \frac{1}{2}\quad \frac{b}{2}(d_1 + d_2) + d(s_1 + s_2)$$

where

$$s_1 = \frac{b}{2} + \quad d + \frac{b}{2m}\qquad \frac{ms}{m - s}$$

and

$$s_2 = \frac{b}{2} - \quad d - \frac{b}{2m}\qquad \frac{ms}{m + s}$$

where $m : 1$ is the longitudinal slope of the ground.

The same explanation and formula will hold good for banking also.

14.4.2 Partial Cutting and Partial Banking

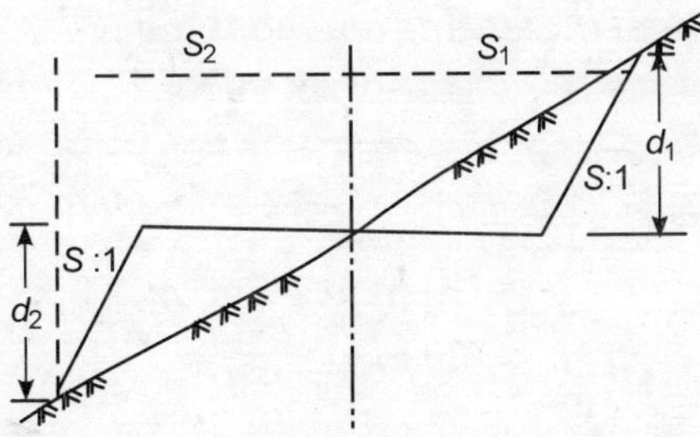

Figure 14.6

The ground has a longitudinal slope of $m : 1$ and the formation is such that a part of it is in cutting and a part in banking. Let d_1 be the depth of cutting and d_2 be the height of banking and d be the depth of cutting at the centre of formation. From Fig. 14.6

$$\text{Area of banking} = \frac{b}{25} - d \quad (s_2 - md) - \frac{b}{5} \; (b - md)$$

$$= \frac{1}{2} \; \frac{b}{2} - md \quad d_2$$

$$\text{Area of cutting} = \frac{1}{2} \; \frac{b}{2} - d \quad (s_1 + md_1) \; - \frac{b}{2s} \; + \; \frac{b}{2} - md$$

$$= \frac{1}{2} \; \frac{b}{2} + md \;\; d_1$$

where
$$s_1 = \frac{b}{2} + \; d + \frac{b}{2m} \quad \frac{ms}{m - s}$$

and
$$s_2 = \frac{b}{2} + \; \frac{b}{2m} - d \quad \frac{ms}{m + s}$$

14.4 METHODS OF CALCULATING EARTH WORK

The earth work quantities can be calculated by either of three formula e.g., mid-sectional area method, mean sectional area method or presmoidal formula method. The following illustrative examples will be found useful.

Example 14.1 *The formation width of a road is 10 m, which is to be constructed on a hill having a slope of 9 :1. The depth of cutting at the centre is 5 m and side slopes of the cutting and banking is 1:1. Calculate the quantity of earth work in horizontal length of 30 m.*

Solution

From Fig. 14.7, we have

$$b = 10 \text{ m}, \; d = 5\text{m}, \; m = 9, \; s = 1$$

$$L = 30 \text{ m}$$

$$ms = \frac{b}{2} + \; d + \frac{b}{2m} \quad \frac{ms}{m - s}$$

$$= \frac{10}{2} + \; 5 + \frac{10}{18} \quad \frac{9 \times 1}{9 - 1}$$

$$= 5 + \quad 5 + \frac{5}{9} \quad \frac{9}{8}$$

$$= 5 + \frac{50}{9} \times \frac{9}{8}$$

$$= 5 + \frac{25}{4} = \frac{45}{4} = 11.25 \text{ m}$$

$$s_2 = \frac{b}{2} + \frac{b}{2\,m} \quad \frac{ms}{m - s}$$

$$s_2 = \frac{b}{2} + d - \frac{b}{2\,m} \quad \frac{ms}{m - s}$$

$$= \frac{10}{2} + \quad 5 - \frac{10}{18} \quad \frac{9 \times 1}{9 + 1}$$

$$= 5 + \frac{40}{9} \times \frac{9}{10} = 9 \text{ m}$$

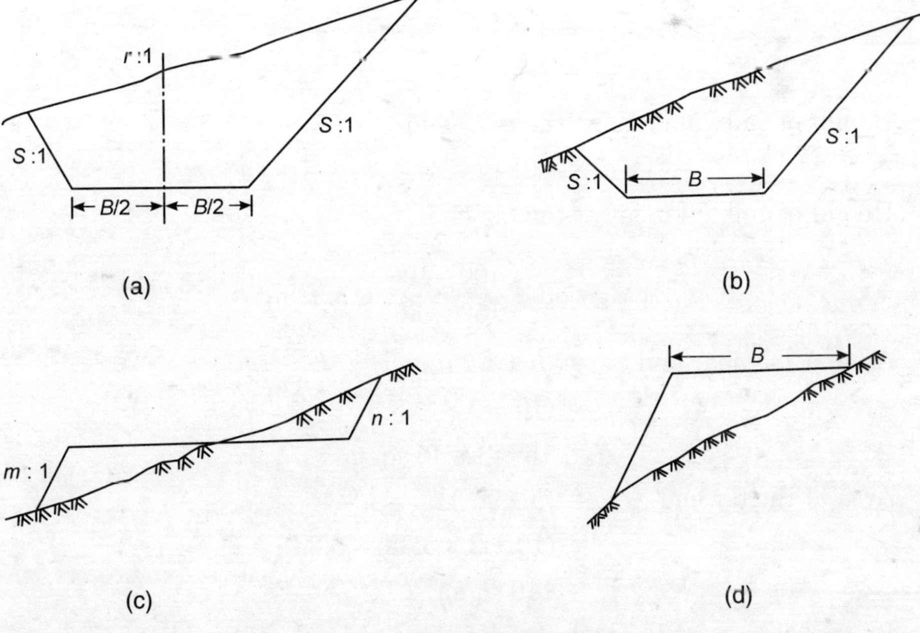

Figure 14.7

Area of banking

$$= \frac{10}{2\times1} - 5 \ (9 - 9 \times 5) - \frac{10}{2\times1} \ (10 \ - 9 \times 5)$$

$$= (5 - 5)\ (-36) - \frac{10}{2} \times - 35$$

$$= 165 \text{ sq. m}$$

Earth work in banking = 165 × 30 = 4950 cu. m.

Example 14.2 *Calculate the quantity of earthwork in embankment, given the following data:*

Height of bank at the near end = 3 m

Upward gradient of formation = 1 in 150

Downward gradient of the ground = 1 in 30

Horizontal length of embankment = 100 m

Width of formation = 10 m

Side slope of bank = 1 in 2

Solution

Here $b = 10$ m

$s = 2$

$m = 3$

Height of embankment at the near end,

$d = 30$

Height of embankment at the far end

$$d_1 = 3 + \frac{100}{150} + \frac{100}{30} = 5.33 \text{ m}$$

Area at the near end $= (b + sd)\ d$

$= (10 + 2 \times 3) \times 3$

$= 16 \times 3 = 48$ sq. m

Area at the far end $= (b + sd_1)\ d_1$

$= (10 + 2 \times 5.33) \times 5.33$

$= 110$ sq. m

By the average end area method volume of earthwork

$$= \frac{(A_1 + A_2)}{2} \times L$$

$$= \frac{(48+110)}{2} \times 100$$

$$= 79 \times 100 = 7900 \text{ cu.m}$$

Example 14.3 *Calculate the quantity of earthwork for a road in banking having a length of 100 metres, on a uniform ground. The width of formation is 10 m and the side slopes are 2 :1. The height of banks at the two ends is 1.00 m and 1.6 m.*

Solution

Volume of earthwork $= (bd + sd^2)\, L$

Here $b = 10$ m

(Mean depth) $d = \frac{1.0+1.6}{2} = 1.3$

$s = 2$

$L = 100$ m

$= (10 \times 1.3 + 2 \times 1.3^2) \times 100$

$= (13 + 3.38) \times 100$

$= 1638$ cu m.

Example 14.4 *Find out the area of the banks of an embankment 10 m wide and 100 m long having a side slope of 2.11. The height of embankment at the two ends is 2.5 m and 3.5 m.*

Solution

Mean height of the embankment

$$d = \frac{2.5+3.5}{2} = 3 \text{ m}$$

Width of slope at the mid-section

$$= \sqrt{2^2 + 1^2 \times d}$$

$$= \sqrt{5} \times 3 = 3\sqrt{5} \text{ m}$$

Area of two slopes $= 100 \times 3\sqrt{5} \times 2$

$= 600\sqrt{5}$ sq. m.

= 1342 sq.m.

Example 14.5 *Calculate the quantity of earthwork for a road in embankment given the following data :*

Length of embankment is 300 m

Width of embankment is 10 m

Side slopes of embankment are 2 :1

The formation level has a downward gradient of 1 in 150 upto 120 m and then the gradient changes downward to 1 in 100 from 120 m to 300 m. The R.L. of formation at the start is 107. The R.L. of ground are :

0	30 m	60 m	90 m	120 m	150 m	180 m	210 m	240 m	270 m	300 m
105.0	105. 6	105.44	105.90	105.42	104.3	105. 0	104.10	104.62	104. 62	103.3

Solution

Here $b = 10$ m and $s = 2$

R.L. of formation at the different points can be calculated from the given gradient of the formation.

The R.L. at the start = 107.00 m

The R.L. at 30 m = 106.80 m

at 60 m = 106.60 m

at 90 m = 106.40 m

at 120 m = 106.20 m

Now the gradient changes

The R.L. at 150 m = 105.90 m

at 180 m = 105.60 m

at 210 m = 105.30 m

at 240 m = 105.00 m

at 270 m = 104.70 m

at 300 m = 104.40 m

Calculations of the earthwork will be computed in the following tabular form.

Length of bank	Height of banking m	Mean height d of banking	Area of Central portion m^2	Area of triangles m^2	Total sectional area m^2	Length L m	Quantity of earth work cu. m.
–	2.0	–	–	–	–	–	–
30 m	1.2	1.60	16.00	5.12	21.12	30	633.6
60 m	1.16	1.18	11.80	2.78	14.58	30	437.4
90 m	0.50	0.83	8.30	1.38	9.68	30	290.4
120 m	0.78	0.64	6.40	0.82	7.22	30	216.6
160 m	1.60	1.19	11.90	2.83	14.73	30	441.9
180 m	0.60	1.10	I1.00	2.42	13.42	30	402.6
210 m	1.20	0.90	9.00	1.62	10.62	30	318.6
240 m	0.38	0.79	7.90	1.25	9.15	30	274.4
270 m	0.70	0.54	5.40	0.58	5.98	30	179.4
300 m	1.10	0.9	9.00	1.62	10.62	30	318.6

Total Earth Work – 3,513.6 cu.m

PROBLEMS

14.1. A cutting is to be made for a line of uniform gradient passing through the first and last points. Calculate the gradient and the volume of earthwork width of formation is 10 m and cutting slope is 1 ½: 1.

14.2. Calculate the volume of earth work by prismoidal formula in a road embankment with the following data :

Chainage	0	30	60	90	120
Ground level	60.1	60.4	61.02	60.5	61.2

Formation level at chainage 0 in 60.3 m, width is 10 m and side slope 2 :1. The gradient of formation 1 in 100.

14.3. A cutting is to b: made through the ground where cross slope varies considerably. At p the depth of cutting is 4 m at the centre line and cross slope is 8 : 1. At Q the depth is 3 :1 m and cross slope is 12 : 1 and at R 4.3 m and 10 : 1 PQ and QR are 50 m apart the formation width is 10 m and side slope $1\frac{1}{2}$: 1. Calculate the volume of cutting.

15

Waterlogging of Roads

In this Chapter you will study,

- Causes of Waterlogging • Harmful Effects of Waterlogging • Anti-Waterlogging Measures • Precautions to be Taken While Constructing Roads on Waterlogged Areas • Embankment on Marshy Land

GENERAL

Waterlogging is the effect or damage caused to a sub-soil due to excessive sub-soil moisture. The problem of waterlogging has taken a serious turn in our country on account of increase in the irrigated areas and shift on the emphasis from protective irrigation to intensive irrigation for increasing the agricultural production. A large area of roads passing through agricultural fields is greatly affected by waterlogging. Waterlogging is mostly due to the creation of irrigation channels in the locality of roads without considering adequate and proper drainage of seepage water. *A land is said to be waterlogged when the soil pores are effectively saturated due to lack of aeration of the soil.*

15.1 CAUSES OF WATERLOGGING

Waterlogging is the result of rising of water table due to the following reasons:

1. Due to increased inflow of water in the sub-soil of embankment or sub-grade of the road surface from the adjoining areas.

2. Due to constant inflow of seepage of irrigation channels from the adjacent upper region.

3. Due to inflow of seepage water from the adjacent tanks, water reservoirs etc.

4. Due to percolation of water from improper surface drainage or sub-surface drainage along the road-side.

5. Due to lateral flow of sub-soil water raising the water-table on the upstream side due to impervious obstruction.

6. Due to poor natural drainage i.e., flat terrain with depression leads to detention of storm water which adds to raising the water-table and hence waterlogging.

7. Due to underground ridges and barriers perched water-table aquefers.

15.2 HARMFUL EFFECTS OF WATERLOGGING

Water–logging raises the water-table of the area which in turn helps the moisture to rise due to capillary action and thereby reducing the bearing capacity of soil. As a result of this the sub-grade is deformed and unequal or excessive settlements take place which make the road surface undulated and unfit for traffic movement. Waterlogging produces some deterimental salts such as sodium sulphates, magnesium sulphates and sodium carbonates etc. which are very dangerous to the embankment of roads.

15.3 ANTI-WATER-LOGGING MEASURES

The anti-waterlogging measures may be preventive as well as curative.

15.3.1 Preventive Measures for Waterlogging

Waterlogging may be prevented by the following methods :

1. Reducing percolation from canals, by lining the canals, lowering the full supply level (F.S.L.), providing intercepting seepage or parallel drains etc.

2. Reducing the sub-soil flow of the ground water by providing artificial seepage drains, providing natural seepage drains, pumping sub-soil water by *well point system*.

3. Reducing percolation of irrigation water by providing intercepting drain and by improving surface drainage of roads and embankments.

15.3.2 Curative Measures for Waterlogging

Curative measures constitute extraction of excessive water from the sub-soil of roads and embankments. Actually speaking when the land adjoining a roadway or railway or an embankment has become waterlogged, it becomes really difficult to eradicate water and prevent waterlogging. Water is mainly extracted from the waterlogged areas by well point system. Under this system vertical perforated pipes are driven at short distances along the road-side or embankment. The top of these vertical pipes is connected by a horizontal pipe, through which the water is sucked up by pumps. This will help in lowering the water-table and hence waterlogging.

15.4 PRECAUTIONS TO BE TAKEN WHILE CONSTRUCTING ROADS ON WATER-LOGGED AREAS

Waterlogged areas pose a serious problem of constructing roads particularly highways and expressways. Under the same conditions of traffic and pavement thickness, the deformation of the road surface is directly proportional to the moisture content of the sub-soil or sub grade of the road with higher moisture content of sub-grade, there will be more undulations in the road surface due to excessive and unequal settlements. Due to the formation of salt such as sodium sulphates, magnesium sulphates and sodium carbonates, the pavement surfaces disintegrade.

While constructing roads on waterlogged areas the following precautions should be taken to prevent any damage to the road :

1. The sub-grade should be thoroughly compacted.

2. The thickness of crust should be designed properly so as to rectify unequal settlements of sub-soil.

3. Roads in cutting should be avoided as far as possible.

4. Deep and narrow pits should be constructed and should be joined with some adjacent stream or nallah. These barrow pits will suck moisture from beneath the road and help in lowering the water-table and hence water-logging.

5. For the construction of bitumen roads, a thin layer of bitumen should be sprayed over the sub-grade to seal the pores of soil and prevent rising of moisture by capillary action. Due to this moisture, the binder gets stripped off, thereby reducing the strength of the road.

15.5 EMBANKMENT ON MARSHY LAND

Construction of embankment on marshy land used to pose a real big problem. The reason being that the bearing capacity of the soil is almost negligible and to get a dry compact embankment is real task. Before starting the construction of an embankment on a marshy land, the pre-requisite is that the sub-soil water must be drained off completely either by well point system or by constructing lateral drains. After the water has been drained off to a reasonable limit, the construction of embankment is started.

A fascine consisting of bundles of green wood of diameter 2.5 cm to 4.0 m tied at intervals into bundles of 15 to 18 cm diameter and placed at a distance of 1 to 1.2 m apart, is placed over the base of the embankment. The green wood and the rope should be such that it is not easily spoiled in the wet soil. The fascine which is just like a raft is filled with sand covered with the branches of trees. Now over this green wood raft or fascine, filling for embankment is started. For the stability of this bank, the water-table should be kept as low as possible.

PROBLEMS

15.1. What is waterlogging and what are its ill-effects.

15.2. What are the protective and creative measures of waterlogging?

15.3. Elucidate the various causes of waterlogging.

16

Highway Maintenance

In this Chapter you will study,

• Maintenance Systems • Classification of Maintenance • Causes of Failure of Pavement Surfaces • Analysis of Flexible Pavement Failures • Analysis of Rigid Pavement Failures • Prevention of Skidding • Evaluation of Road Surfaces • Overlays • Flexible Overlays for Over Cement Concrete Road • Maintenance of Bituminous Surfaces • Repair of Highway Drainage

GENERAL

Maintenance is defined as preserving and keeping the serviceable conditions of a structure as normal as possible, so as to maintain its original condition as constructed or as subsequently improved. Maintenance is an important activity which helps in providing better service facilities, longer life and better appearance. Highway location, design and construction have a bearing on the maintenance cost. At the time of alignment studies, if proper consideration is given to drainage problems, soil conditions, directness of route, landslide problems etc., it will generally reduce maintenance cost and maintenance problems. The maintenance of a highway will include (i) maintenance of surfacing, (ii) maintenance of shoulders and footpaths, (iii) maintenance of underground and surface drainage, and (iv) maintenance of culverts and other structures.

16.1 MAINTENANCE SYSTEMS

All types of roads require maintenance. The type and extent of maintenance will depend on the serviceability standards, availability of fund and the priorities. On account of diversified operations which are interlinked, an appropriate system of management should be there for effective implementation of maintenance programme. Following are the various factors involved in the management systems:

1. Routine surveys are to be conducted to assess the maintenance needs.

2. To fix up priorities of maintenance depending on the availability of funds.

3. To fix up minimum serviceability standards for different type of roads depending upon the type, density and volume of traffic.

4. To workout different alternatives to meet up the maintenance requirements.

5. To phase out the availability of materials, maintenance gangs and optimum use of tools and equipments.

6. To decide various factors influencing the maintenance needs for example, sub-grade, drainage, environment traffic etc.

16.2 CLASSIFICATION OF MAINTENANCE

The highway maintenance are divided into three categories:

(a) Routine maintenance

(b) Periodic maintenance

(c) Special repairs.

(a) Routine Maintenance. After knowing the possible needs for a minimum serviceability standard and to improve the life of a road, routine maintenance is essential. The proverb 'stitch in time saves nine', holds good here for increasing the serviceability of road. Since the road surface is subjected to jerks and vibrations due to moving traffic and is exposed to the diverse atmospheric conditions, the wearing and process is very quick. The routine maintenance consists of :

(i) Cleaning and upkeeping of side drains and cross drainage works. This is more important for city or urban road and is also very important for all type of roads before and during the rainy season.

(ii) Upkeep of carriageway–The entire network of roads is divided geographically and each section of the road/roads is allotted to different engineering departments for maintaining the carriageway by repairing patches and potholes, edges and junctions.

(iii) Maintenance of shoulders, foot paths, pedestrian crossings, road signs, signals etc. by conducting routine checks.

(b) Periodic Maintenance. Periodic maintenance is different than the routine maintenance. Periodic maintenance is done periodically at regular intervals of time. The time period of periodic maintenance depends on the serviceability standard to be maintained, type of surfacing, type and intensity of traffic, weather conditions, availability of funds. Periodic maintenance is needed for all type of roads irrespective of mode and standard of construction. Even if a road is not being used routinely it still requires maintenance. Maintenance schedule is prepared and is followed strictly for maintaining standards and increasing the life of roads.

(c) Special Repairs. Due to natural calamities or due to structural failures, some special type of repairs to be done. This may consist of overhauling the entire road length or a portion of it, re-surfacing or re-conditioning of the old surfaces particularly of W.B.M. road, changing or modifying camber and gradients, relaying of shouldering, widening of carriage, *overlaying* over the existing pavement due to increased traffic load etc. comes under the category of special repairs. Repairing cracks and joints of the rigid pavements and removing un-evenness of flexible payments can also be termed as special repairs.

16.3 CAUSES OF FAILURE OF PAVEMENT SURFACES

Following are the general causes of the failure of road surfaces:

(i) Poor quality of material used either in the surfacing or base course. The material generally get crushed under the wheel of the traffic and disintegrated.

(ii) Constructional deflects and lack of quality control. Due to poor supervision and ignoring standard specifications will lead to pavement failures.

(iii) Inadequate surface and sub-surface drainage is also responsible for the structural failure of roads. Stagnation of water on the road surface and its subsequent penetration into the sub-grade will cause failure.

(iv) If the wheel load is increased beyond the designed capacity, the road surface will give way.

(v) Settlement of base, sub-grade or embankments.

(vi) Natural calamities such as heavy snow fall, frost, heavy rains and floods, earthquake etc. lead to pavement failure.

A flexible pavement failure is by the formation of corrugation, pot holes, ruts, cracks, depressions and by general settlements. Pavement failure is a complex problem and several factors are responsible for this. The aging and oxidation of bituminous film lead to the deterioration of surfaces and subsequent failure as the water will have a direct access to the foundation. The rigid pavement of cement concrete develops cracks due to repeated loads and fatigue stresses. A rigid pavement failure is observed by the development of structural cracks and discontinuity of surface. Moderate irregularities in the supporting layers beneath the cement concrete layer are sustained due to inherent bending properties of cement concrete but excessive bending stresses cause structural failure.

16.4 ANALYSIS OF FLEXIBLE PAVEMENT FAILURES

After analysing the various types of failures structural and otherwise, it has been observed that the failure of flexible pavements is due to :

1. Failure of sub-lease or base
2. Inadequate stability or strength
3. Loss due to bending properties
4. Inadequate wearing surface

1. Failure Due to Sub-base or Base. The failure of sub-base or base is attributed to soft sub-grade, excessive water content, excessive wheel load, unequal settlements and up-heavals. The sub-base or base failure may be due to inherent weakness of the material, improper compaction or crushing of the base material.

2. Inadequate Stability or Strength. Poor mix proportioning or inadequate thickness are the primary seasons for lack of stability. Soft varieties of stone aggregates cause instability of the base. Lack of supervision and quality control during laying and compactor will cause weakness of base.

3. Loss of Bending Properties. Due to excessive wheel load repetition the aggregates move laterally. Due to this lateral escape of aggregate materials the composite structure of the base course is disturbed which results in loosening of aggregates. This loosening of aggregates can be seen on the bituminous surface by alligatory or map like cracks. Due to this permanent deformations occur.

4. Inadequate Wearing Surface. The wearing course strength depends on sub-grade, base or sub-base strength and vise-versa. If this thickness of wearing course is inadequate or due to the absence of wearing coat, the base or sub-base is exposed to direct damaging effect of the atmosphere such as rains and frost action. This shows that not only the base or sub-base is responsible for the strength and stability of the wearing surface but the wearing surface is also responsible for the strength of the base.

16.5 ANALYSIS OF RIGID PAVEMENT FAILURES

The analysis of the failure of rigid pavements show that failure in cement concrete road is due to the following reasons :

(i) Inferior or soft aggregates which get crushed under the wheel load particularly under iron wheels.

(ii) *Poor construction and workmanship.* The strength of cement concrete depends on proper proportioning, mixing, water cement rates and compaction. Poor workmanship, improper curing, poor construction of joints and their filling will ultimately result in pavement failure.

(iii) *Poor surface finish.* Improper finishing of the pavement surface will lead to disintegration of concrete, honey combing and ultimate failure.

(iv) Curving is one of the most important factors which attribute to the overall strength of concrete. Concrete pavements are exposed to the atmosphere and variation of temperature due to day and night temperature will cause shrinkage and warping if sufficient and proper curving is not done.

Following defects are observed due to the above reasons:

(a) Formation of cracks

(b) Spalling of joints

(c) Slippery and un-even surface

*(d) Mud pumping.

16.6 PREVENTION OF SKIDDING

For traffic safety on highway it is essential that the pavements have skid resistance property. The skid resistance or the friction of the pavement surface may be measured using pendulum type friction recorders or some other instrument. On the basis of statistics we can say that 40% of accidents occur due to skidding of vehicle. Skidding cannot be prevented completely by any method but can be reduced. Following are some of the reasons of skidding :

(a) Water (b) Clay

(c) Sand (d) Dust

(e) Oil and Grease.

---

* Mud pumping is the process of mud ejection through the joints and cracks.

Investigation show that sufficient degree of skidding resistance can be built into the road surface and can be maintained through proper construction and maintenance.

Skidding is of three types viz., straight, sideways and impending. Straight skidding occurs where the brakes are suddenly applied. The vehicle moves or skids in the direction of the travel. Sideway skidding occurs on curves due to lack of super-elevation and inadequate coefficient of resistance. Impending skidding is encountered when the braking is applied. Gradual skidding can be measured by

(i) Pendulum type skid friction recorders

(u) Skid testing device attached to a test vehicle and

(iii) Dynamic skid testor lowered by another vehicle.

If the roads are properly maintained and kept face from dust and water, sufficient amount of skidding can be reduced. Rough surfaces of W.B.M. gravel and cement concrete offer sufficient skid resistance. Presence of water film and the smooth surfaces of aggregates influence the skid resistance property. Wheels of the vehicles do the polishing of aggregates and make their surfaces smooth. Excessive bitumen bleeding and smooth aggregates particularly when wet produce skid prone surface. Renewal of wearing surface, spreading of grit on the excessive bleeding surfaces is the only remedy for preventing skid. At the time of construction care should be taken to see that the wearing surface is not made too smooth and excess use of bitumen is avoided. Proper maintenance and keeping the surface dry and free from dust can prevent sufficient amount of skidding.

16.7 EVALUATION OF ROAD SURFACES

For effective planning of maintenance and strengthening of pavement surfaces, it is essential to evaluate the pavement for composition, sub-grade support, traffic volume and density. There are various methods for evaluating the pavements.

For structural evaluation of flexible and rigid pavements plate bearing tests are conducted. It is assessed by the load for a specific deflection of the plate. Investigations show that the performance of flexible pavements is closely related to the elastic deflection of materials measurement of deflection under the wheel loads serves as an indeed to carry a specified traffic load. *Benkelman beam* devised by A.C. Benkelman of U.S.A. is generally used for measuring the deflection. The Benkelman beam is a simple and easy to operate method of measuring deflections.

(ii) Flexible overlay over an existing rigid pavement of cement concrete

(iii) Rigid over of cement concrete over the existing rigid pavement of cement concrete and

(iv) Rigid cement concrete overlay over the existing flexible pavement of bitumen.

16.8.1 Method of Constructing Overlays

The overlay should be properly planned and designed before launching the project. The overlay designed is different than the design of pavements. For the design of overlays the Indian Road Congress have given certain guidelines. The guidelines are measurement of deflection by Bankelman's Beam. The deflection measurement are very sensitive to temperature, the I.R.C. has recommended that looking to Indian climate, the deflection measurements are related to standard temperature of 35° C. Reading are taken when the temperature is above 30°C. The correction of deflection measurement is 0.0065 mm for each degree of temperature greater or less than 35° C. Deflection also vary with the climate and moisture content of sub-grade. Correction factors for dry and wet sub-grade vary from 1.2 to 2.0 for clayey and sandy soils I.R.C. recommends that total design thickness of the pavement the expected design traffic volume is determined. The thickness of the overlay is found by subtracting the existing thickness of pavement.

As discussed in Art. 16.7, the rebound deflection values D_0, D_1 and D_t are computed. The outlay thickness may be determined by Ruiz's equations

$$h_0 = \frac{R}{0.434} \log_{10} \frac{D_c}{D_a}$$

where

h_0 = overlay thickness

R = deflection reduction factor depending on the overlayed material (usual value for bituminous overlays range from 10 to 15 and the average values may be taken as 12)

D_a = allowable deflection which depends upon the pavement type and the designed life and is equal to 0.75 to 1.25 mm. The I.R.C. suggests the following formulae of overlay thickness.

$$h_0 = 550 \log_{10} \frac{D_c}{D_a}$$

When bituminous concrete or bitumen macadam surface is provided as the overlay surface, an equivalently factor of 2.0 is suggested.

When laying overlay, the existing surface should be properly prepared by cutting diagonal grooves at least 60 mm deep and 900 to 1200 mm apart. The grooves should be properly cleaned and conditioned to receive the overlay layers which are subsequently compacted. If the thickness required of the overlays is more than 100 mm two layers of overlay should be used. Each layer of overlay should be properly compacted.

16.9 FLEXIBLE OVERLAYS OVER CEMENT CONCRETE ROADS

Flexible overlays can also be provided over rigid pavements of cement concrete slab roads. Following are the I.R.C. recommendations for providing flexible overlays over cement concrete surfaces :

1. For heavy traffic of 1500 C.V.D. 7.5 cm Bituminous macadam over 4 cm asphaltic concrete 15 cm granular or layer under 4 cm asphaltic concrete.

2. For 501 to 1500 commercial vehicles/day 7.5 cm bituminous macadam under 2 cm premixed carpet with seal coat.

or

7.5 cm built up spray under 2 cm premixed carpet with seal coat.

or

7.5 cm granular layer under 4 cm asphaltic concrete with seal coat.

The concrete surface should be properly recokened and cleaned before applying overlay.

It is suggested that 3–5 cm groove depth at regular intervals of 45–60 cm should be made on the road surfaces. The grooves should be inclined at 45° to 60° to the edge of the road, crossing each other so as to form a mesh type of pattern. The grooves then should be cleaned with a wire brush. Hot bitumen should then be spread so as to wet the surface and then the overlay should be laid.

For area of high intensity of rainfall and unfavourable drainage conditions, following are the recommendations:

(i) For 1500 C.V.D.

10 cm bituminous macadam under 4 cm asphaltic concrete and seal coat.

(ii) For 501—1500 C.V.D.

7.5 cm bituminous macadam under 4 cm asphaltic concrete and seal coat.

For areas of heavy rain, poor drainage facilities, high plasticity sub-grade and very heavy commercial vehicle traffic.

(i) 5 cm coated macadam with 40 mm single size aggregate mixed with 2.5–3.0% bitumen under 7.5 cm bituminous macadam with 4.0 cm asphaltic concrete as wearing surface.

or

(ii) 11 cm bituminous macadam with 4 cm asphaltic concrete as a wearing course.

16.9.1 Maintenance of Surfacing

Whatever type of road surface it may be, the maintenance should be a preventive maintenance. For important highways and busy lanes, greater is the importance of preventive maintenance. Defects in the road surface should be corrected as soon as they appear so that they should not become too serious to be remedied quickly. This will reduce the maintenance cost and keep the road surface in good running condition. Surface maintenance is more important than any other maintenance because it is only the surface which is being used by the traffic. About 50% or more of the maintenance budget is spent on the maintenance of road surface only. Most of the maintenance work is done departmentally by employing work-charge staff on muster-roll as the maintenance work is so diverse, so subject to variation that it does not lend itself to competitive bidding.

16.10 MAINTENANCE OF BITUMINOUS SURFACES

Maintenance of bituminous surfaces consists in repairing pot holes and ruts, surface treatment, resurfacing etc.

16.10.1 Repair of Pot Holes, Ruts and Patches

Patch repair may be needed for the damaged portions of a road, due to pot holes etc. Sometimes patching is also required to remove inequalities in shape and surface and removing waviness in order to make the surface smooth riding. Pot holes are formed due to poor gradation of surface materials. Corrugation or waviness is caused due to incorrect gradation of aggregates, excessive bitumen in the mix, traffic excessiveness or overloads. Corrugations are caused due to unequal settlement of sub-soil or moisture in the bitumen mix.

Pot holes are cut in square or rectangular shape and the loose material is removed. These are the cleaned by a wire brush. Bitumen is heated and applied on the cut portion of the pot holes. After applying bitumen, patch mix, consisting of bitumen and grit or stone ballast, is filled and thoroughly compacted. The surface of the patch should be a little higher than the normal surface of road so as to allow for compaction by the traffic.

Sometimes pot holes occur on more than one-third of the surface ; it is preferable to resurfacing the entire road surface than to repair the patches. (The method of repainting or resurfacing has been discussed in the chapter on Construction of Roads.)

16.10.2 Repair of Base-course

Flexible pavements may cause structural failure due to inadequate thickness of base, weak sub-grade and sub-base. Excessive moisture or unequal settlement may also result in the failure of base. Poor consolidation of the base, sub-base or sub-grade may cause failure too. Constant percolation of surface water may also be responsible for the failure of base. Before undertaking the repair of base course, investigations by trenching or visual inspection should be done. Repair work should not be started unless the cause of the failure is not determined.

If the failure is due to improper surface drainage which helps in the percolation of water to the base course, proper drainage should be provided so as to facilitate easy removal of surface water. This will help in avoiding unnecessary expenditure.

However, if it is decided to increase the base thickness, the whole surface is loosened by scarifying to the full depth. All loose materials are removed, sub-grade is rolled and a new base or sub-base of the required thickness is laid and rolled. Now the new base is covered with suitable surface treatment or a premixed carpet.

16.10.3 Repair of Bleeding Surface

If the surfacing consists of excessive bitumen, the surface will become slippery during rainy season and will bleed during hot weather. Bleeding will cause unnecessary inconvenience to the traffic as the bitumen will stick to the tyres, corrugation or rutting in the surface of the road. Bleeding normally occurs just after the construction of the road. Repair of bleeding surfaces is also called *Surface Treatment.* Bleeding can easily be corrected by spreading a layer of dry coarse sand in a thickness varying from 5 mm to 10 mm and rolling the surface.

16.10.4 Repair of Cement Concrete Roads

The maintenance of rigid pavements consists (i) in filling and sealing joints and cracks in pavement surfaces, (ii) patching of damaged areas, (iii) repairing of blow-ups.

(i) *Sealing of joints and cracks.* Joints and cracks will allow water to percolate to the base and sub-grade. Prevention of infiltration of surface water to reach sub-grade under pavement and collection of water in cracks and joints, will help in improving the life of the road, its serviceability and reducing the maintenance cost. It is for this reason that the joints and cracks should be properly sealed and filled. Before filling the joints, the joints should be cleaned. Sometimes compressed air is used to clean the joints before filling or sealing. For better and proper sealing, the old sealing material should be removed. The sealing material consists of bitumen and sand, cork, impregnated fibre board or a ready-made sealing compound.

(ii) *Patching of damaged areas.* Patching of damaged concrete surfaces is a very difficult job and cannot be done satisfactorily. Small patches and holes are usually filled with bituminous materials. The method of filling and repairing is the same as already explained.

(iii) *Repairing of blow-ups.* Due to longitudinal expansion, the transverse joints are buckled up i.e., the ends of the slab are raised up. This is known as blow-up. These blow-ups are sometimes too severe and the pavement remains in a raised position unless broken down. Blow-ups are a traffic hazard and should be repaired as soon as they appear on the surface. The blow-ups are repaired by removing the damaged portion of the pavement and replaced with a patch of concrete or bituminous materials. Sometimes the blow-up is temporarily repaired by road-side soil which has the advantage of allowing the pavement to settle down to its original position under the traffic load itself. After that the permanent patch is applied.

16.10.5 Repair of Shoulders and Foot-Paths

Shoulders and foot-paths constitute a very important part of a highway for the following three reasons, viz., (i) provide additional space on the carriage-way for the traffic to be used in emergency, (ii) protects the edges of the road thereby providing stability to the road surface, and (iii) provide walk for the pedestrians, cyclists etc. thereby reducing congestion on the road. Unless the importance of shoulders is understood properly the importance of its maintenance will remain half-way. It is desirable that there should be a colour contrast between the shoulder and the road surface. While maintaining shoulders the function of shoulders should be kept in mind. Shoulder and berms should be repaired as soon as they are damaged.

Foot-paths are raised platforms on one side or both sides of the road, for the use of pedestrians. Foot-paths are important feature of the urban roads. Foot-paths reduce congestion on the carriageways and thereby reducing chances of accidents. Therefore it is very important that foot-paths should be properly maintained and repaired.

16.11 REPAIR OF HIGHWAY DRAINAGE

Proper maintenance of highway drainage system means removal of water from highway surface and appurtenances as quickly as possible so as to keep the highway in good running condition. For efficient surface drainage the road surface should be smooth and should have proper gradients and camber. The water should be drained off smoothly to the side drains through the shoulders. The shoulders should be flush with the pavement so that water while flowing from the road surface should not enter the joints and thereby harming the sub-grade or base.

The side drains should be cleaned and properly maintained. The drains should have sufficient capacity to take the run-off effectively.

In urban roads sub-surface drainage (underground drains) also contribute a lot to the maintenance problems. The maintenance of sub-surface drains consists of maintaining and cleaning of manholes, inspection chambers, lamp holes etc. Periodic inspection of these appurtenances of the sub-surface drainage will reduce maintenance problems and hazards.

PROBLEMS

16.1. What is maintenance and repair of highways? Also elucidate the importance of maintenance.

16.2. What are the essential features of maintenance of roads? Elucidate in brief.

16.3. How will you justify that the maintenance of shoulders directly affects the functioning of highway ?

16.4. What are the causes of pot-holes and ruts and how are they repaired in a bituminous road ?

16.5. What are the points to be considered while maintaining surface and sub-surface drainage of highways ?

16.6. State the causes of failure of flexible pavements in our country. How varies and corrugated are formed ?

16.7. State the methods of strengthening of existing pavements. How overlays are designed ?

16.8. Elucidate the method of deflection by Benkelman's beam for designing overlays.

17.1 DESIGN CONSIDERATIONS

The purpose of pavement design is to produce a soil-structure system that will carry any traffic smoothly and safety with minimum cost. The performance of flexible pavements depends to a large extent on the strength properties of sub-grade as the slab action in layers is almost negligible, when subjected to vehicular load on the edge of the pavement. Pavement base failure as well as structure failure takes place due to shear. The sub-grade failure is accompanied by small amount of surface movement outside the pavement.

The increased axle load and the growth of traffic warrant has much importance in design, construction and maintenance of shoulders, as is necessary for pavement structures. The shoulders provide lateral support to the pavement as well as safety to vehicles. If the proportion of commercial vehicles is more than other traffic as is the case in almost all developing countries, the damage to the pavement is more from the heavy axle. In many situations the heaviest 10 per cent axle load cause more damage to a greater extent than all other combined traffic.

Bituminous base has better load spreading property. Moreover, the price of bitumen has increased more than six times in the last 10–12 years, therefore for the construction of new roads for very heavy traffic too, bituminous base should be avoided if possible. The study of the confinement effect in the pavement structure has been subject of interest to the highway engineer. In the flexible pavement design the elastic modulii of different components of the pavement, is an important parameter, its variation under different confinement conditions will help in developing the design methodology.

Terzaghi mentioned that in granular soils the modules of elasticity increases with depth. This amounts to sensivity to internal pressure. Similarly Sowers and Vesic also observed that variation in elasticity modulus are significant as most of the elastic theories of pavement analysis are based on single layer, double layer and three layer concepts. Kachroo observed that the elastic modulus of pavement material depends on the level of stress developed which is a variable in pavement layers under the same loading conditions.

17.2 APPLICATION OF ELASTIC THEORY IN PAVEMENT DESIGN

Non-homogeneous materials are used in pavement layers and their properties are sometimes dependent on weather and moisture conditions. Many researches have been made in many countries of the world including India and the eminent highway engineers are of the opinion that more complicated pavement design approaches are merely of theoretical interest and lack practical applications.Very limited information is available on the characteristic behaviour of pavement materials. It is, therefore, the elastic theory which gives a promising design approach for flexible pavements. The linear elastic theory is modified to deal with non-linear material characteristics of the pavement components.

17.3 PAVEMENT DESIGN COMPONENTS

The pavement design rigid as well as flexible is governed by the following components :

1. Wheel load
2. Impact
3. Position and distance of wheel load from the edge of pavement
4. Type of wheel viz., pneumatic and iron wheel
5. Repetition of wheel load
6. Rainfall
7. Temperature
8. Geometry of road
9. Soil strength and
10. Drainage conditions.

The wheel load and the number of repetition of the load are the two most important factors governing the pavement design. The wheel load is half the axle load. The axle loads has been standardised in different countries for the purpose of design of pavements. The Indian Road Congress in collaborations with Road Research Institute has adopted 8.16 tonnes as a standard axle load. More advanced countries have adopted higher axle loads such as U.K. has opted for 10.17 tonnes, U.S.A 8.2 to 10.9 tonnes, France 13.0 tonnes and so on.

Repetition of wheel load is a very important component causing pavement failure. It is difficult to assess the number of repetition of wheel load for a pavement life. The standard procedure to deal with this problem is to express the traffic in terms of an equivalent number of standard axles. The standard axle load is 80 kN.

17.4 TYPE OF PAVEMENTS

Pavements can be classified as flexible pavements and rigid pavements.

Flexible pavements are those pavements which can resist tensile stresses and having very little resistance to deformation under the wheel loads. The example of the flexible pavements are earth roads—stabilized and unstabilized, bituminous roads of all types such as surface painted, bituminous concrete premixed carpets etc. and water-bound macadam roads. The rigid pavements are those which provide great resistance to deformation under the wheel load. Plain and reinforced cement concrete roads are the examples of rigid pavements.

The essential differences between the rigid pavements and flexible pavements are:

(i) Temperature variations in different weathers do not produce stresses in the flexible pavements, but heavy and dangerous stresses are developed in the rigid pavements due to temperature variations.

(ii) The flexible pavements yield to occasional excessive wheel loads for which they are not designed but rigid pavements at once rupture due to excessive stresses caused due to occasional wheel loads.

(iii) If the sub-grade is of varying strength, the rigid pavement will not adjust to the differential settlements and will act as a beam or slab at the points of support whereas the flexible pavements will not adjust to the irregularities due to differential settlement.

17.5 DESIGN CRITERIA

As have been said earlier the pavement design is relatively a new science and in the initial stage some stress-strain assumptions were made on the characteristics of soil and distribution of the wheel load through the thickness of pavement. Most of the design methods are based on past experience. Some are empirical methods based on soil classification tests, soil strength tests etc. Some of the design methods are based partly on theory and partly on experience such as C.B.R. (California Bearing Ratio) method and Waterguard method for rigid pavements. The recently developed method is Group Index Method of U.S.A., which gives approximate thickness of pavement.

Early methods of design of the pavements were based on the assumption that the load through the pavement is spread on the sub-grade through a cone. The pressure on the sub-grade thus calculated was compared with the actual bearing capacity of the soil and the thickness of the pavement thus determined. On the basis of theoretical and practical assumptions, the following factors have been vitally considered for a rational design :

1. The characteristics of the natural sub-grade which underlies the road.
2. The intensity and nature of traffic (to be assumed).

3. The amount of moisture present in the sub-soil and its drainage conditions.

4. The climatic conditions of the locality in respect of rains, snow, landslide etc.

17.6 FLEXIBLE PAVEMENT DESIGN METHODS

The flexible payments consist of number of layers. In the design process, it is to be ensured that under the application of load none of the layers is overstressed. This means that at any instance no section of the pavement structure is subjected to excessive deformation from depression. The top layers are subjected to maximum stresses and the intensity of stresses decrease as their distance from the top layers increases. As the top layers of the flexible pavements are subjected to maximum intensity of stress, superior grade of the material should be used for the top layers.

No rational design method for the design of flexible pavement have been evolved uptil now. The design process and service behaviour of pavements can be only predicted theoretically by mathematical laws. The design methods are either theoretical, empirical or semi-empirical and are based on the experience gained in the past.

The empirical methods are based on physical properties of strength of subgrade, soil classification, stress deformation and stress-strain functions. The design method are based on soil classification such as Group Index method, methods based on soil strength like CBR method, stress deformation characteristics of elastic layered system like Burmister method and so on. Following are some of the standard methods of flexible pavement design:

1. Empirical methods
   (a) Group Index Method
   (b) California Bearing Ratio Method
   (c) California Resistance value or Stabilometer-method
   (d) Mc Lead Method
2. Some-empirical methods; Triaxial Method
3. Theoretical Method; Burmister Method.

Empirical methods are based on physical characteristics or strength parametres, semi-empirical method on theory and empirical equations, whereas the theoretical method is based on theoretical approach using two-layer elastic theory.

There are several methods of designing flexible pavements but the following two methods are commonly used in our country.

Group Index Method. Group Index is a number which characterises the nature of soil. The Group Index Method was first introduced by D.J. Steele. The Group Index is the inverse measure of the qualities of a sub-

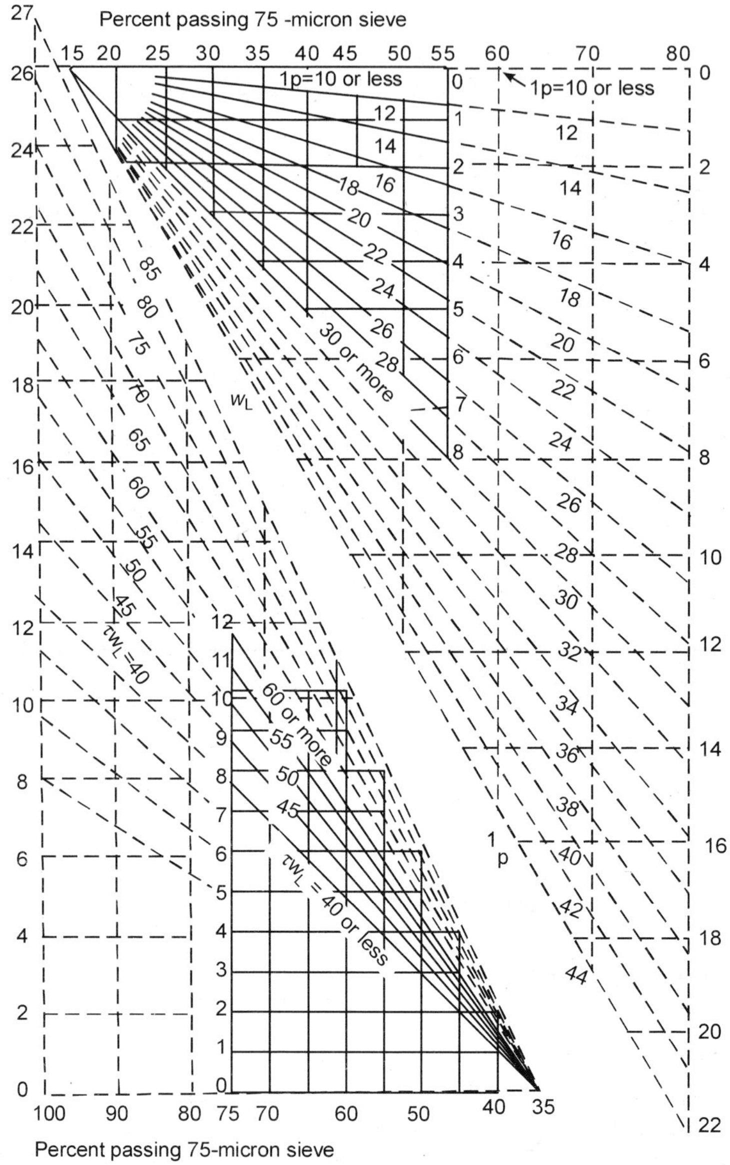

Figure 17.1 *Group Index*

grade. More the Group Index number, lesser the bearing capacity of soil i.e., more the G.I. of a soil, the poorer the soil is, and hence more the thickness of the pavement and sub-base. The actual thickness of the base and the surfacing will depend on the intensity of traffic. The group index is a function of the amount of material passing the 75-micron I.S. Sieve, the liquid limit and the plastic limit is given by the following equation:

Group Index (G.I.) = $0.2\,a + 0.005\,ac + 0.1\,bd$

where

a = that portion of percentage passing 75-micron sieve greater than 35 and not exceeding 7.5 expressed as a positive whole number between 0 and 40.

b = that portion of percentage passing 75 micron sieve greater than 15 and not exceeding 55 expressed as a positive whole number (0–40)

c = that portion of the numerical liquid limit greater than 40 and not exceeding 60 expressed as positive whole number between 0 and 20.

d = that portion of the numerical plasticity index greater than 20 and not exceeding 30 expressed as a positive whole number between 0 and 20.

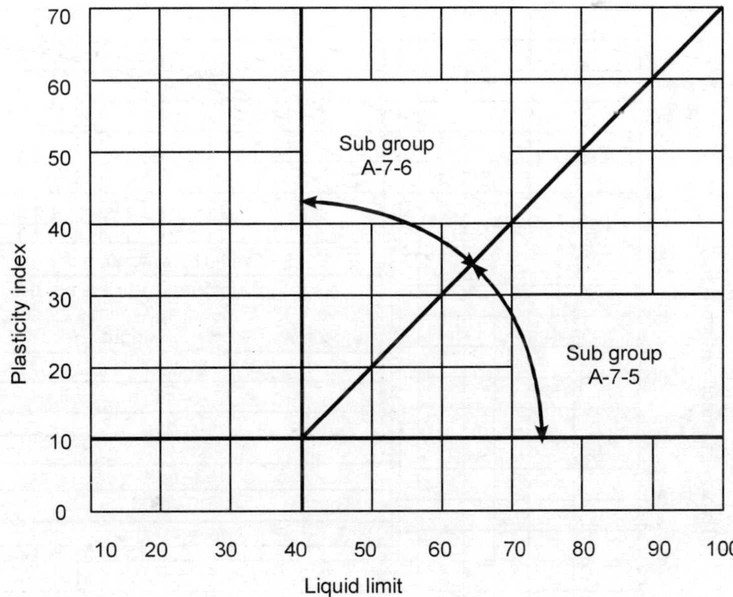

Figure 17.2 *Group Index*

Figure 17.1 shows a chart of Group Index numbers from which G.I. can be conveniently obtained as the sum of readings on the vertical scale of the chart.

As has been said earlier, higher the group index the poorer is the sub-grade and greater is the thickness of pavement. The thickness of the sub-base will increase with the increase in the group index.

The thickness of surfacing and the base will depend on the traffic volume and intensity. If the number of commercial vehicles passing on the road per day is less than 50, it is termed as light traffic, if the number of commercial vehicles is between 50 and 300, it is termed as medium traffic and if the number of commercial vehicles passing on the road is more than 300 per day, it is termed as heavy traffic.

Curve *A* gives the thickness of sub-base to be provided, curve *B* gives the overall thickness of the road crust for light traffic, curve *C* for medium traffic and curve *D* for heavy traffic.

C.B.R. Method. The California Bearing Ratio (C.B.R.) method is the most commonly used method of designing flexible pavements. The C.B.R. is

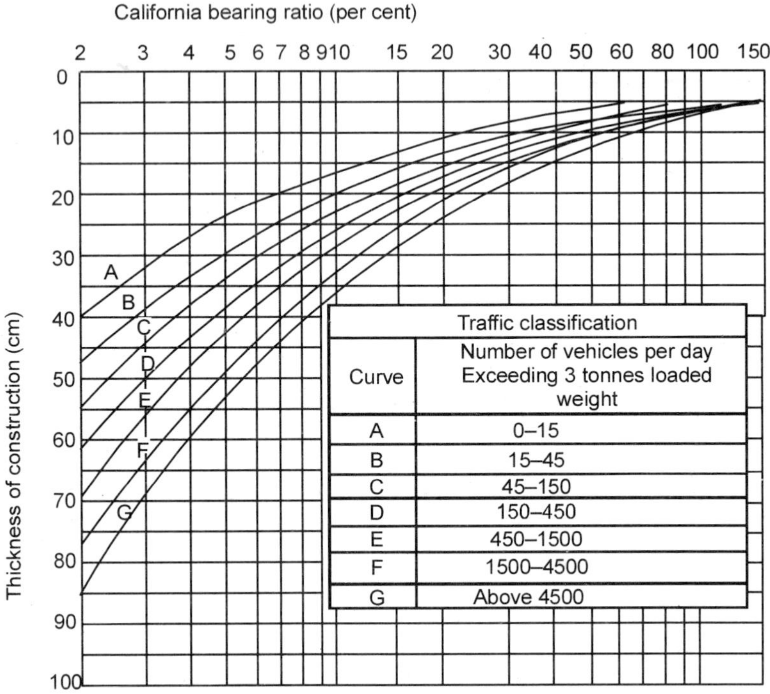

Figure 17.3

defined as the ratio of test load required to force a cylindrical plunger of 3 m^2 or 19.355 cm^2 cross-sectional area into a soil mass at the rate of 0.05 cm or 0.127 cm per minute, to the load required for corresponding penetration of the plunger into a standard sample of crushed stone. The later load being known as *Standard Load.*

$$\text{C.B.R.} = \frac{\text{Test load}}{\text{Standard load}} \times 100$$

17.6.1 C.B.R. Test

The C.B.R. Laboratory test is conducted in the following manner:

A sample of soil to be evaluated is placed in a steel mould with attached collar which is detatchable. The diameter of the mould is 150 mm. The mould is placed on a base plate which has a loading frame with 50 mm diameter plunger to penetrate a pavement component material at the rate of 1.25 mm/ minute. The load required to penetrate the plunger from 2.5 mm to 50 mm are recorded on the dial gauge. These loads are expressed as percentages of the standard load values in respective deformation values to obtain CBR value. The specimen in the mould is subjected soaking and swelling for 4 days.

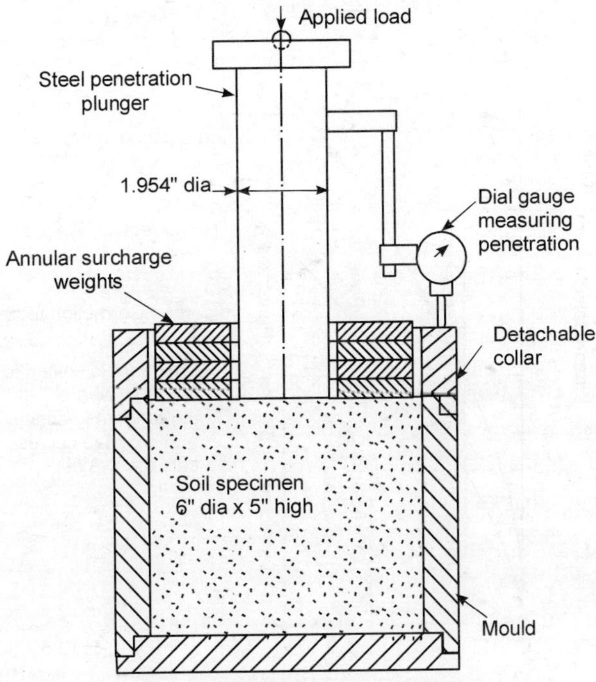

Figure 17.4 *C.B.R loading frame*

Following compaction of the sample when still in the mould it is subjected to a surcharge weight equal to the expected load of the pavement. Now the specimen and the mould assembly placed in the loading frame. For 0.0, 0.5, 1.0, 1.5, 2.0, 2.5, 3.0, 4.0, 5.0, 7.5, 10.0 and 12.5 mm penetration of the plunger load values are recorded, the observation show the resistance of a material against penetration of a standard plunger. The load penetration graph is plotted. These curves are based on the principle that a meterial of a certain C.B.R. value requires a certain minimum of construction over it. For standard crushed stone loads of 6.895 MN/m^2 and 10.343 MN/m^2 cause 2.5 mm and 5.0 mm penetration. A correction is made for curves concaving upwards. Some typical values of C.B.R of Indian soils are :

	Type of soil	*C.B.R value*
1.	Black cotton or hearing soils	1 to 2
2.	Alluvial silt	2 to 5
3.	Silty clay	3 to 5
4.	Sandy clay	4 to 6
5.	Well graded sand	8 to 20
6.	Muram	8 to 20
7.	Gravel	8 to 20

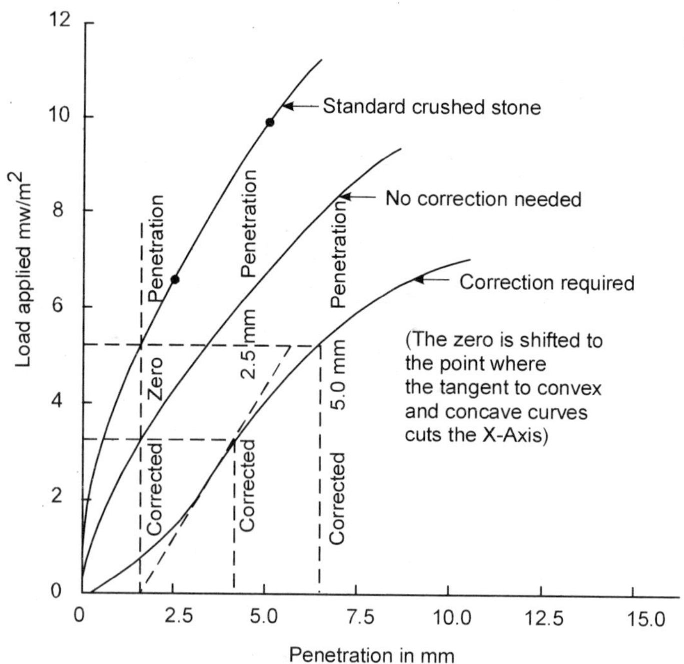

Figure 17.5 *Load penetration curve*

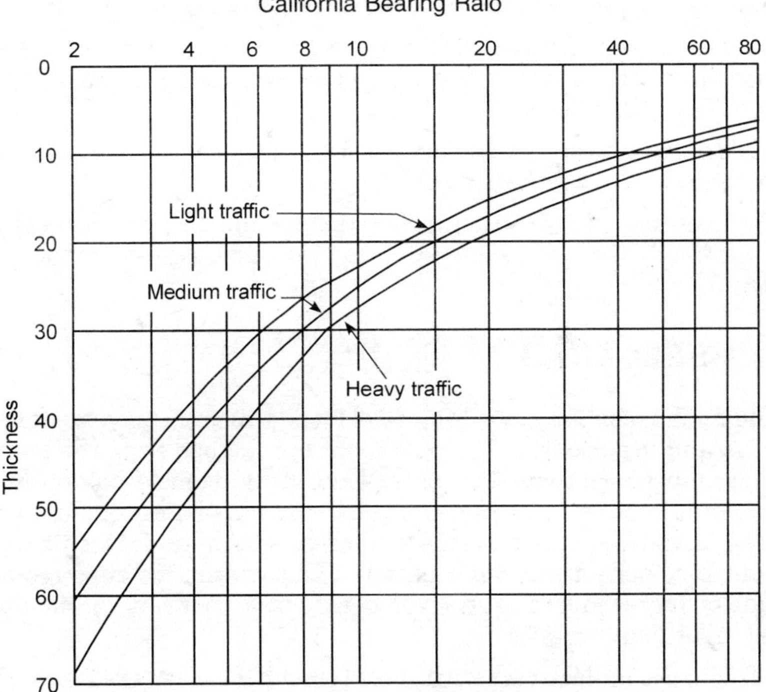

Figure 17.6 *C.B.R graph*

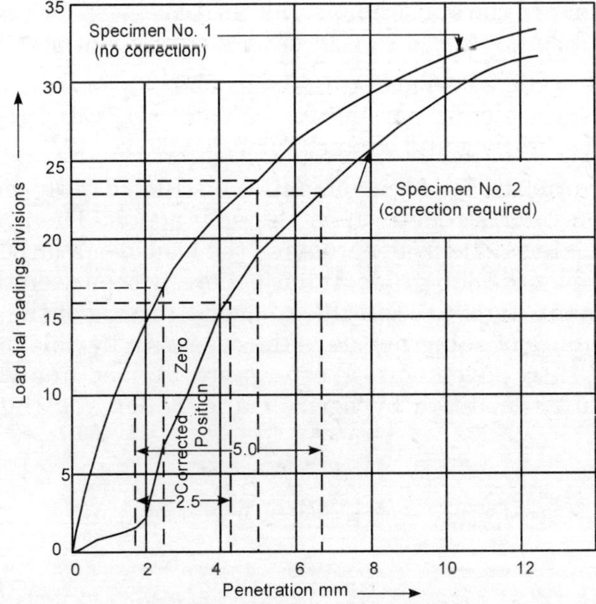

Figure 17.7 *Comparison of stresses*

Following are some of the drawbacks of C.B.R. test:

1. It is essentially an arbitrary strength test and cannot be used for the evaluation of other properties such as cohesion, angle of repose or internal friction, shearing strength etc.

2. The test procedure must be followed strictly if results are to be comparable with standard result.

3. Presence of coarse material initiate the results. For this purpose only material passing through 20% mm sieve is used for the preparation of specimen.

17.7 WHEEL LOAD

For the design of pavements, the wheel loads which the pavement is expected to withstand are decided. The pavement is designed from the point of view of maximum wheel load. The gross load of the vehicle has no importance and is not taken into consideration. Another important consideration is the effect of repetition of loads on the pavement which induce failure by causing commutative permanent deformations. To counter this effect an allowance is made by increasing the static wheel load by 10 per cent for medium traffic and 20% for heavy traffic.

In India, the bullock cart with steel-tyre wheels is a very common mode of transportation. A design curve proposed by N. Mohan Rao of Road Research Institute, New Delhi, gives pavement thickness for known C.B.R. values is shown in Fig. 17.7. Since the stresses caused due to bullock cart are very high particularly in the surface layers of the pavement for bullock carts the top 10 cm thick layer should consist of very strong material.

In India the static wheel load is taken as 2268 kg for commercial vehicles with contact surface of 12 cm and for bullock cart traffic the static wheel load is taken as 1270 kg and contact surface 4.5 cm.

Rigid Pavements. Rigid pavements constructed with cement concrete will depend on their strength upon the slab action. The cement concrete being a rigid material, helps in spreading the load over a large area and that too uniformly over the sub-grade. It has generally been seen that the failure of rigid pavements tend to occur due to over stressing of the concrete itself than by the failure of sub-grade as is the case with flexible pavements. For the design of rigid pavements, Westergaard derived analytical formulae which were later modified by himself and Teller and Sutherland. The formulae are :

$$f_{(\text{centre})} = 0.275(1+\mu)\frac{P}{h^2}\; 4\log_{10}\frac{l}{b}$$

$$+\log_{10}\{12(1-\mu^2)\} - 54.54\,c_2\;\frac{l}{c_1}^{\;2}$$

$$f_{(\text{edge})} = 0.529\ (1 + 054\ \text{m})\ \frac{P}{h^2}\ 4 \log_{I0}\ \frac{1}{b} + \log_{10}\ \frac{b}{2.54}$$

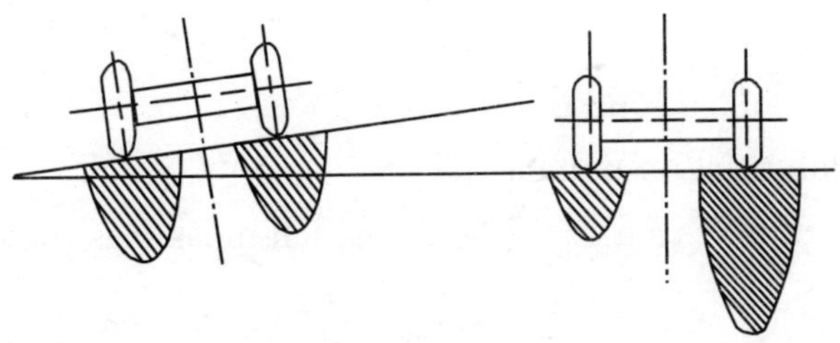

(a) Wheel Load

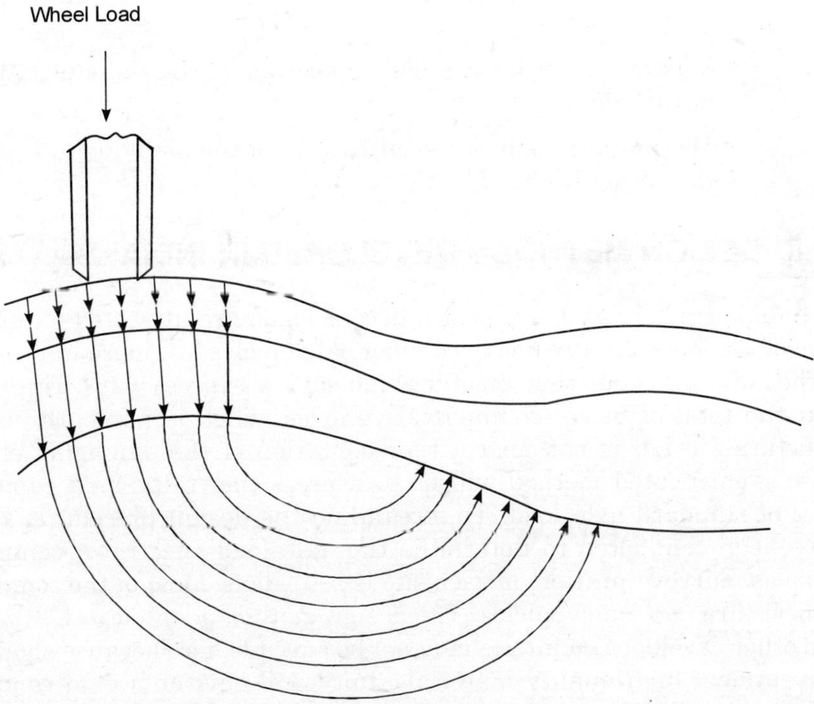

(b) Deflection of sub-grade under wheel load

Figure 17.8

$$f_{(\text{corner})} = \frac{3P}{h^2}\left[1-\left(\frac{a\sqrt{2}}{l}\right)^{1.2}\right]$$

where

P = Wheel load in kg

h = Slab thickness in cm

l = Radius of relative stiffness which measures the stiffness of slab.

$$l = \left[\frac{Eh^3}{12\left(1-\mu^2\right)k_s}\right]^{1/4}$$

μ = Poisson's ratio for concrete (0.1 – 0.35)

$f_{(\text{centre})}$ = Maximum tensile stress at the bottom of the slab due to loading in the centre (kg/cm^2).

$f_{(\text{edge})}$ = Maximum tensile stress at the bottom of the slab due to loading at the edge.

$f_{(\text{corner})}$ = Maximum tensile stress at the top of the slab due to loading at the corner of the slab.

17.8 DESIGN METHODS DEVELOPED IN INDIA

In India, commercial vehicles of different loads are attracted on different type of roads. Generally heavy commercial vehicles are more in proportion in the industrial belts than on other highways or express ways. This implies that the total of heavy commercial vehicles which is presently used for structural design is not correct representation of the damaging effect. A more sophisticated method will be to express the traffic as a cumulative total of standard axle load. To streamline the design procedure, surveys have been conducted to determine the axle load spectra of commercial vehicles. Survey duration on each site is 2 – 3 days. Most of the commercial vehicles are two axle vehicles. The design method adopted in U.K., U.S.A. and other developed countries may not be suitable here because shortage of conventional high quality materials, untypical environmental conditions, changes in compaction techniques etc. Development of a rational and scientific approach to deal with varying situations require development of an analytical based approach and structural design procedure.

For analytically based structural design and rehabilitation studies a number of computer programmes are available. These programmes give either an elastic or linear elastic solution permanent deformation of the road structure are beyond the passive of these computer programmes. DEFPAV is a computer system of prediction of pavement deformation is capable of predicting the lateral profile of pavements after any number of axle loading. The programme is also capable of bearing used to estimate load equivalence factors for different environments and loading condition.

17.9 DESIGN CURVES

Design curves giving the appropriate thickness of construction required above a material with a given CBR have been developed for different wheel loads and traffic condition after carrying out experiments on sub-grade, sub-base or base materials for satisfactory and un-satisfactory pavements. The curves have been extended for the design of runways and are widely used in our country, though they are abused also for being arbitrary and ad hoc.

To design a pavement, the CBR value for the sub-grade material is determined and the corresponding total thickness of the pavement is found from the curve. The total thickness of the pavement so obtained is sub-divided into base, sub-base and the surfacing by knowing the CBR values of sub-base, base and the surfacing material and determining from the thickness of construction required above the respective layers. The chief requirement is that each layer should be of superior quality and stronger than the layer below it. CBR design curves have been developed and used for the design of flexible pavements but can be used for determining the total thickness of concrete pavements also.

The Indian Road Congress has recommended a CBR design chart for use in India. Seven different curves based on number of commercial vehicles passing on that road are given in Fig. 17.9.

17.9.1 Design Method

For a flexible pavement design by C.B.R. method the following procedure should be adopted:

(i) Evaluate the C.B.R. value for the sub-grade.

(ii) Select the appropriate design curve.

(iii) Calculate the anticipated traffic based on growth rate and the design life of road.

(iv) Calculate the thickness of base, sub-base and surfacing.

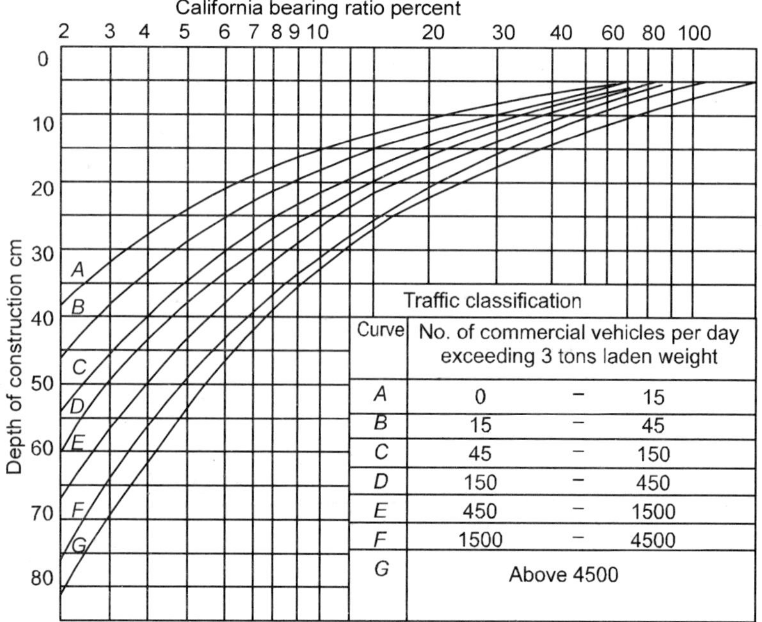

Figure 17.9 *C.B.R. design curves*

17.9.2 I.R.C. Recommendations

I.R.C. (I.R.C.-37 – 1970) has recommended the following points on C.B.R. design method.

1. I.R.C. recommended that in-situ tests should not be conducted. Tests should be conducted in the laboratory on re-moulded samples of soil. Standard test procedure should be strictly followed on prepared and compacted samples.

2. For the design of new roads the soil samples of sub-grade should be compacted at OMC and the same should be achieved in the field also, otherwise the soil sample should be compacted to dry density expected to be achieved in the field.

3. The test samples should be soaked in water for four days before testing. In places where the water table is deep and rainfall is less than 50 cm, soaking of the test sample is not necessary. Wherever possible the most adverse condition of moisture effecting the sub-grade should be determined by conducting actual field studies at different locations along the alignment.

4. At least three samples should be tested at each location of similar conditions of density and moisture content. If the variation in the test results is too large or exceeding specified limits, then average of the samples should be taken. The maximum specified limits for C.B.R. values are 3 for values upto 10.5 for values upto 10 – 30 and 10 for values upto 60%.

5. An estimate of the expected traffic during the design life period, be made, keeping in view the existing traffic and the annual growth rate. Pavement for major roads should be designed for a period of 10 years and the following formula should be used

$$A = P(1 + r)^{(n+10)}$$

where

A = No. of heavy vehicles per day at the end of the design period (vehicles ladden weight > 3 tonnes)

P = No. of vehicles at the last count

r = Rate of growth annually

n = No. of year of anticipated time of construction

The value of P should be taken on a seven day average of heavy vehicles. A growth rate of 7.5% may be assumed in the absence of any reliable data.

6. When the sub-base course contains large proportion of material of 20 mm size or more, CBR values for these materials will not be valid for the subsequent layers. Thin layers of surfacing such as a seal coat should not be counted towards the thickness of the pavement.

7. The results be compared by the following design formulae

$$t = \sqrt{P}\left(\frac{1.75}{\text{CBR}} - \frac{1}{p\pi}\right)^{1/2}$$

and

$$t = \left(\frac{1.75P}{\text{CBR}} - \frac{A}{\pi}\right)^{1/2}$$

where

t = total thickness of pavement

P = wheel load in kg

A = contact area cm^2

p = tyre pressure kg/cm^2

Note. These equations are valid for CBR value for sub-grade to be less than 12%.

Example 17.1 *Design a suitable thickness for a state highway to be newly constructed on a sub-grade of CBR value 5% and wheel load 4200 kg with 6 kg/cm² tyre pressure. Medium to light traffic of 200 vehicles are anticipated.*

Solution Using the design curve of C.B.R. for C.B.R. value 5% and wheel load 4200 kg, thickness of pavement is 38 cm by using graph *D*.

Using the design formula

$$t = \sqrt{P}\left[\frac{1.75}{\text{CBR}} - \frac{1}{p\pi}\right]^{1/2}$$

We have

$$P = 4200 \text{ kg}$$

$$\text{CBR} = 5\%$$

$$p = 6 \text{ kg/cm}^2$$

$$t = \sqrt{4200}\left[\frac{1.75}{5} - \frac{1}{6\times\frac{22}{7}}\right]^{1/2}$$

Solving the equation we have

$$t = \mathbf{36.2 \text{ cm.}}$$

Hence a total thickness of 38 cm is recommended.

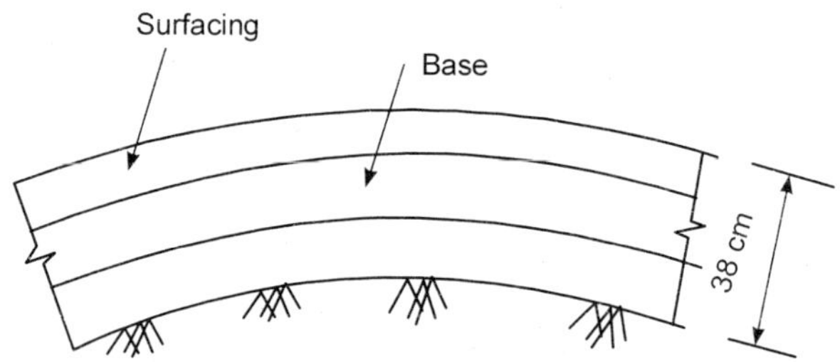

Figure 17.10

Example 17.2 *Design a bituminous surface pavement for an average of 1000 vehicles per day of weight 3 tonnes the C.B.R. value of subgrade is 5. Two available road aggregates when compacted have C.B.R. value of 20 and 60 respectively.*

Solution For the given traffic intensity of 1000 vehicles I.R.C. recommended graph E to be used. The total thickness of the pavement comes out to be 43 cm.

Thickness above the sub-base

CBR – 20 = 19 cm

Thickness above the base CBR – 60 = 10 cm

Hence the pavement will be

(a) Surfacing = 10 cm

(b) Base = 19 cm

(c) Sub-base = 43 – 19 – 10 = **24 cm.**

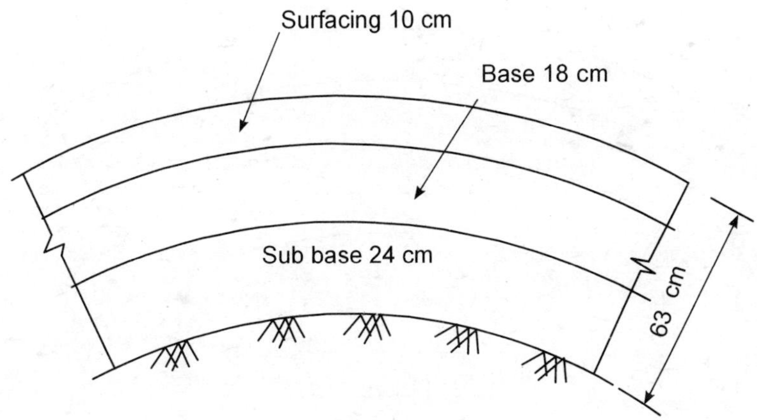

Figure 17.11

Example 17.3 *A double lane highway is to be constructed for the present traffic load of 1000 vehicles per day average with an estimated growth rate of 7.5%. The road is to be designed for a life time of 10 years as per I.R.C. guidelines. The C.B.R. value of sub-grade is 3. Locally available material is gravel whose C.B.R. is 25.*

Solution Traffic at the end of 10 years

$$= 100 \quad 1+\frac{7.5}{100}^{\;10}$$

$$= 2062 \text{ vehicles per day for both the lanes}$$

$$\text{Vehicles per lane} \quad = \frac{2062}{2} = 1031$$

IRC modified CBR curve is to be used.

For CBR value 5 on *F* curve the total thickness comes out to be 62.5 cm.

For CBR 25 gravel the thickness above the base comes to be 17 cm.

The surfacing to consist of 17 cm thick bituminous layer and the base will be 62.5 – 17 = 44.5 cm. The base is to be divided into two layers as 44.5 cm layer is difficult to be compacted.

It is proposed to provide a premixed carpel at 7.5 cm thickness (equivalent to 2 × 7.5 = 15 cm) with 2 cm thick surfacing or seal coat.

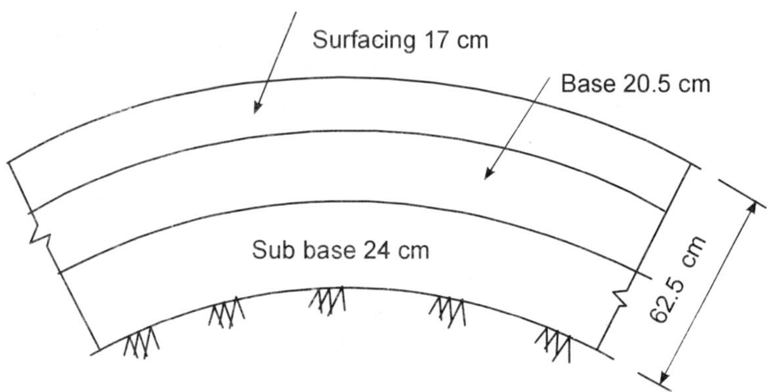

Figure 17.12

17.10 CALIFORNIA RESISTANCE VALUE OR STABILO-METER METHOD

Most design empirical formula procedures attempt to estimate the thickness of the pavement necessary to distribute the load or applied wheel pressure to the sub-grade, the stabilometre method requires the pavement to be designed to resist any lateral and upward movement of soil particle which may take place as shown in Fig. 17.2. The three factors considered to be of major importance in developing the flexible pavement thickness and which are taken into account by this procedure are :

1. *The structural quality of the sub-grade soil.* This quality is measured by means of a *stabilometer* and expansion pressure tests. It is expressed in terms of resistance value *R*.

2. *The traffic conditions.* The design procedure considers that the required thickness over any particular lays is proportional to the average tyre pressure, the square root of the effective tyre imprint area, and the logarithm of the number of load repitition. In utilizing these considerations the magnitude and number of wheel loads estimated for the design year are corrected into equivalent numbers of 2268 kg wheel loads.

3. *The tensile strength of the pavement layers.* The required thickness of a given layer or group of layers is considered to be inversely proportional to the 5th root of its tensile strength. The tensile strength is expressed in terms of a cohesive value.

17.10.1 Stabilometer Test

Prior to using the actual design procedure it is recommended to conduct a series of tests on the soil samples in order to obtain basic relative date. The step in the test is to prepare three specimen of 10 cm (4") diametre and 6.25 cm (21/2") height. Soil used in these specimen is restricted to material passing 20 mm sieve. The specimens are covered or encased in rubber membrane which

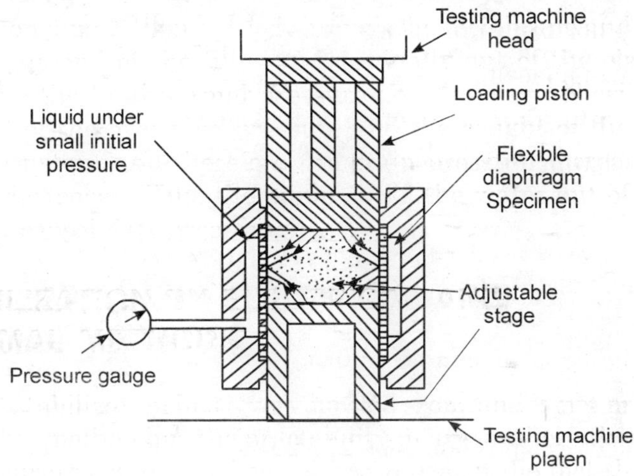

Figure 17.13 *CBR test*

will act as a inner wall. The soil in the first sample is mixed with slightly more moisture than is expected to produce saturation of the specimen voids after compaction. Lateral pressure is applied on the specimen by fluid pressure through the membrane and the vertical pressure by loading head placed on the loading machine. Each specimen is tested in Hveen Stabilometer. The Siabilometer test is a triaxial test except that the results are influenced primarily by the resistance due to internal friction and only slightly by the portion of the total resistance that may be due to cohesion. The stabilizer test specimen are kept at a temperature of 60°C. The lateral pressure is increased from 0.35 kg/cm^2 to 7.0 kg/cm^2 and the vertical pressure 2722 kg in increments of 454 kg.

The bulk density of the specimen is calculated after releasing the pressure. The temperature is maintained at 60°C for two hours before it is placed in the cohesiometer. The loading is done by lead shots at the rate of 1800 gm per minute. The breaking weight of lead shots is noted. The strength of soil as measured by stabilometer is called its resistance value R which is calculated from the following equation

$$R = \frac{22.2}{\dfrac{P_h \times D}{P_v - P_h} + 0.222}$$

where

P_v = vertical pressure

P_k = horizontal pressure

D = displacement

Here the value of P_v is token to be 28 kg/cm^2.

The cohesionmeter value c will be

$$c = \frac{L}{d(0.2\,H - 0.0176\,H)^2}$$

where

d = diameter of sample = 10 cm

H = height of sample = 6.25 cm

The value of c comes out to be 211.2533.

17.10.2 Design Procedure

The actual design procedure is based on three concepts :

1. There must be sufficient thickness of overlying material to provide a weight which will present expansion of the underlying material after displacement thereby preventing it from taking up more moisture. The thickness estimate is determined on the basis of the expansion pressure test results.

2. There must be sufficient cover thickness of material to prevent the plastic deformation of the underlying material. This thickness is determined from the stabilometer test.

3. There must be sufficient thickness of cover material to resist strength loss due to accumulation of moisture while the highway is in service. This cover thickness is determined from the exudation pressure test.

Based on performance data it was established by Hveern and Carmany that pavement thickness varies directly with R value and logarithm of load repetition. It varies inversely with the fifth root of C value. The expression thickness is given by the empirical equation

$$T = \frac{K\,(T_1)\,(90 - R)}{C^{1/5}}$$

where

T = Total thickness of pavement, cm

K = Numerical constant equal to 0.166

T_1 = Traffic Index = 1.35 $(\text{EWL})^{0.11}$

R = Stabilometer Resistance Value

C = Cohesiometer value

The equivalent wheel load is the sum of accumulated sum of the products of number of constants and axle loads. The various constants on yearly basis for the number of axles are

Number of Axles	*Equivalent wheel load (EWL) constants*
2	330
3	1070
4	2160
5	4620
6	3040

17.10.3 Equivalent Value *C*

The cohesiometer value for each layer of pavement is obtained from the test results. Following relation is used for calculating C value

$$\frac{t_1}{t_2} = \frac{C_2}{C_1}^{1/5}$$

While designing a pavement as the thickness of the pavement is not known, the pavement is assumed to consist of any one material like the gravel base course of known C-value. Individual thickness of each layer is converted in terms of gravel equivalence by using the above formula where t_1 and t_2 are thickness of the two layers and C_1 and C_2 are their corresponding cohesiometer values.

Metric equivalence of C-values for some materials are given below

Gravel base course	15
Soil cement base course	120 – 230
Bituminous concrete	60 – 62
Open graded bituminous size	22 –30

17.11 DESIGN PROCEDURE OF CBR RESISTANCE VALUE

The following design procedure is adopted :

1. For different moisture contents different pavement thickness are calculated as per R-value required.

2. Calculate pavement thickness values to balance the sub-grade expansion pressure by dividing the expansion pressure by the average density of the sub-grade or pavement. Pavement thickness as per expansion pressure at different moisture contents are calculated.

3. Pavement thickness satisfying expansion pressure and R-values at different moisture contents are determined by plotting graphs.

4. The exudation pressure of sub-grade soil ground at various compacting moisture contents are plotted against the pavement thickness based on moisture contents.

17.12 Mc LEAD METHOD

The Canadian Development of Transport under the direction of Mc Lead made an extensive investigation of the stability of existing airfields in Canada using plate bearing tests corelated with in-situ CBR tests, North Dakota cone tests, Housel penetrometer tests and Triaxial compression tests. The tests were conducted on the surface of roads, base course and sub-grade at large number of test locations. This investigation resulted in a definite design method. The test result of Mc Lead indicated that for any given size of bearing plate, the load carried by the first application is about 115 percent of that which will the supported at ten repetitions for any specified total deflection.

On the basis of tests the following corelation was established :

$$T = \log_{10} \frac{\rho}{s}$$

where

T = required thickness of gravel
ρ = applied wheel load for the same contact area and deflection
s = subgrade support for the same contact area and deflection
k = base course constant

The value of k varied with the size of bearing plate and can be obtained following graph.

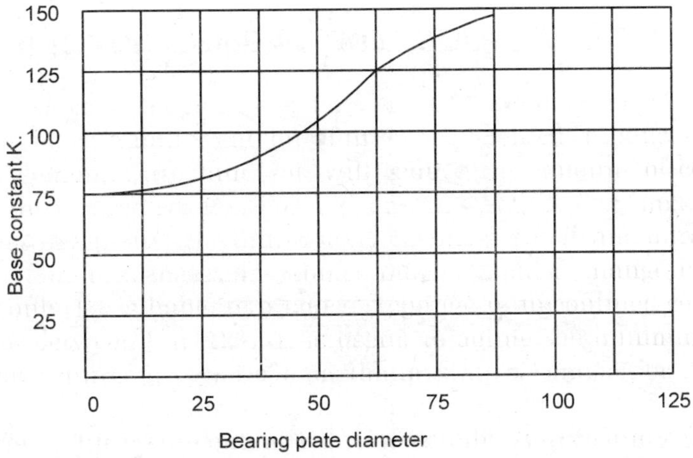

Figure 17.14 *Bearing plate and Base constant*

The sub-grade support s for the design of highway pavement is calculated from the support measured or calculated for 30 cm diameter plate at 0.5 cm deflection and ten repetitions.

17.13 TRIAXIAL METHOD

The method was first proposed by Palmer and Barber and deals with theoretical stresses in ideal masses using Boussineq's equation. The design equation was further developed by Kansas.

A lateral pressure of 1.4 kg/cm^2 is applied in the test for determining the estastic modulii for various materials. The lateral pressure is arbitrarily assumed as the lateral confinement of soil layers. The design equation is

$$\Delta = \frac{3Pa^2}{2\sqrt{(a^2+h^2)}}$$

where

Δ = Deflection under the designed load

= 0.254 cm

$$P = \frac{p}{ka} \quad \frac{\text{Wheel load}}{\text{Area of bearing plate}}$$

Substituting the value of P, we have

$$\Delta = \frac{3p}{2\pi E\sqrt{(a^2 + h^2)}}$$

$$\therefore \qquad (a^2 + h^2) = \frac{3p}{2\pi E\Delta}^2$$

or

$$h = \sqrt{\frac{3p}{2\pi E\Delta}^2 - a^2}$$

For incompressible pavements becomes the required thickness of pavement. If the modulus of elasticity is different for the subgrade and the pavement, a stiffness factor has to be introduced and the equation is modified as

$$T = \sqrt{\frac{3\rho}{2\pi E\Delta}^2 = a^2} \times \frac{Es}{Ep}^{1/3}$$

Some more modifications for traffic coefficient and saturation are also suggested.

17.14 BURMISTER METHOD

Burmister conceived a pavement as a two layer flexible system. This is a basic theory which utilizes the assumptions close to actual condition in a flexible road. Stress and deformation values obtained by these assumptions are dependent on the modulii of deformation of the different layers. Since most of the typical flexible roads are normally composed of layers whose modulii of decrease with the depth, the net effect is that the Burmister equations predict stresses and deflections in the subgrade that are considerably less than those obtained from the Boussinesq equations. Refering to Fig. 17.6, the left side of the figures as if the load is considered to be resting directly on the sub-grade whereas on the right hand side pavement of thickness h_1. and modulus E_1 is inserted and the stresses calculated on the basis of this. The data are plotted in the form of bulbs of pressure for the two systems. A bulb of pressure is a surface obtained by connecting points of equal stresses on the various horizontal planes at various depths. The pressure on any one point on the surface of the bulb is the same as any

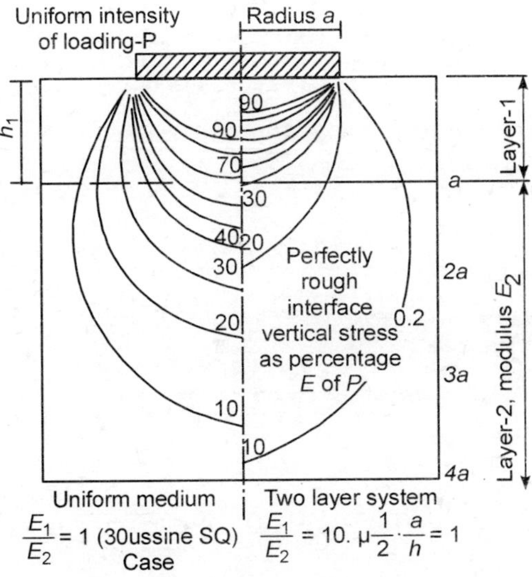

Figure 17.15 *Burmister chart*

other point. Pressures of points inside any given bulb are greater than those on the bulb, while pressures, outside the bulb are less. This figure also illustrates the main function of a pavement, which is to reduce to an acceptable level the pressures applied to the sub-grade. As can be seen the stresses in the sub-grade at a depth h_1 are considerably influenced by the insertion of the strong pavement material. With the pavement inserted, the natural stress at the interface and directly below the centre of the applied load is estimated to be approximately 30 percent whereas without pavement the stress at depth h_1 is approximately 70 percent of the applied unit load.

If E_1, E_2 and E_3 are the elastic modulii of sub-grade, sub-base and base course and if $E_3 > E_2 > E_1$ the effectiveness of the reinforcing action of the pavement layers is logically utilized. The deflection factor F is introduced in two layered system which is dependent on ratio of modulii and depth.

The displacement equation of Burmister is $\Delta = 1.5 \dfrac{P}{E_1} \cdot F$

where

p = uniform pressure on the sub-grade

E_1 = modulus of elasticity of the sub-grade

F = deflection factor

(F is a dimensionless factor and depends on the ratio of modulii of elasticity of sub-grade pavement E_1/E_2 obtained from the graph).

Example 17.4 *Design a suitable pavement (flexible) by Burmister's two layer theory for wheel load of 40 kN and tyre pressure of 0.5 MN/m². The modulus of elasticity of the bituminous concrete of which the pavement is to be made, is 150 MN/m². A normal sub-grade of 30 MN/m². Assume any other data not given. Burmister displacement graph can be used.*

Solution

$$\text{Tyre pressure} = \frac{40 \times 10^3}{\pi a^2} = 0.5 \times 10^6$$

$$\therefore \quad a^2 = \frac{40 \times 10^3}{\pi \times 0.5 \times 10^6}$$

$$a = \sqrt{\frac{4}{\pi \times 50}}$$

$$= 15.96 \text{ or say } 16 \text{ cm}$$

Thickness of paveemnt = 32 cm

$(2a = 2 \times 16 = 32 \text{ cm})$

$$= \frac{\text{Modulus of elasticity of pavement}}{\text{Modulus of elasticity of sub-grade}}$$

$$= \frac{150MN}{30MN} = 5$$

$$\text{Deflection of pavement} = F \times \frac{1.5pa}{E}$$

Here

E = modulus of elasticity of sub-grade

F = displacement factor = 0.43 from the Burmister displacement graph

p = Tyre pressure = 0.5×10^6

Hence $$\Delta = \frac{0.043 \times 1.5 \times 0.5 \times 10^6 \times 16}{30 \times 10^6}$$

$$= 0.173 \text{ cm}$$

As the displacement is less than 0.5 cm hence 32 cm pavement thickness is safe

Example 17.5 *Design a flexible pavement for 41 kN wheel load for a maximum deflection of 0.5 cm and tyre pressure of 0.5 MN/m². Assume any other data not given and use Burmister deflection graph. For 0.5 cm deflection the pressure developed on a 30 cm diameter bearing plate is 125 MN/m² and on 0.5 cm diameter plate 400 MN/m².*

Solution

$$\text{Tyre pressure} = \frac{\text{Wheel load}}{\pi a^2}$$

$$0.5 + 10^6 = \frac{4\times 10^3}{\pi a^2}$$

$$a^2 = \sqrt{\frac{41000}{\pi\times 5}}$$

$$a = 16.1 \text{ cm}$$

For single layer and rigid circular plate of 30 cm diametre ($a = 15$)

$$\Delta = 1.18\,\frac{p.a}{E}\times F$$

$$0.5 = \frac{1.18\times 1.25\times 15}{E}\times F$$

$$F = 1 \text{ for single layer}$$

$$\text{Hence } E \text{ for sub-grade} = \frac{1.18\times 1.25\times 15\times 1}{0.5}$$

$$= 44.2 \text{ kg/cm}^2$$

$$\text{Deflecuon} = \frac{1.5\,p.a}{E} = \frac{1.5\times 0.5\times 16.1}{44.2}$$

$$= 0.183$$

Hence pavement thickness for 0.183 cm deflection will be

$$2.1\times a = 21\times 16 = 33.6 \text{ cm.}$$

Example 17.6 *Design a flexible pavement on the basis of two-layered theory for a wheel load of 40 kN, tyre pressure 0.5 MN/m² and maximum deflection 0.5 cm. The results of the plate load bearing test on 75 cm diameter plate on the sub-grade and aggregate (granular) base are :*

Stress developed for 0.25 cm deflection on the subgrade is 0.07 MN/m² and stress developed on base course for the same deflection is 0.14 MN/m².

Solution For sub-grade

$$\Delta = \frac{1.18\, p.a}{E_1}$$

$$p = \text{for sub-grade is } 0.07 \text{ MN/m}^2$$

$$a = \frac{75}{2} = 37.5 \text{ cm} \quad \text{or } 0.375 \text{ m}$$

$$\Delta = 0.25 \text{ cm} = 0.0025 \text{ m}$$

Hence

$$E = \frac{1.18 \times 0.14 \times 37.5}{0.0025} \times 12.4 \text{ MN/m}^2$$

For the base course

$$0.0025 = \frac{1.18 \times 0.14 \times 37.5}{E_1} \times F$$

$$= \frac{1.18 \times 0.14 \times 37.5}{12.4} \times F$$

Hence

$$F = 0.50$$

Assuming the thickness of the base course to be 15 cm, the ratio of base course to the radius of the bearing plate is

$$\frac{15}{37.5} = 0.4 \qquad \text{(i)}$$

For the ratio of base course and the radius of the bearing plate to be 0.4 and deflection factor 0.5, we have from the graph

$$\frac{E_2}{E_1} = \frac{1}{100}$$

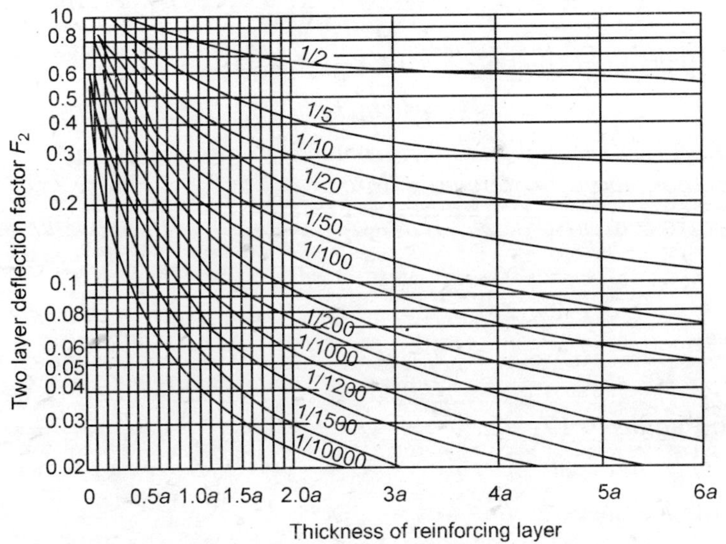

Figure. 17.16 *Deflection graphs*

$$a = \sqrt{\frac{\text{Wheel load}}{\mu \times \text{tyre pressure}}}$$

$$= \sqrt{\frac{40 \times 1000}{\mu \times 0.5 \times 1000000}} = \sqrt{\frac{4}{3.14 \times 50}}$$

$$= 0.16 \text{ cm}$$

$$= 16 \text{ cm} \qquad \text{(ii)}$$

Now deflection under the wheel load

$$\Delta = \frac{1.5 p \cdot a}{E_2} \times F$$

$$\frac{0.5}{100} = \frac{1.5 \times 0.6 \times 10^6 \times 0.16}{12.4 \times 100} \times F$$

Solving the equation for F, we have $F = 0.52$

Now $\frac{\text{Pavement thickness}}{\text{Radius}} = 0.4$ from (i)

Hence pavement thickness = 0.4 × 16 from (i)

= 6.4 cm

Example 17.7 *Design a flexible pavement by stabilometer and cohesiometer value method, with the following data :*

Mioisture content %age	*R*	*Pressure kg/crn²*	
		Exp.	*Exd.*
15.0	*65*	*0.10*	*48*
18.5	*40*	*0.085*	*35*
23.0	*18*	*0.04*	*20*

Traffic Value = 10

C value = 90 gm/cm²

Solution We have

$$T = \frac{K\,(T_1)(90 - R)}{C^{1/5}}$$

where

T = design thickness

K = a constant = 0.166

T_1 = traffic index = 10

R = stabilometer value = 65

C = cohesiometre value = 90 gm/cm^2

Therefore

$$T = \frac{0.166\,(10)\,(90 - 65)}{90^{1/5}}$$

$$= 16.9 \text{ cm}$$

Similarly for R = 40, the thickness of pavement will come out to be 33.8 cm and for R = 18, the thickness will be 48.5 cm.

Example 17.8 *Design a flexible pavement using the following data :*

CBR (soaked subgrade) = 3%

CBR (W.B.M. base) = 90%

Design life = 10 years

No. of heavy vehicles per day with 10% expected increase per year = 200

Solution No. of expected vehicles at the end of 10 years

$$= 200 \left(1+\frac{1}{100}\right)^{10}$$

$$= 200 \times (1.1)^{10}$$

$$= 200 \times 2.59$$

$$= 518 \text{ vehicles per day.}$$

So we will design the pavement for the maximum anticipated traffic I.R.C design curve E will be used.

Depth for 3% CBR on E curve = 57 cm

Depth for 90% CBR on E curve = 7 cm

Hence depth of sub-base = 57 – 7 = 50 cm.

If moorum is used for the sub-base the thickness of sub-base = 57 – 20 = 37 cm. Depth of W.B.M. course = 20–7 = 13 cm.

Note : For moorum sub-base the CBR is assumed to be 20% when it is soaked.

To summarise the above, the actual design thickness is determined on the following concepts :

1. The design thickness of the overlaying material should be sufficient to present expansion of the underlaying material by preventing it from absorbing more moisture from the top.
2. The design thickness of cover surface could be sufficient to prevent plastic deformation.
3. The cover material should be sufficient to resort accumulation of moisture. This thickness is determined on the basis of exudation pressure. The exudation pressure is taken as the design thickness and also to prevent loss of strength.

17.15 RIGID PAVEMENT DESIGN

Rigid pavements are those which contain sufficient rigidity and high modulus of elasticity to be able to budge over weaker spots in the sub-grade. Because of these properties of rigidity and flexural strength, the wheel loads are distributed over a large area and so deflection are small and unit pressures on the sub-grade is very low. Normally it is not necessary to provide sub-base for a rigid pavement. But where sub-grade condition could lead to non-uniform support for the pavement base, sub-base is sometimes used. The sub-base prevents damage from

1. Frost action

2. Poor damage
3. Mud pumping
4. Swell and shrinkage

The most common and popular rigid pavement is the cement concrete pavement. Following are the advantages and disadvantages of cement concrete roads:

Advantages

1. Provide a smooth riding surface.
2. Provide clean and dust free surface.
3. Provide long and maintenance free life.
4. Provide less operation cost surface.
5. Provide surface for withstanding heavy and impact loads.
6. Provide good visibility during day and night.

Disadvantages

1. High initial cost.
2. No additions and alteration possible.
3. Produce noise and glare.
4. Joints are a source of headache.
5. Surface becomes slippery during rainy season.

Stress developed in cement concrete roads. The stresses in cement concrete roads are due to variety of reasons viz. changes in temperature, wheel or impact load, changes in moisture contents in subgrade. These sources induces stresses in the concrete pavements resulting in deformations. To analyse the various stresses i.e., tensile, compressive and flexural, produced due to deformation, is a complex and tedious problem.

17.16 DESIGN FACTORS

The following design factors are to be considered for the design of cement concrete pavements:

1. Traffic Factor. The traffic factor consists of volume and character of volume anticipated to use the road which includes maximum wheel load contract area, tyre pressure, wheel configuration, repitition of load and impact Indian Road Congress have specified maximum axle load of 8200 kg and tanden axle load of 14500 kg. Load on one set of dual tyres is one half of the axle. In order to cater heavy loads it is common to add a tanden axle, having tanden spacing between 10 m to 1.2 m. The maximum safe laden weights of vehicles or combination should not exceed the value given by

$$W = 465(24 + 3.2L) - 14.6L^2$$

where W = safe weight in kg.

and L = distance between the extreme axles in metres.

The Ministry of Shipping and Transport have further recommended an increase of 2.5% in gross vehicle weight certified by the manufacturer. However a pilot study conducted at few places indicates a much higher axle load than stipulated. The pavement design procedure followed is based on the number of commercial vehicles of more than 3 tonne laden weight. The axle load should also be taken into account during the course of design. This aspect can be accounted for from the equivalence factors, which express the damage caused by known axle load as a ratio of the damaged caused by a standard 8.2 tonne axle. ASSHO Road test have revealed that 10 tonne axle would do 2.3 times as much damage as 8.16 tonne axle load while 16.3 tonne axle load would do 15 times as much damage as 8.16 tonne axle.

2. Temperature Differential Factor. Temperature differential is a function of the variation of minimum and maximum temperature during the day and during the year. The daily and annual mean temperature cycle of concrete pavement help in determination of maximum spacing of contraction and expansion joints. The most important environmental factor is the determination of temperature differential between top and bottom layer of the rigid pavement. Indian Road Congress have formulated the following temperature differentials for different zones.

Name	*Temperature variation in slabs*				
	10 cm	*15 cm*	*20 cm*	*25 cm*	*30 cm*
Punjab, U.P., Rajasthan, M.P., Haryana and coastal areas	10.2°C	12.3°C	13.1°C	14.1°C	15.8°C
Bihar, West Bengal, Orissa, Assam and coastal areas	14.4 C	15.6°C	16.4°C	16.6°C	16.8°C
Maharashtra, Karnataka, Andhra Pradesh, Chennai and coastal regions	14.75°C	17.3°C	19.0°C	20.3°C	21.0°C
Kerala and South Chennai	13.2°C	15.0°C	16.4°C	17.6°C	18.8°C
Coastal areas bounded by hills of Southern India	12.8°C	14.6°C	15.6°C	16.2°C	17.0°C

17.17 RIGID PAVEMENT DESIGN THEORIES

Stresses are induced in concrete pavements or rigid pavements due to bending or its prevention cement concrete pavements should be designed and controlled on the basis of flexural strength of concrete i.e., 40 kg/cm^2 (4 .0 MN/m^2) and compressive strength 38–42 kg/cm^2 (3.8 – 4.2 MN/m^2) is assumed to have modulus of elasticity as 3×10^5 kg/cm^2 (3×10^4 MN/m^2) Poissons ratio as 0.15 and coefficient of thermal expansion as 7.5×10^{-6}/C°.

1. Westergaurd Theory of Load Stresses. The pioneering work of rigid pavement design can be attributed to Professor H.M. Westergaurd way back in 1925. His theory was based on thin plates resting on an elastic foundation to determine the stresses due to interior and edge loading. He assumed that

1. Concrete slab ascts as a homogeneous thin elastic plate.

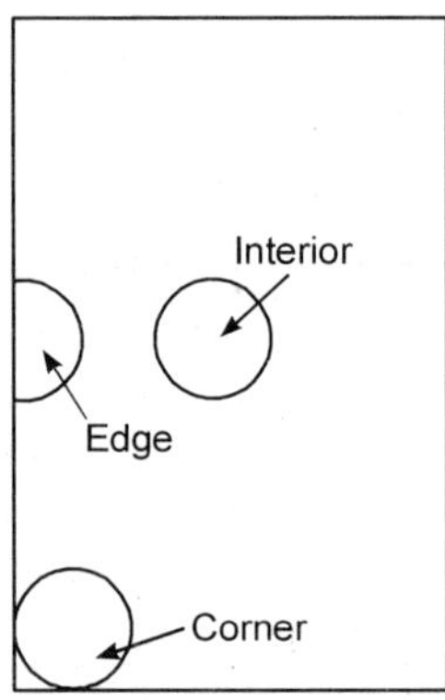

Figure 17.17 *Load stress diagram*

2. The reaction of the sub-grade is proportional to the deflection and the sub-grade has no shear strength.

3. The reaction of the sub-grade is equal to the deflection × modulus of sub-grade reaction.

4. The load is circular in shape in the middle as well as corner of the slab as shown in Fig. 17.17.

Westergaurd Stress Equation for Edge Loading

$$\sigma_e = \frac{0.512p}{h^2} [4 \log^{10} (l/b) + 0.359]$$

Corner Loading

$$\sigma_c = \frac{3p}{h^2} \; 1 - \left(\frac{a\sqrt{2}}{l}\right)^{0.6}$$

Interior Loading

$$\sigma_i = \frac{0.3162}{h^2} - [4 \log10\ l/b + 1.069]$$

where σ_e, σ_c and σ_i are stresses due to edge, corner and interior loading kg/cm^2

h = Thickness of slab

p = Wheel load

a = Radius of wheel load distribution in cm

b = Radius of resting section in cm

l = Radius of relative stiffness in cm

2. Gerald Pickett Equation. Gerald Pickett presented two formulae for two types of corners viz., protected and unprotected. A protected corner is that in which provision is made to transfer at least 20% of loading from one slab to another by some mechanical device such as dowel bar etc.

For Protected Corners

$$\sigma_c = \frac{3.36p}{h^2} \quad 1 - \frac{\sqrt{a/l}}{0.925 + 0.22\ a/l}$$

For Unprotected Corners

$$\sigma_{uc} = \frac{4.42p}{h^2} \quad 1 - \frac{\sqrt{a/l}}{0.925 + 0.22\ a/l}$$

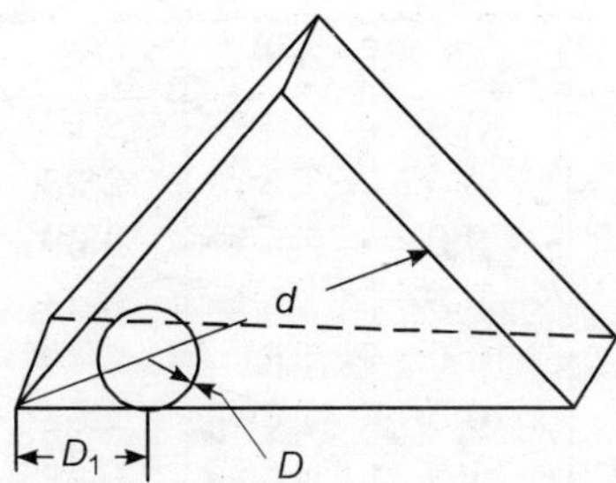

Figure 17.18 *Goldback theory*

3. Goldback's Equation. The assumption of this equation is that rigid pavement act as a cantilever beam with the load at the corner and the bending moment is assumed to be acting on a plane diagonally across the corner and is distributed uniformly over the cross section. The bending equation is

$$f = \frac{My}{I}$$

where

f = unit stress in kg/cm^2

$M = p \times x$

$y = h/2$ (distance of the extreme fibre from the neutral axis)

h = thickness of slab

I = Moment of inertia = $2x \times h^2/12$

Stress due to corner load

$$\sigma_e = \frac{px \cdot h/2}{2x \cdot h^2/12}$$

4. Bradbury Equation. Bradbury in 1938 presented an equation on the options tnat the sub-grade provide only partial support to the pavement slab at the edges. His equation is

$$\sigma_e = \frac{p}{h^2} \times Q$$

where

σ = Stress in kg/cm^2

p = Wheel load in kg

h = Thickness of slab in cm

Q = Stress coefficient depending upon l/b and $a\sqrt{2/l}$ ratios

l/b	Q_i	Q_e	l/b	Q_i	Q_e	l/b	Q_i	Q_e
1.0	0.34	0.21	8.0	1.48	2.27	14.0	1.79	2.84
1.5	0.55	0.61	8.5	1.52	2.33	14.5	1.81	2.86
2.0	0.72	0.89	9.0	1.55	2.39	15.0	1.83	2.90
2.5	0.85	1.12	9.5	1.58	2.44			
3.0	0.94	1.30	10.0	1.61	2.49			
3.5	1.02	1.45	10.5	1.63	2.54			
4.0	1.10	1.58	11.0	1.66	2.59			
4.5	1.17	1.73	11.5	1.68	2.63			
5.0	1.22	1.81	12.0	1.71	2.68			
5.5	1.28	1.90	12.5	1.73	2.72			
6.0	1.33	1.99	13.0	1.75	2.75			
6.5	1.38	2.07	13.5	1.77	2.80			
7.0	1.41	2.14						
7.5	1.45	2.21						

$\frac{a\sqrt{2}}{l}$	Q_c	$\frac{a\sqrt{2}}{l}$	Q_c	$\frac{a\sqrt{2}}{l}$	Q_c
0.00	3.00	0.25	1.69	0.50	1.02
0.05	2.50	0.30	1.54	0.55	0.90
0.10	2.25	0.35	1.40	0.60	0.79
0.15	2.04	0.40	1.27		
0.20	1.86	0.45	1.14		

5. Kelly Equation. In 1939 Kelly forwarded his equation for corner loading based on the tests carried by Teller and Southerrland

$$\sigma_c = 3p/h^2[1-(a\sqrt{2l})^{1.21}$$

All the symbols have the similar notions.

6. Spranglar Equation

$$\sigma_c = \frac{3.2p}{h^2} 1 - \frac{a\sqrt{2}}{l}$$

17.18 RIGID PAVEMNT DESIGN METHODS

(a) Portland Cement Association (PCA) Method. This method of design of rigid pavement or cement concrete pavement to be more precise, is based on Pickett's and Ray's extension of Westergaurd theory of distribution of static load applied over a circular area of surface of a homogeneous, isotropic and elastic slab resting over a uniform sub-grade of elastic nature. The charts and graphs developed by Pickett and Ray were computerized and the design thickness so obtained are realistic and practical.

(b) Losberg Method. This method is based on the assumption that the thickness of the rigid slab, resting on an elastic foundation, when loaded by a static load distributed over a small circular area, the maximum radial bending moment along a critical failure line does not exceed the bending moment which would rather produce continuous circumferential crack at a given radial line around the loaded area. From the ultimate negative bending moment slab section is determined.

(c) Load Classification Number Method. This method, also known as LCN method, originally developed in Great Britain by its Public Building and Works Ministry after carrying out extensive series of loading tests on existing pavements of different composition and sub-grade supports. The purpose of these tests was to ascertain or determine the load carrying capacity of a pavement through these tests standard load classification concept was developed to express the capacity of a pavement to carry an aircraft in terms of a single number known as LCN.

17.19 L.R.C. RECOMMENDATIONS FOR THE DESIGN OF CONCRETE PAVEMENTS

I.R.C. has recommended the following design procedure :

1. The design wheel load may be taken as 4800 kg with a tyre inflation pressure of 5.3 to 6.3 kg/cm^2.

2. The traffic volume may be projected for 20 years.

3. The traffic should be classified as under and adjustment in the thickness of slab be made.

Type	*No. of vehicles of weight more than 3 tonnes passing per day*	*Adjustment of thickness*
A	0 – 15	– 5 cm
B	15 – 45	– 5 cm
C	45 – 150	– 2 cm
D	150 – 450	– 2 cm
E	450 – 1500	0 cm
F	1500 – 4500	+ 2 cm
G	> 4500	+ 5 cm

4. The mean daily and annual temperature variation cycles for different states and zones should be collected as given in the Table on page 468.

5. The modulus of sub-grade K value is deterimined for 0.125 cm deflection on a 75 cm diameter standard plate.

6. The flexural strength of cement concreted to be used should be taken as 40 kg/cm^2.

7. The length and width of the slab should be decided on the basis of lane width and spacing of joints.

8. A trial thickness of the slab is assumed on the basis of past experience and standard practices, for the purpose of calculating stresses. The warping stress at the edge regions is calculated and this value is subtracted from the allowable flexural strength of concrete (40 kg/cm^2) to find the residual strength in the pavement to support the edge load. The equation for warping stress is

$$\sigma = \frac{E_{et}}{2} \frac{C_x + \mu C_y}{1 - \mu^2}$$

where

σ = Warping stress in kg/cm^2 at the interior

E = Modulus of elasticity of cement concrete in kg/cm^2

e = Thermal coefficient of concrete per °C

t = Temperature differential of the region

C_x = Coefficient based on L_y/l

C_y = Coefficient based on L_x/l in the desired direction (right angle to the direction of L_x/l)

μ = Poisson's ratio

9. The load stress in the region is calculated by the following formula.

$$\sigma_c = 0.529 \frac{p}{h^2} (1 + 0.54\ \mu)\ (4 \log_{10} l/b + \log_{10} b - 0.4048)$$

where

σ_c = Load stress

p = Wheel load

h = Thickness of pavement

l = Relative stiffness radius in cm

b = Radius of resting section

u – Poissons ratio of concrete = 0.15

The available factor of safety in edge load stress with respect to the residual strength is found. If this value of safety factor is less than 1.0 or much more than 1.0, another trial thickness of the slab is assumed till the factor of safety is nearly equal to 1.0 but not less than 1.0.

10. Cover load stress is checked as per Westergaurd equation

$$\sigma_c = \frac{3p}{h^2}\left(1 - \frac{a\sqrt{2}}{l}\right)^{1.2}$$

If this stress value is less than the allowable flexural stress in concrete, the thickness is sufficient and *vice versa*.

11. Spacing of joints should be adjusted depending on the length of slab and joint thickness. For 25 mm wide joints, the maximum recommended spacing of expansion joints is 140 m. Maximum spacing of contraction joints may be kept at 45 m in unreinforced slabs.

$$= \frac{3 \times 4100}{20^2} \; 1 - \left(20\sqrt{2 \times 91.4}\right)^{1.2}$$

$$= 15.4 \text{ kg/cm}^2$$

(b) *Bradury's method*

Strees at the edge

$$\sigma_c = ph^2 \times Q_i$$

where Q_i is a coefficient depending upon l/b. From the table on page 438 for 91.4/18.4 the value of $Q_i = 1.205$.

Therefore

$$\sigma_a = 4100 \times \; 20^2 \times 1.205$$

$$= 12.7 \text{ kg}$$

Stress at the corner

$$\sigma_c = \frac{p}{h^2} \times Q_c$$

$$= \frac{4100}{20^2} \times 1.54$$

$$= 15.8 \text{ kg/cm}^2$$

(c) *Picket's Method.* Stress on the unprotected corner

$$= \frac{4.2p}{h^2} \; 1 - \frac{\sqrt{a/l}}{0.925 + 0.22\; a/l}$$

$$= \frac{4.2 \times 1400}{20^2} \; 1 - \frac{\sqrt{20}/91.4}{0.925 + 0.22 \times 20/91.4}$$

$$= 43.1 \times (1 - 1.48)$$

$$= 22.3 \text{ kg/cm}^2$$

Stress on protected corner

$$= \frac{3.36p}{h^2} \; 1 - \frac{\sqrt{a/l}}{0.935 + 9.22\; a/l}$$

$$= \frac{3.36 \times 4100}{20^2} \quad 1 - \frac{\sqrt{20/91.4}}{0.935 + 9.22 \times 20/91.4}$$

$$= 17.92 \text{ kg/cm}^2$$

Example 17.10. *Calculate the thickness of concrete pavement for a wheel load of 4100 kg and 10% additional load for impact, the tyre pressure is 5 kg/cm^2, modules of sub-grade pressure is 6 kg/cm^3, Poisson's ratio of concrete is 0.15, modulus of elasticity of concrete is 3 × 10^5 kg/cm^2, flexural strength = 50 kg/cm^2. Assume factor of safety = 2.*

Solution

$$\text{Total load} = 4100 + \frac{10}{100} \times 4100$$

$$P = 4100 + 410 = 4510 \text{ kg}$$

Allowable flexural strength = 50/2 = 25 kg/cm^2

Trial thickness h

$$= \sqrt{3p/\sigma_c}$$

$$= \sqrt{\frac{3 \times 4510}{25}}$$

$$= 23.5 \text{ cm}$$

By Westergaurd formula, corner stress

$$= 3p/h^2 \quad 1 - \frac{a\sqrt{2}}{l}^{0.6}$$

where

$$a = \sqrt{\frac{P}{p \times \pi}}$$

$$= \sqrt{\frac{4510}{5 \times 3.14}}$$

$$= 17.1 \text{ cm}$$

and

$$l = \frac{Eh^3}{12(1-\mu^2)h}^{1/4} \quad (\text{assume } h = 20 \text{ cm})$$

$$= \left[\frac{3\times10^5\times20^3}{12\,(1-0.15^2)^6}\right]^{1/4}$$

$$= 68.4 \text{ cm}$$

Therefore corner stress

$$= \frac{3\times4510}{20^2}\left[1-\left(\frac{17.1\sqrt{2}}{68.4}\right)^{0.6}\right]$$

$$= 47.2 \times 0.45$$

$$= 21.3 \text{ kg/cm}^2$$

As the allowable stress is 25 kg/cm^2, hence assumed thickness of slab as 20 cm is reasonable and adequate.

Example 17.11 *Calculate the thickness of cement concrete pavement for a total wheel load of 4150 kg, tyre pressure 5 kg/cm^2, modulus of elasticity 2 × 10^5 kg/cm^2, modulus of sub-grade reaction k = 2.75 kg/cm^3 and Poisson's ratio = 0.15.*

Solution Bradbulry's corner formula

$$\sigma_c = 3p/h^2\,[1-(a/l)^{0.6})]$$

For h = 15 cm, l = 71 cm and a = 21.2 cm

$$\sigma_c = \frac{3\times4150}{15^2}\left[1-\left(\frac{21.2}{71}\right)^{0.6}\right]$$

$$= 28.2 \text{ kg/cm}^2$$

As the value of σ_c is more than the permissible limit, assume another thickness

$$h = 17.5 \text{ cm} \qquad \text{then} \quad l = 79.2 \text{ cm}$$

$$\sigma_c = \frac{3\times4510}{17.52}\times\left[1-\left(\frac{21.2}{79.2}\right)^{0.6}\right]$$

$$= 22.34 \text{ kg/cm}^2$$

This value is within the permissible, limit and hence safe.

The design can be repeated by Kelly's Formula

$$\sigma_c = \frac{3p}{h^2}\left[1-\left(\frac{a\sqrt{2}}{l}\right)^{1.2}\right]$$

Assume h = 17.5 cm and l will be 79.2 cm

$$\sigma_c = \frac{3 \times 4150}{17.5^2} \times 1 - \frac{21.2\sqrt{2}}{79.2}^{1.2}$$

$= 28.1$ kg/cm^2 which is higher than the permissible value of 25 kg/cm^2

Now assume h = 20 and the l will be 89 cm

$$\sigma_c = \frac{3 \times 4510}{20^2} \times 1 - \frac{21.2\sqrt{2}}{89}^{1.2}$$

$= 25.5$ kg/cm^2 which is also higher.

Now assume h = 22.5 cm then l = 97

$$\sigma_c = \frac{3 \times 4150}{20^2} \times 1 - \frac{21.2\sqrt{2}}{97}^{1.2}$$

$= 18.7$ kg/cm^2

which is safe

Example 17.12 *Design a dowel bar for a cement concrete road for a 20 cm thick concrete pavement, given the following data:*

Wheel load = 4100 kg

Joint thickness = 2 cm

Permissible flexible strength of dowel bar = 1400 kg/cm^2 and shear strength = 1000 kg/cm^2

Permissible bearing stress in concrete = 100 kg/cm^2

E for concrete = 3 × 10^5 kg/cm^2

k value for sub-grade = 8 kg/cm^3

$\mu = 0.15$

Assume any other data not given.

Solution Assume diameter of dowel bar to be 2.5 cm

$$\text{Length of dowel bar} = 5 \times d \quad \frac{f_s}{f_c} \times \frac{l + 1.5\,\delta}{i + 9.8\,\delta}^{1/2}$$

$$= 5 \times 2.5 \quad \frac{1400}{100} \times \frac{l + 1.5 \times 2}{l + 8.8 \times 2}$$

$$= 12.5 \quad 14 + \frac{3l}{l + 17.6}$$

$$= 40.5 \text{ cm}$$

Total length of dowel bar = 40.5 + 2 = 42.5 cm

p in shear $= 0.785\ d^2 \times fb$

$= 0.785 \times (2.5)^2 \times 1000$

$= 07.85 \times 6.25 \times 1000$

$= 4900$ kg

p in bending $= \dfrac{2d^2 f_s}{l + 8.8\delta}$

$$= \frac{2 \times 2.5^2 \times 1400}{40.5 + 8.8 \times 2}$$

$= 754$ kg

p in bearing $= \dfrac{f_c\ l^2 d}{12.5\ (r + 1.5\ \delta)}$

$$= \frac{100 + 10.5^2 \times 2.5}{12.5\ (40.5 + 1.5 \times 2)}$$

$= 755$ kg (aproximately)

Assume the load transfer capacity of dowel bar to be 40% of the wheel load, the capacity of the dowel bar will be

$$4100 \times \frac{400}{100} = 1640 \text{ kg}$$

If the factor of safety is assumed to be 2 then load transfer capacity will be

$$= \frac{1640}{2} = 820 \text{ kg}$$

which is much more than the value of p in bending, shear or bearing.

PROBLEMS

17.1. Differentiate between flexible pavements and rigid pavements as far as their design criteria is concerned.

17.2. What are the design criteria for the design of flexible pavements. Illucidate briefly.

17.3. Discuss those properties of cement concrete which are of vital importance in the design of concrete pavements.

17.4. Discuss the merits and demerits of the Group Index Method of flexible pavement design.

17.5. Design a suitable pavement for light traffic and calculate the Group Index for the following characteristics of a sub-grade soil :

Percentage passing through 75 micron sieve = 80, liquid limit = 30%, plasticity index = 15.

17.6. Elucidate any one method for the design for a flexible pavement of a highway.

17.7. A vehicle weighing 12000 kg with 60% of its weight is on the rear axle with dual tyres. Calculate the thickness of the pavement to be provided to sustain this load assuming the following data :

Permissible stress = 25.0 kg/cm^2

Modulus of elasticity of concrete = 2.2×10^5 kg/cm^2

Poisson's ratio μ = 0.15

Modulus of sub-grade reaction k = 2.75 kg/cm^3

Compare the results by using Bradburry, Kelly and Pickett's equation.

17.8. Using the following data, compute the total thickness of a flexible pavement. A section of the highway is passing through a water logged area with water table at 80 cm with adverse drainage conditions. The CBR value of the subgrade is 3% the total average daily traffic is 600 PCU.

17.9. Design a flexible pavement with the following data :

CBR value for subgrade = 3%

CBR value for mooram = 20%

CBR value for W.B.M. = 90%

No. of heavy vehicles per day = 200

Design life = 20 years

Annual rate of increase = 10%.

17.10. Design total thickness of a flexible pavement using 5 cm thick bituminous concrete surfacing, with the following data :

CBR value for compacted subgrade = 10%

CBR value for poorly graded gravel = 25

CBR value for broken stone ballast = 95%.

17.11. Design the thickness of a concrete pavement to support a maximum load of 4100 kg with an allowance of 10% for impact the tyre pressure may be assumed to be 5 kg/cm^2, using the following data:

K value for subgrade = 2.75 kg/cm^3

Modulus of elasticity for concrete = 2.1×10^5 kg/cm^2

Poission's ratio = 0.15

Flexural strength of concrete = 1000 kg/cm^2

Factor of safety = 2.

17.12. Design the thickness of a base course for a single wheel load of 22500 kg, 14 kg/cm^2 tyre pressure. Assume the following data :

1. 75 cm diameter plate load test, 0.8 kg/cm^2 pressure and 0.125 cm deflection.
2. Plate load test on 15 cm diameter lease and 75 cm diameter steel plate 2.1 kg/cm^2 pressure and 0.125 cm deflection.
3. Use charts and assume any other data not provided.

17.13. Calculate the thickness of a concrete pavement using Westergaurd corner stress equation with the following data :

E for concrete = 2.5×10^5 kg/cm^2

Poission's ratio = 0.15

K value for sub-grade reaction = 4 kg/cm^3

Wheel load = 4500 kg

Tyre pressure = 6 kg/cm^2

BIBLIOGRAPHY

1. Burmister, theory of displacement in layered surfaces—Highway Research Board U.S. 1962.
2. The AASHO Road Test – 61A – 1961.
3. DSIR 'Soil Mechanics for Road Engineers' H.M.S.O. London – 1970.
4. I.R.C. Standard Specifications and Code of Practice for the Construction of Concrete Roads I.R.C. – 1978.

5. S.K. Khanna, M.G Arora, B.R. Marwaha – Monograph for wheel load stresses in concrete pavements I.R.C. – 1970.
6. A.C.I. Recommended Practices for Design of Concrete Roads – American Concrete Institute – 1988.
7. I.R.C. Guidelines for design of rigid pavements IRC – 58 – 1974.
8. Kadyali LR – Highway Engineering.
9. S.K. Khanna – Highway Engineering.
10. Kelly F.F. – Structural Design of Concrete Pavement – 1945.

requirements and specific development. This is the basis of the eighth five year plan.

Main concentration on the development of new roads system for connecting interior villages is needed. For main roads the aim is to provide roads of fairly good condition connecting all districts and tehsil head quarters, ports, important towns and places of tourist interests, religious places etc. Provision of bridges, fly overs, by passes for places which cause considerable delay to traffic should receive priority in respect of high traffic volume roads, should be made. Construction of new roads to provide additional connections or short cuts should receive the least priority, unless these are needed for a specific project or special need or unless they are self financing through toll-tax or other means.

Modernisation of the high traffic density roads which carry bulk of country's total road transport movement, will bring about a marked improvement in road transport and should be given the highest priority.

Greater attention needs to be paid to the strengthening and widening of weak and narrow bridges on roads of various categories, to meet the increased axle load to be carried by the roads.

The policy of state construction of roads should be restricted to only rural roads. This restriction should not be applied to express ways, state highways and major district roads where properly designed surface should be provided to cater to the designed traffic volume.

As far as possible during the eighth plan periods, all major villages with a population of 5000 should be connected by fair weather roads.

18.2 RURAL ROADS AND VILLAGE CONNECTIVITY

Under the minimum need programme all villages having a population of 500 and above should be connected by fair weather roads. Generally speaking this seems to be a sound principle, it does tend to create some distortions, particularly in districts where the village population is scattered in small hamlets. The approach of the eighth five year plan should be to connect as much population as possible. So the yard stack should be not the number of villages but the percentage population of a district connected by roads.

Master plan on the basis of rural roads in each district should be prepared developing an optimum network of village road for connectivity. It is essential to identify nodal departments for planning and budgeting of rural roads matching with the increased traffic volume. This will help in automatic development of rural areas.

For road development, several innovative measures are underway including mobilization of private sector capital for the construction and maintenance of roads, a central road fund can be used the objectives of the 8th plan. For expressways development efforts must be made to mobilize the private sector resources by developing toll roads. Other sources like funds of Agriculture Produce Marketing Committees, Industrial Development Corporations, Tourism Corporations, Sugarcane cess funds, Salt cess funds, mineral cess fund need to be tapped to finance the specific road to meet their requirements.

Roads construction and maintenance technology needs to be modernised. Mechanisation for improvement of high density roads should be encouraged. An efficient information system and a strong data base of planning of future roads need to be developed. Research and development for properly carrying out a large programme of rural road work should receive due importance.

Roads should be planned to open up new areas for education, health, family welfare, postal services, commercial activities etc.

18.3 ROAD PATTERNS

For the planning road system in urban areas special care and precaution are to be taken for providing accident free traffic flow. The following patterns are usually adopted for planning of roads in nearly developed cities or innovating the existing roads:

(i) Minimum Travel distance pattern
(ii) Rectangular pattern
(iii) Star and Circular pattern
(iv) Radial and Grid pattern
(v) Star and block pattern
(vi) Hexagonal pattern.

(i) Minimum Travel Distance Pattern. In this system the entire city is divided into nodal points around a central portion by forming sectors. Each

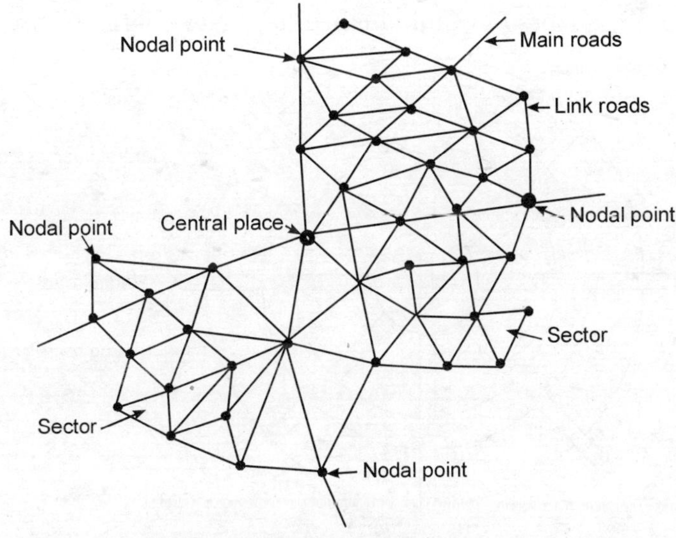

Figure 18.1 *Minimum travel distance*

sector is again sub-divided in such a fashion that from each of the nodal centres, the distance to the central portion is minimum as shown in Fig. 18.1.

(ii) Rectangular Pattern. It is also called block pattern. This system is not very convenient from the traffic point of view. In this pattern the

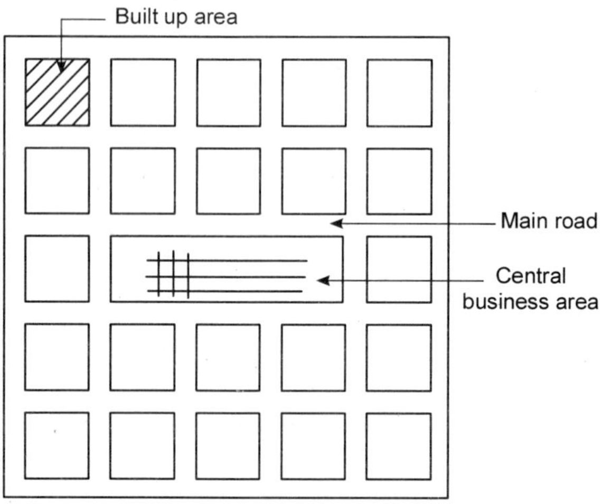

Figure18.2 *Rectangular pattern*

entire area is divided in rectangular blocks surrounded by lanes and bylanes on each side with central marketing and business area. All important central services are located in the central place. The main roads are around the central place. This is also called Chandigarh pattern.

(iii) Star and Circular Pattern. This is also called Radial and circular pattern and is adopted in Cannaught Place, New Delhi. In this pattern the

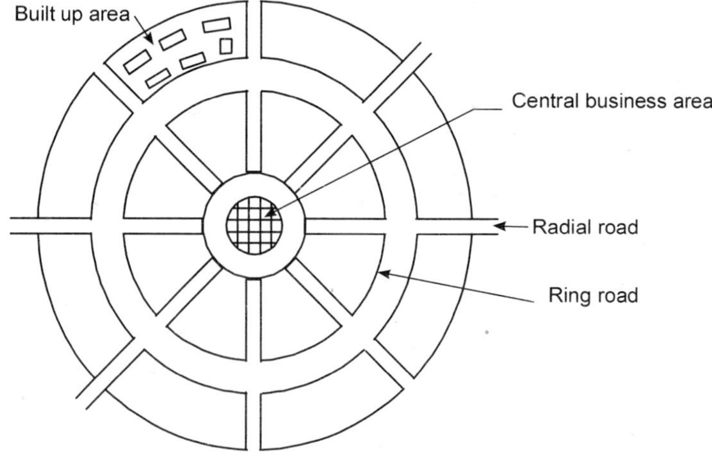

Figure 18.3 *Circular pattern*

entire area is divided into segments with central circular business area. Almost all the business activity is located in the centre or on the periphery. The business centre is connected by radial roads which in term are connected by ring roads.

(iv) Radial and Grid Pattern. This system is very much similar to the star and circular pattern. The only difference is that the ring roads

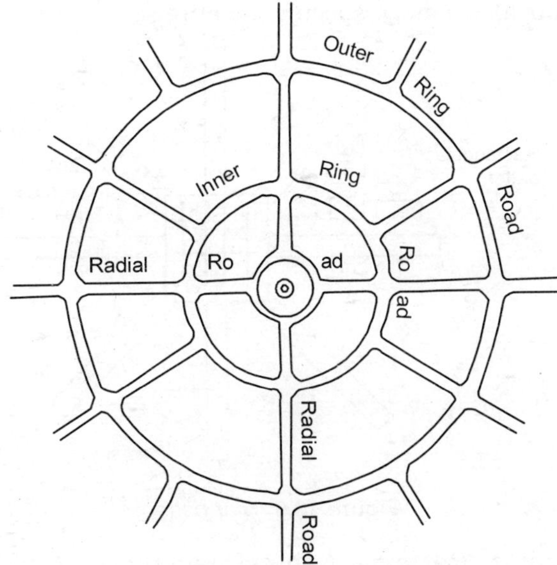

Figure 18.4 *Radial pattern*

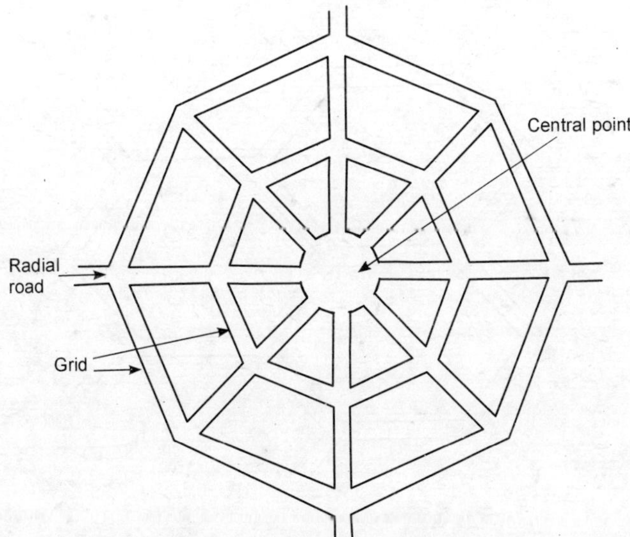

Figure 18.5 *Grid and star pattern*

and radial roads form the segments with built up area in between and business centre at the centre thereby forming a grid around the business centre as shown in Fig. 18.5. It is based on Nagpur plan.

(v) Star and Block Pattern. It is a combination of star and circular pattern and block pattern. In this system the ring roads are avoided and the radial roads enclose the blocks with a central business area. This system is not very popular as it has no specific advantage.

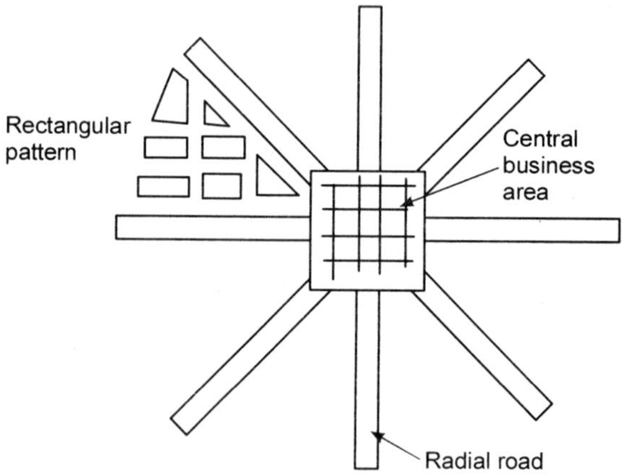

Figure 18.6 *Star pattern*

(vi) Hexagonal Pattern. In this system the entire area of planning is divided into hexagonal zones. Each zone will have a separate marketing

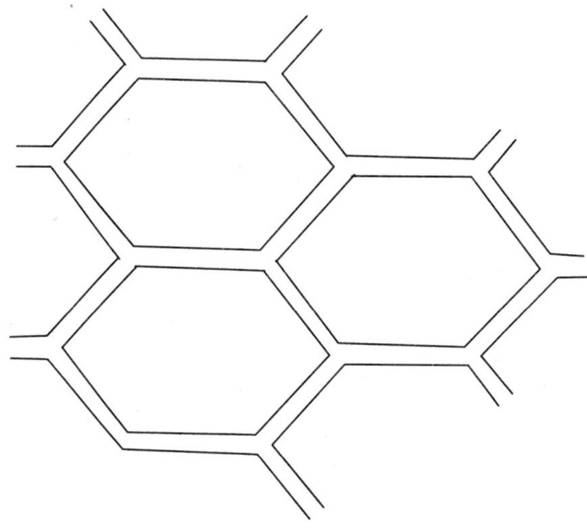

Figure 18.7 *Hexagonal pattern*

zone and central services surrounded by hexagonal pattern of roads. Each hexagonal segment is independent in almost all respects.

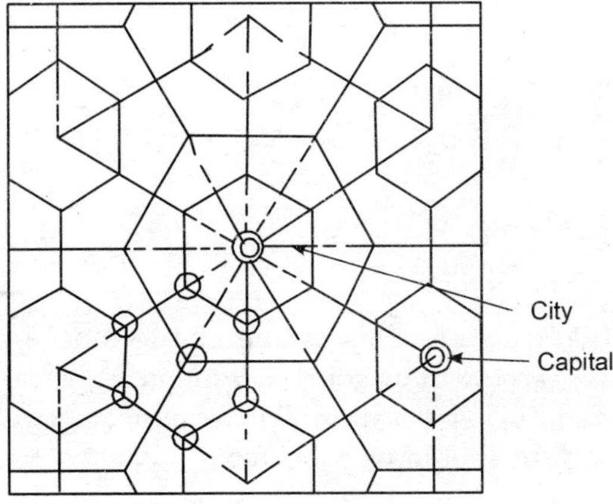

Figure 18.8 *Star and grid pattern*

18.4 METHOD OF PLANNING A ROAD SYSTEM

The highway planning for a country or an area includes

(i) Assessment of total road length

(ii) Location of nodal points to be connected invariably

(iii) Preparation of plans or master plan

(iv) Phasing of budget

(v) Fixing up priorities.

Thus for launching a project of highway planning, field survey is to be conducted for determining the total length of roads required, the volume of traffic, nature of traffic etc. These are fact finding surveys and are called planning surveys. The planning surveys include the following :

(i) Economic studies

(ii) Financial studies

(ii) Traffic studies

(iv) Engineering studies.

(i) Economic Studies. The economic studies consist of collecting data regarding service to be provided by each road system to the population and produce of the area. Details regarding the existing facilities available. The economic justification of each system is to be worked out. The following details should be collected by conducting surveys :

(a) Distribution of population of each area, village and town with classified groups.

(b) Population growth strategy.

(c) Industrial produce and the marketing centres with specific distances and classified groups.

(d) Agricultural and industrial development of the area and future trends.

(e) Per capita income.

(f) Existing road facilities.

(ii) Financial Studies. This is one of the most important aspect of highway planning. The various sources which can be tapped, the financial institutions, the funding agencies etc. The revenue to be generated. Following informations and data should be collected :

(a) Sources of income and estimated toll tax or revenue tax.

(b) Resources from other sources.

(iii) Traffic Studies. Before planning a road, traffic studies should be conducted by conducting the following studies :

(a) Traffic volume of vehicles per day, annual average, design hourly traffic volume and peak density. This data is precisely collected manually.

(b) *O* and *D* studies (Origin and Destination)

(c) Traffic flow pattern

(d) Other transport facilities

(e) Accidents—causes and patterns and cost studies

(f) Future trend and growth of volume of expected traffic and its pattern.

(iv) Engineering Studies. The engineering studies consist of the following:

(a) Topographical surveys

(b) Soil survey or analysis of soil its bearing capacity, porosity etc.

(c) Alignment of existing roads.

(d) Road life, its crust, etc.

(e) Drainage and cross drainage.

(f) Bypasses, links, bridges, crossings etc.

(g) Possible future development of the area.

All details of the topography of the land, soil conditions, construction and future extension, maintenance problems, availability of raw materials etc. are to be collected factually. Factual data collection is known as *Fact Finding Surveys.*

All the above information collected on the basis of fact finding surveys are tabulated and plans are prepared showing :

(i) Almost all details of existing road network, drainage structures, rivers, buildings, monuments, towns and villages, agricultural lands, commercial and industrial areas etc.

(ii) Distribution of population groups, population concentration etc.

(iii) The existing network of roads with traffic flow.

(iv) Agricultural and Industrial produce quantities of the proposed area.

18.5 MASTER PLAN

It is the development plan of an area, indicating all possible alignments and the final proposed alignment. The master plan is to be based on the basis of various studies conducted viz. Economic studies, Financial, and Engineering studies. Sometimes a target of road length is fixed for a particular area based on Nagpur Plan. The Nagpur plan is based on population, industrial and agricultural produce and future developments to be achieved. The master plan should also indicate phasing of the road development plan by fixing up the priorities for construction of different links. Of all the proposed alignments the best is to be selected which is a very difficult problem.

Saturation System. This system of selection of best alignment is based on optimum utility of the unit length of road. This system of highway planning is being used in most of the western countries. The utility of a road is based on population served and the traffic handled. The agriculture and industrial produce of the area to be served is of utmost importance. So far selecting the best alignment population and productivity units should be calculated by the following methods:

1. Population Unit. The proposed alignment will serve town and villages of varying population. These villages and towns are grouped or clubbed together thereby forming population units. For example, villages with population.

Less than 500 utility unit	=	0.25 population units
501 – 1000 utility unit	=	0.50 population units
1001– 2000 utility unit	=	1.0 population units
2001 – 5000 utility unit	=	2.0 population units

All the villages and towns to be served by the proposed alignment, are assigned utility units based on their population. These utility units are marked on the map showing the proposed alignment.

2. **Productivity Unit.** Total agricultural and industrial produce of the area under consideration is worked out by actual data collection. The weight of agricultural produce and industrial produce are assigned some units for example 1000 tonnes agricultural product may be taken as a one productivity unit. Similarly industrial produce can also be assigned similar units. These productivity units are marked on the map clubbing together different areas and locations.

Utility of Alignment. Total of population units and productivity units of each alignment is worked out and is divided by the total length of the proposed alignment to obtain utility per unit length of road. Each alternative alignment is considered based on utility units and the best is selected. This is the simplest and most reliable method but the limitation is the variation in assigning weightages to population, industrial and agricultural products. So while assigning utility units for population and the products special care, based on experience and expertise, should be given for a balanced proposal. Based on this the project can also be phased out and priorities can be allotted for financial considerations.

18.7 PLANNING BASED ON NAGPUR PLAN

First attempt for a proper and long term planning of road development in our country, was made in the Nagpur Conference in 1943. After the completion of Nagpur Road Plan, second twenty year plan, 1961–81 was drawn and the third twenty year plan 1981–2001 has also been approved by the Planning Commission. Following two formulae have been approved under the Nagpur Plan. The first formula is for calculating length of National Highways, State Highways and Major District roads, whereas the second formula is for calculating the length of village roads and minor district roads.

Formula I :

Total road length of National Highway, State Highway and Major District Roads in a particular area in kilometres will be

$$= \frac{A}{8} + \frac{B}{32} + 1.6\,N + 8\,T + D - R$$

where

A = Agricultural area in km^2

B = Non-agricultural area in km^2

N = Number of towns and villages with a population of 2001–5000

T = Number of towns with a population more than 5000

D = Development allowance of 15% or road length calculated to be provided for agricultural and industrial development during the next 20 years

R = Existing length of railway track in kilometres.

Formula II :

Total length of minor district roads and village roads

$$= 0.32\,V + 0.8\,Q + 1.6\,P + 3.2\,S + D$$

where

V = Number of villages with population less than 500

Q = Number of villages with population ranging between 501 – 1000

P = Number of villages with population 1001 – 2000

S = Number of villages with population 2001 – 5000

D = Development allowance of 15 percent for the next 20 years.

The above empirical formulae are based on actual data collection and experience. All possible aspects of population ranges, number of towns and villages and future development of the area has been taken into account. The formulae are based on *grid and star* pattern and hence sometimes called as *star and grid formulae.*

Analysis of the Formulae. From the above formulae it is clear that in addition to road length based on agricultural and non-agricultural areas, specific road length for towns and villages of various population groups, have been allotted. Towns and villages with population more than 5000 have been allotted 1.6 km of the first formula and 3.2 km of the second formula. Villages with population less than 500 have been allotted 0.32 km of second formula. A length of 1/8 km of first formula is provided for one km^2 of agricultural area. That is, that grids of the first category of roads (NH, SH and MDR) are 16 km apart such that an area of 16 × 16 km^2 is provided with 32 km of road length or 1.0 km road length is available for 8 km^2 agricultural land, that is why the term 1/8 is used in the second formula.

Example 18.1 *Calculate additional length of metalled and un-metalled roads to be provided for a town having the following data :*

(*i*) *Total area = 40,000 km*^{2}

(*ii*) *Agricultural area = 12,000 km*^{2}

(*iii*) *Length of railway track = 160 km*

(*iv*) *Length of existing metalled road = 450 km*

(*v*) *Length of existing un-metalled road = 600 km*

(*vi*) *Number of towns and villages to be connected with various population groups are*

Population	*>5000*	*2001–5000*	*1001–2000*	*501–1000*	*<500*
No. of towns and villages	*10*	*50*	*150*	*200*	*600*

Solution According to Nagpur Plan or Star and Grid formula. Total length of metalled roads required

$$= \frac{A}{8} + \frac{B}{32} + 1.6\,N + 8T + D - R$$

Here

$$A = 12{,}000 \text{ km}^2$$
$$B = 40{,}000 - 12{,}000 = 28{,}000 \text{ km}^2$$
$$N = 50$$
$$T = 10$$
$$D = 15\% \text{ of Road length}$$
$$R = 160 \text{ km}$$

$$= \frac{12000}{8} + \frac{28000}{2} + 1.6 \times 50 + 8 \times 10 + D - 160$$

$$= 1500 + 875 + 80 + D - 160$$

$$= 2455 + \frac{2455 \times 15}{100} - 160$$

$$= 2455 + 368.25 - 160$$
$$= 2663.25 \text{ km}$$

Additional metalled road required

$$= 2663.25 - 450$$
$$= 2213.25 \text{ km}$$

Total length of un-metalled road

$$= 0.32\,V + 0.8\,Q + 1.6\,P + 3.2\,S + D$$

Here

$$V = 600$$
$$Q = 200$$
$$P = 150$$

$$S = 50$$

$$D = 15\% \text{ of road length for future development}$$

Substituting the above data in the grid formula, we have

$$0.32 \times 600 + 0.8 \times 200 + 1.6 \times 150 + 3.2 \times 50 + D$$

$$= 192 + 160 + 240 + 160 + D$$

$$= 752 + \frac{15 \times 752}{100}$$

$$= 752 + 122.8$$

$$= 874.8 \text{ km}$$

Additional un-metalled road length required

$$= 874.8–600 \text{ km}$$

$$= \mathbf{274.8 \text{ km}}$$

18.7 THIRD TWENTY YEAR ROAD DEVELOPMENT PLAN FOR THE YEAR 1981–2001

This twenty year plan was finalised in Lucknow (U.P.) is sometimes known as Lucknow Plan. The final documents of this plan were published in the year 1984. Following are the salient features of this plan on which the plan is based:

1. Long term master plan for overall road development of the area should be prepared at the District, State and Block or taluka level.

2. Roads should also be built in less developed or industrially backward areas.

3. The future road development should be based on the revised classification of roads namely primary, secondary and tertiary roads.

4. The overall road density in the country should be increased to 82 km per 100 km^2 area by the end of this plan.

5. The National Highway Grid should form square grid of 100 km side.

6. Express ways should be constructed along the major traffic corridors to provide fast travel.

Methods of Calculating Road Length

1. Primary Roads. The length of the National Highways is calculated on the basis of 100 km square grid system i.e., for each 10000 sq. km area the road length should be 100 + 100 = 200 km. For each one square km area the length of National Highway should be

$$100/10{,}000 = 1/50 \text{ km}$$

Hence, Total length of National Highways anticipated by 2001 will be

$$32{,}87{,}782 \text{ sq km}/50$$

$$= 65{,}760 \text{ km approximately.}$$

Apart from this 2000 km expressways to be developed for fast traffic movement.

2. Secondry Roads. The secondry roads consist of State Highways and Major District Roads.

(*i*) *State Highways.* At least one national and one state highway should pass through a town. According to 1981 census there are 3364 towns all over the country. Hence the area of one square grid consisting of national highway and state highway should be the total area of the country divided by number of towns 32,87,782/3364

$$= 977\ 3 \text{ km}^2 \text{ i.e., a grid of } 31.26 \text{ km}$$

The length of National highways and State highways will $2 \times 31\ 26 =$ 62.52 km for each square grid. Hence total length of national highways + State highways will be equal to

$$62.52 \times 3364 = 2{,}10{,}360 \text{ km}$$

$$\text{Length of State highway} = 2{,}10{,}360 - 66{,}000$$

$$= 1{,}44{,}360 \text{ km}$$

Length of State highway in a particular state will be area of state in $\text{km}^2 \times$ 62.52 or

Area of State in $\text{km}^2/25$.

(*ii*) *Major district Roads.* Total length of MDR required by a state will be area of the stale in $\text{km}^2/12.5$ or

$90 \times$ No. of towns in the state.

(*iii*) *Rural roads and other district roads.* These are also known as *Tertiary roads.* As envisaged by the third twenty year plan, total length of teritary roads shall be 21,89,000 km. The length of National highways, State highways and Major district roads have been calculated and hence the length of village roads and other district roads can be calculated by subtracting the length of NH + SH + ODR from the proposed length of the roads as envisaged in the third plan. The proposed length of all types of roads in a state are given below :

Assam	85,280 km
Andhra Pradesh	1,74,856 km
Arunachal Pradesh	56,400 km

Bihar	2,12,000 km
Gujarat	1,14,800 km
Haryana	34,800 km
Himachal Pradesh	62,000 km
Jammu and Kashmir	54,000 km
Karnataka	1,45,000 km
Kerala	1,30,000 km
Madhya Pradesh	3,42,000 km
Maharashtra	2,20,000 km
Mizoram	26,000 km
Manipur	23,000 km
Orissa	1,66,000 km
Punjab	52,000 km
Rajasthan	2,10,000 km
Ultar Pradesh	3,56,000 km
West Bengal	1,16,000 km

18.8 APPROPRIATE TECHNOLOGY FOR RURAL ROADS

Most of our villages are un-connected even by fair weather roads and majority of ihc villages are connected by fair weather roads only. Majority of the trible areas are not even accessible during rainy season and they are completely cut off from rest of the country. The benefit of modern technology in agricultural farming, improved seeds and manures, tools and equipments cannot be provided without proper accessibility. Development of rural roads is also necessary for achieving the targets of integrated rural development. According to the third road development plan 1981–2001 the length of rural roads should increase from 1.3 million kilometre to 2.2 million kilometres for providing access to all villages having a population of 500 and above by 2001. Construction of such a huge network of roads is a challenging job. In the face of continued financial constraints, highway engineers are hard pressed to evolve the most economical design low cork construction techniques and maximum use of locally available materials to match overall socio-economic needs.

The choice of appropriate construction technology is to be based on technical feasability, economic viability and social desirability. The appropriate technology must be labour intensive to generate much needed employment opportunity. In the Indian context, employment of rural poor must be the deciding factor of adopting the appropriate technology.

For rural road construction improvised and hand tools should be encouraged. Animal driven rollers, cart mounted water bowsers fabricated from empty oil drums etc. should be used. Use of walktype vibratory compactors should be encouraged. Use of hand operated bitumen sprayers and power rollers should be encouraged.

The labour intensive technology or the appropriate technology for village road construction has the following advantages :

1. It creates more employment opportunities.

2. The construction is cheap and economical and hence permits provision for expansion of network to larger areas.

3. Large share of the funds allotted for the work remain in the villages thereby improving the life style of the rural masses.

In order to achieve the objectives of rural road constructions, the following measures are suggested:

1. Identify the technique of construction based or rural work force and locally available materials.

2. Organise such set up for employing large number of local persons.

3. Increase efficiency of individual labourers by supplementing with right type of hand tools and providing training with improvised machinery.

4. Giving incentives to labour for productivity.

18.9 ROLE OF TRANSPORTATION IN RURAL DEVELOPMENT

For the rapid and effective development of rural areas where more than 76% of our population live, road network and letter transportation facilities can play a very important role. Only with the improvement in transportation facilities fertilizers and other inputs for agriculture and cottage industries could reach the rural population easily and the produce can be sold easily in the nearby places for better renumerative price resulting in faster economic development growth and decreased wastages. With improved transportation facilities, education, health care and other social needs of the villages are effectively met.

PROBLEMS

18.1. Discuss the various features of highway planning in our country based on twenty year road plan.

18.2. Elaborate the importance of road transportation in rural development.

18.3. Explain how road lengths required for different states by the year 2001 are determined using Third Road Development Plan concept.

18.4. From the following data for a district, calculate the road length required based on Nagpur Plan and 20 year plan and compare the results :

Population of villages/Towns	*No.*
< 500	90
501 – 1000	150
1001 – 2000	200
2001 – 5000	180
5001 – 10,000	40
10,001 – 20,000	25
20,001 – 50,000	10
> 1,00,000	3

Area of the district = 12,000 km^2
Agricultural area = 4,800 km^2
Under developed area = 2,200 km^2

Pune 75, Nagpur 76

18.5. Explain why saturation system is considered a rational method to decide the final road network and for phasing the road development plan ?

18.6. Explain how master plan is prepared and the road development plan, is phased ?

BIBLIOGRAPHY

1. Vaish V.R.,'Project Management' Lok Udyog Vol. IX No. 4 – 1975.
2. Chahraboite N.K. 'Cost and Time over runs in New Projects' Lok Udyog Vol. IX – 1975.
3. Merani N.V. 'Initial Planning for Civil Engineering Projects and Special Problems affecting execution in Urban Areas' IRC 8 – 1986.
4. Report of the National Transport Policy Committee (Pande Committee), Planning Commission, 1980.
5. Chief Engineers Conference of Road Development (1943) Postwar Road Development (Nagpur Plan) I.R.C.

6. Report of the Committee on Transport Policy and Co-ordination (Trilok Singh Committee) 1966.
7. Nagaraj B.N., B.P. Chandrasekhar and S. Raghwa Chari. 'Delineation for Transportation Planning' IRC : 42–1981.
8. Dahui M.Q., Tarun Das and A. Kumar, 'Transport Model to Generate Inter regional Flow' Seminar Paper at I.I.T. Kanpur – 1978.
9. Kadiyali L.R. – Highway Engineering – 1983.
10. Report on Research Project-R-3 P.W.D. (B&R) Rajasthan,1979.
11. Road user Cost Studies – Central Road Research, Institute, 1982.
12. Traffic census on Non-Urban Roads IRC : 9–1979.
13. Report on Research Project R – 3 U.P. P.W.D. Research Institute, 1980.
14. Highway Research Board Iterim Report of the working group – Vol. 43 – 2. 1982.
15. Report of Chief Engineers on Road Development Plan for India (1961 – 81) M.O.S.T., 1958.

19

Highway Construction Management

In this Chapter you will study,

• Tenders and Contracts • Construction Programming • C.P.M. and PERT • Gantt-Bar Chart • Milestone Charts • Work Division Structure • PERT and CPM Network • Estimate of Time • Critical Path • Critical Path Method • Procedure for Preparation on CPM Network

GENERAL

Highway construction and management is an important aspect of highway construction and planning of major highways. The basic objective of the construction management is to have an economical construction, honouring time schedule, optimum use of available machinery, achieving cost benefits and providing maximum service to the road user. The construction planning should also aim at providing comfort and minimum disturbance to the adjoining areas. The management of construction consists of scheduling the work, controlling the quality control, managing finances, correlating the other infrastructural activities etc. In short it can be said that highway construction management is the process of getting the work done at the cheapest rate within the scheduled time period and maintaining quality control. As the highway construction involves lumpy investment of public

the work and all measures are taken that work is completed as per proposed schedule.

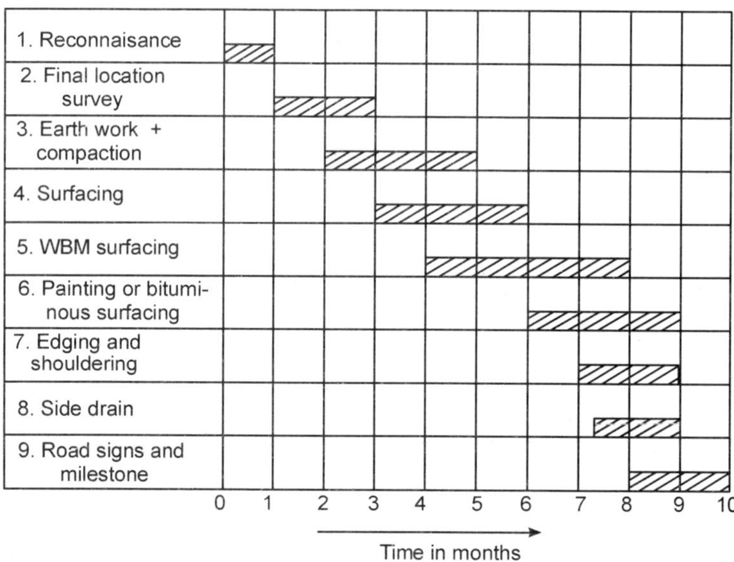

Figure 19.1 *Simple Gantt-Bar Chart*

19.5.1 Shortcomings of Gantt-bar System

Gantt-bar charts has some limitations in practice and it does not help the controller in many matters. Following are the main shortcomings of Gantt-bar chart:

(i) Position of Inter dependencies between various works.

(ii) Non-control on work progress.

(iii) Uncertainties.

(i) Position of Interdependencies between Various Works. In large works, where a job can be started before the completion of its previous job, the bar chart cannot show clearly the interdependencies among the various activities. This is big drawback of bar-chart. The overlapping time between the various activities or works does not show the relation or inter-dependent or complete independent of them.

(ii) Noncontrol of Work Progress. Gantt-bar chart cannot be used as a control device for the control of the work. When the work is in progress, it is absolutely necessary to have the uptodate knowledge of actual quantities of works done. But this chart of Fig. 19.1 cannot give uptodate position of work done. In all the major works, mostly some changes are unavoidable in some works, but Gantt-bar chart cannot give any assistance under such circumstances. The uptodate position of various work completed can be shown by modifying the Gantt-bar chart as shown in Fig. 19.2. In such cases the works actually completed upto a particular date are shown by

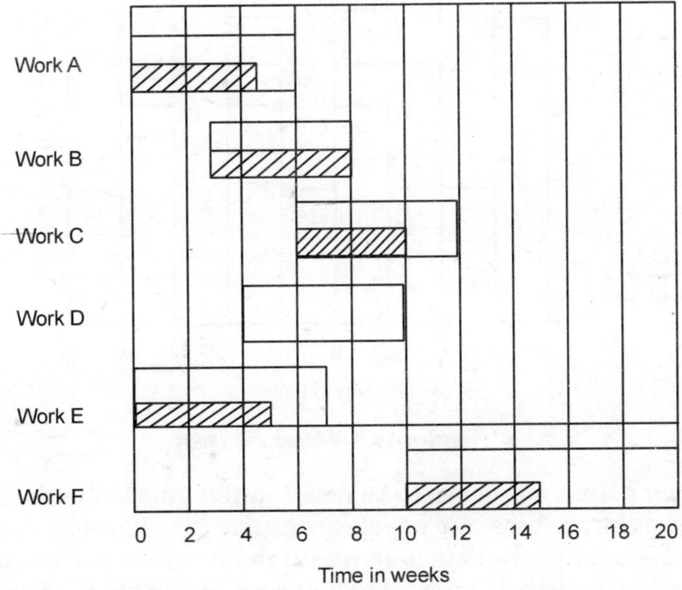

Figure 19.2 *Modified Gantt-bar chart*

partially filling the blank bars before the works. The filling is done daily, weekly or monthly as desired on the requirements. From this modified chart the works which are lagging behind, in schedule or ahead of schedule time can be directly noted.

(iii) Uncertainties. The main shortcoming of the Gantt-bar charge is its inability to illustrate or reflect the uncertainties or tolerances for the completion of various jobs. In the modern times during the execution of various projects the tolerance limits are required, so no proper watch can be on jobs having no tolerance limit of time.

19.6 MILESTONE CHARTS

Due to various shortcomings of Gantt-chart as described in the previous article, to meet the modern requirements for proper control of the work, various improvements are made in it from time to time. The improvement over the Gantt-bar chart is milestone chart or milestone system. Figure 19.3 shows a typical milestone chart. In this system the various sub works or events are shown by squares instead of bars of Gantt bar chart.

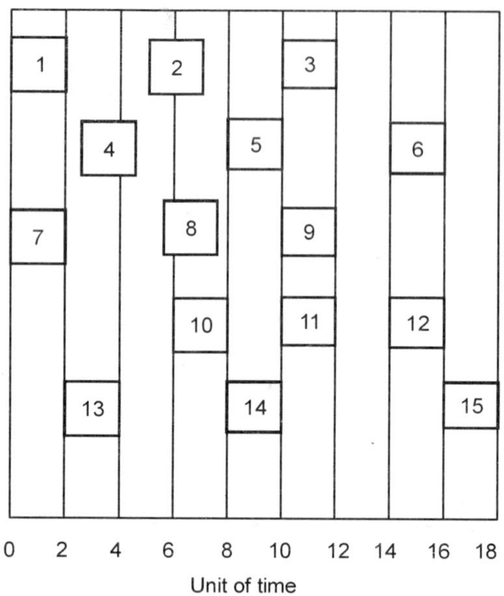

Figure 19.3 *Milestone chart*

The main points which are to be noted in the mile stone chart are (i) the long time works are break up in sub-works and are shown by specific events or milestones against the item in chart, (ii) these milestones indicating events are plotted against the time scale as in case of bar charts and (iii) all the events are serially numbered and such members are centered inside the mile stone indicated by square or circle as the case may be. The milestone chart also has the shortcoming of not showing independencies between various events or works. All the sub-works in the milestone chart are shown by chronological sequence instead of logical sequence.

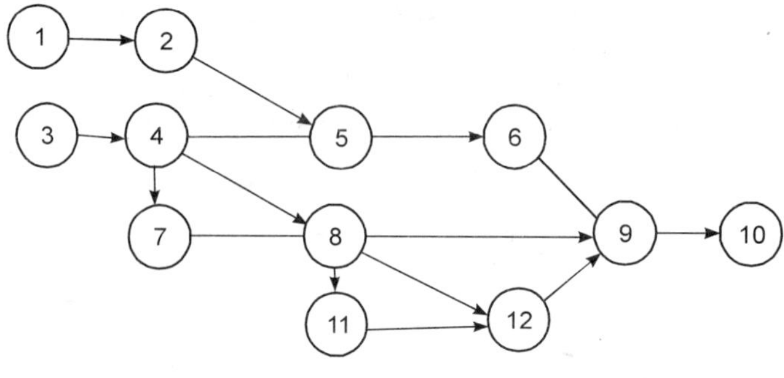

Figure 19.5 *Improved milestone chart*

Figure 19.4 shows the improved milestone chart in the form of a network, having all the sub works and works connected by arrows in a logical sequence.

19.7 WORK DIVISION STRUCTURE

The milestone chart gives the idea of the financial elements of the programme and their interrelationship. For drawing better chart, it is the primary work to divide the work into various possible sub works and prepare its structure. This is known as work division.

The division of the work is continued until it is reduced to elements or components which can be managed and controlled. The number of divisions of the work depends on the experience of the planner and the actual requirements for proper control of the work. Each such division unit is the key point of the work and must be placed at the right position in the chart.

19.8 PERT AND CPM NETWORK

Figure 19.4 shows an improved milestone net work. In the net work there are two basic elements (i) activity and (ii) event. The activity is the time consuming part of the work, whereas event is the beginning or end of a job. The activities are denoted by arrows and the events by circles or rectangles. When all the events and activities are connected logically and sequentially, they form a network. This network is very important in the execution and control of the works.

One of the main difference between the PERT network and CPM network is that PERT network is event oriented, whereas CPM network is activity oriented.

19.9 RULES FOR NUMBERING THE EVENTS

All the events should be numbered in their logical sequences. In case, complicated network has been formed after various additions and deflections, after so many trial networks, the numbering of the events is to be done very carefully.

D. R. Fulkerson has given the following rules which help in numbering the events:

(i) An 'initial' event is one which has arrows coming out of it and none entering it. In any network there will be one such event, Number it '1'.

(ii) Delete all arrows emerging from event 1. This will create atleast one more 'initial'.

(iii) Number these new initial events as '2, 3, 4, ...'.

(iv) Delete all emerging arrows from these numbered events which will create new initial events.

(v) Follow step (iii).

(vi) Continue the process till last event is obtained which has no arrows emerging from it.

19.10 ESTIMATE OF TIME

After deciding the network, the next work while preparing the network is to estimate the time of completion of each activity or job. As a matter of fact due to so many uncertainties, it is not possible to exactly calculate the time required for completion of all activities separately. From the previous experience gains while getting such works executed, the time which is expected to be taken in completion of the activity can be estimated. In complicated and complex projects where so many departments are responsible, the estimate of time for various activities are arrived in consultation with the experts of those departments. For taking into account the various uncertainties, following three types of estimates of time are usually prepared.

(i) Optimum Time Estimate (T_o). The estimate of shortest possible time, when the works can be completed under ideal conditions, is known as optimum time estimate. In this estimate no provisions of any delay or conditions beyond control are taken into account.

(ii) Most Likely Time Estimate (T_L). While preparing this estimate of time provisions of few setbacks, delays and breakdowns, which have occurred while doing such works in the past are taken into account. This time estimate is more nearer to the actual completion time of the activities.

(iii) Pessimistic Time Estimate (T_p). This is the maximum possible time which the activities will take in their completion, under worst possible conditions. While estimating the pessimistic time no provision is made for labour strikes and unrests, acts of God etc. which are beyond human control.

From the above three times, the expected time (T_e) can be calculated by taking average of all the three times above, by the formula,

$$T_e = \frac{T_o + 4\ T_L + T_P}{6}$$

The earliest expected time for each event is entered near the event circle. When an event is connected by more than one set of activity paths, the calculation is not so straight forward.

The latest time by which an event must occur to keep the work schedule is known as the *latest allowable occurrence time.*

For the calculation of the *earliest expected time* for the occurrence of a work, the calculation is to be started from the initial event and ended with the last event. This is known as *forward pass.* On the other hand, the latest allowable time for the occurrence of the work, the calculations are to be started from the last event and ended at the first event. This is known as *backward pass.*

Example 19.1 *Figure 19.5 shows a PERT network. The optimum, most likely and pessimistic time estimates are given on the arrows representing*

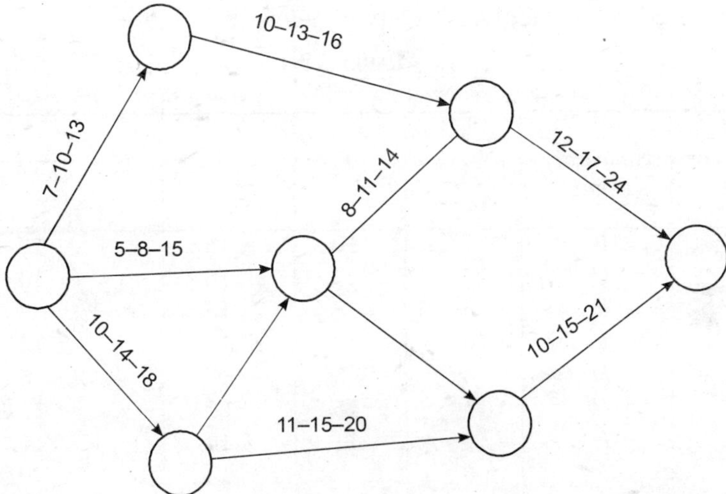

Figure 19.5 *PERT network*

the activities. Using Fulkerson's rule, number the events and work out the expected time for each activity.

Solution. Applying Fulkerson's rule, the first circle on the left hand side is the initial event because no arrows are entering in it. It shall be numbered

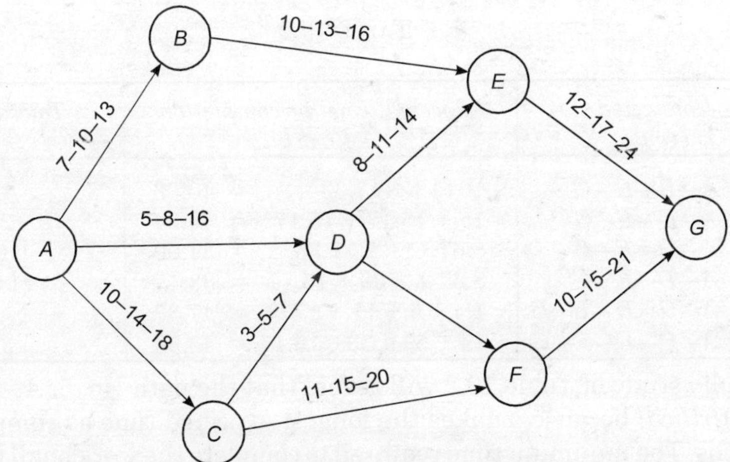

Figure 19 .6 *PERT network*

as *A*. Now deleting all arrows emerging from event *A*, only two circles are left with no arrows entering them. These two circles shall be numbered as *B* and *C* as shown in Fig. 19.6. Now deleting all arrows emerging from *A*, *B* and *C*, and continuing the process events *D*, *E*, *F* and *G* are numbered as shown in Fig. 19.6.

The expected time calculated is given below in Table 19 1.

TABLE 19.1

Predecessor event	*Succesor event*	T_e	T_L	T_P	$T_e = \frac{T_O - 4T_L + T_P}{6}$
A	*B*	7	10	13	10
A	*C*	10	14	18	14
A	*D*	5	8	15	8.67
B	*E*	10	13	16	13
C	*D*	3	5	7	5
C	*F*	11	15	20	15.17
D	*E*	8	11	14	11
D	*F*	6	9	14	9.33
E	*G*	12	17	24	17.33
F	*G*	10	15	21	15.17

19.11 CRITICAL PATH

Now after studying the PERT network of Fig. 19.6, the various paths for the completion of work, expected time to complete the work and tolerance shall be as given in Table 19.2.

TABLE 19.2

Path No.	*Connected events*	*Expected time for completion of work*	*Tolerance*
1.	*A—B—E—G*	10+13 + 17.33 = 40.33	7
2.	*A—C—F—G*	15 + 15.17 + 15.17 = 44.34	2.99
3.	*A—D—E—G*	8.67 +11 + 17.33 = 37.00	10.33
4.	*A—D—F—C*	8.67 + 8 33 + 15.17 = 33.17	14.16
5.	*A—C—D—E—G*	1.4 + 5 + 11 + 17.33 = 47.37	–
6.	*A—C—D—F—G*	14 + 5+9.33+ 15.17 = 43.50	3.83

Carefully study of Table 19.2 will reveal that the path No. 5, *A—C—D—E—G* is *Critical* because it takes the longest expected time as compared to other paths. The minimum time required to complete the work shall be 47.33 because without completing all the connected events, it is not possible to complete the work. The path *A—C—D—E—G* is known as *Critical Path* and any delay in any event on critical path will delay the completion of the work. Therefore it will be the duty of the management to keep strict watch on events *A—C—D—E—G—* connecting the critical path. While studying Table 19.2, it will be seen that other paths have some tolerance limits upto which the events *B* and *F* can be delayed. The method of controlling the events lying on critical path is known CPM (Critical Path Method). The arrow lines of critical path are shown by thicker lines than other arrow lines not-on this path as shown in Fig. 19.6.

19.12 CRITICAL PATH METHOD

In the previous article some idea about CPM is given. The reader should remember that PERT network is event based, whereas CPM network is actively based. But this does not mean that PERT do not take into account the activities or CPM ignores events. As a matter of fact events and activities are interrelated. PERT gives more emphasis on events and CPM on activities. Event is the start or close of an activity and no time is consumed in it.

CPM network mainly differs from PERT network in the following ways:

(i) CPM does not take into account the uncertainties which are involved during time estimation for the execution of the work.

(ii) Times are directly related to cost, with the increase in time cost of the work is increased and vice versa.

(iii) CPM network is prepared on the basis of the activities of jobs or sub-works, instead of events.

Hints for Preparation of CPM Network. Following points help in the preparation of the CPM network :

(i) What activity or activities which should be completed before the start of a particular activity ?

(ii) What activity or activities follow this ?

(iii) What activities can be started or accomplished simultaneously ?

Predecessor and Successor Activities. The activity or activities which immediately come before another activity without any intervening activities are called *Predecessor activities* to that activity. The activity or activities which follow another activity without any intervening activities are called *successor activities* to that activity.

Numbering the Events. The numbering of the events of CPM network is done as per Fulkerson's rule already described in the previous articles.

19.14 PROCEDURE FOR PREPARATION OF CPM NETWORK

Before starting preparation of CPM network, first work is to divide the work into a number of events and activities. This splitting of work into events and activities is a very important point of project planning and require great skill and practical experience. The event is the accomplishment in project planning. It is actually and particular instant of time and is represented by circle, rectangle or square. The start and completion of an activity is an event. Therefore every activity has two events within a project. The activity consumes time and money. It is represented by an arrow and indicates the sequence of the events in a logic manner. Every activity is independent, after the completion of all the activities, the work is completed.

Example 19.2 *If the estimated time for all the activities of the previous example are followed, work out the total project time for the completion of the work.*

Activity symbol	A_1	A_2	B	C_1	C_2	D_1	D_2	D_3	E_1	E_2	E_3	F
Estimated time in days	*10*	*7*	*1*	*2*	*5*	*3*	*4*	*15*	*15*	*3*	*60*	*40*
Acticity time in days	G_1	H	I	J	G_2	K_1	K_2	L_1	L_2	L_2	L_3	
Estimated time in days	*20*	*15*	*20*	*7*	*30*	*10*	*10*	*10*	*7*	*7*	*5*	
Activity symbols	*M*	*N*	P									
Estimated time in days	*20*	*20*	*10*									

Also draw critical path on the network.

Solution CPM network of the project is drawn and all the essential times in days are entered in triangles near the related activity as shown in Fig. 19.7.

Now the total estimated time for completion of work upto all events shall be worked out and entered in rectangles provided near every event in the bottom as shown in Fig. 19.7 as follows :

At event No. 1 as no time is consumed, hence total time shall be zero. At event No. 2 the time for completion of activity A_1 (10 days) shall be the total time. At event No. 3, the total time shall be for completion of activities A_1 and A_2 (10 + 7) days i.e., 17 days. At event No. 4, the total time for work upto this stage shall be estimated time in all activities before i.e., (10 + 7 + 7) = 24 days. Total time upto activities 5 shall be 24 + 2 = 26 days. At event No. 6 the total time shall be the great time of all activities of two paths viz.,

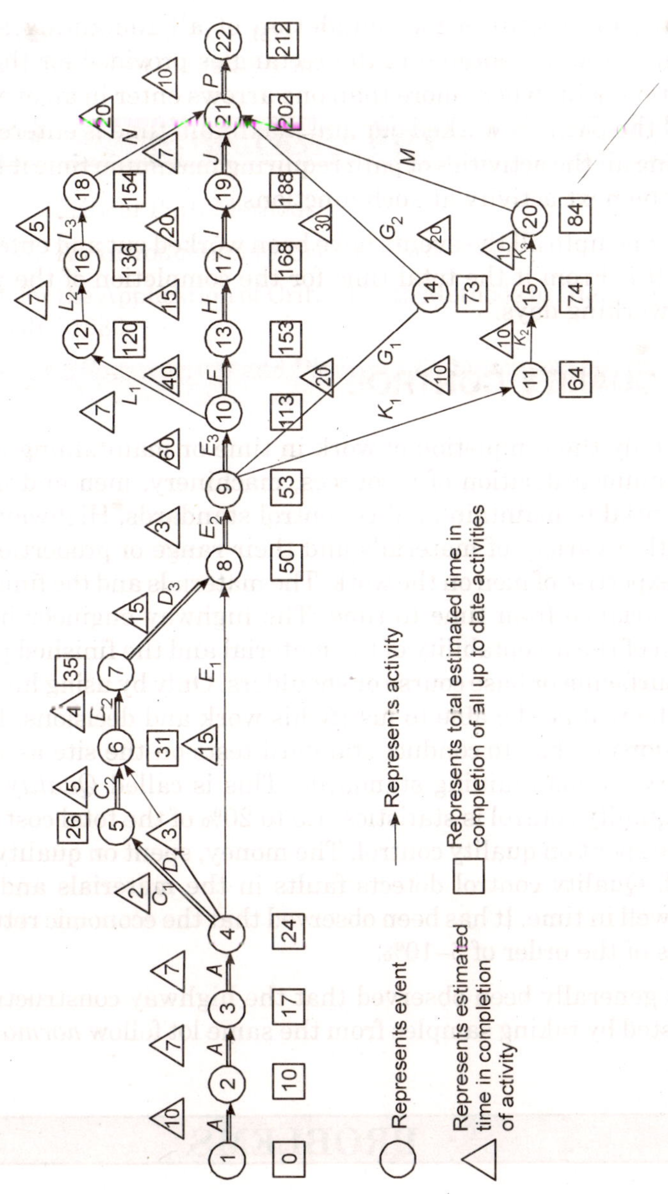

Figure 19.7 *CPM network*

20

Highway Economics

In this Chapter you will study,

• Definitions • Economic Evaluation of Highways • Motor Vehicle Operation Cost • Running Cost of a Vehicle • Time of Travel and its Value • Costs and Benefits • Annual Highway Cost • PERT and CPM Network • Estimate of Time • Critical Path • Critical Path Method • Procedure for Preparation on CPM Network

GENERAL

Highway or road construction is an economic ventures economic analysis are therefore becoming of increased importance to the highway planner and design engineer so as to reach to the most economical design. In situations where construction of a certain highway facility is obviously desirable economically and otherwise, it requires careful economic analysis to determine the comparative advantages of various proposals or features of location, design and construction programme priority. It is, therefore, necessary that each highway proposal should be examined properly to ascertain the economic justification of alternate proposals to ensure the overall soundness of the investment. A developing country like ours has serious shortages of resources needed for economic development. Highway projects involve lumpy investments of exchequer's money. The channelisation of funds for highway projects is at the cost of other sector of economy within the allocation of highway sector, a number of other schemes can be undertaken.

20.1 DEFINITIONS

Road User Services. These are the facilities, advantages, privileges, conveniences provided by the construction of a proposed highway. The road user services should also include economical transportation of goods, safe and comfortable ride.

Annual Cost of Operation. It is defined as the total cost of maintenance of a highway, operating and depreciation cost of vehicles.

Capital Cost. It is the total initial investment on the highway. This includes cost of land, cost of construction and the cost of equipment purchased for its construction exclusively or for kept for its maintenance. This should include cost of planning, designing and overhead cost of supervision and administration.

Maintenance Cost. It is cost of maintaining a highway in a serviceable condition. For maintaining its serviceability, repair work of various items is to be undertaken and this regular repair cost is called maintenance cost. Sometimes, during the years it is felt to innovate the entire pavement by way of overlaying, widening of surface, re-surfacing of the entire pavement to restore the highway to a condition of excellence or service superior to its original conditions. This additional investment is called *Betterment Cost.*

Highway Operating Cost. Cost of fuel for vehicles using the highway, cost of general services provided to road user not directly connected to maintenance of road such as cost of road signs, markings, signals etc. Cost of other physical components is also included in the operating cost.

Salvage Values. It is the value of the property for further use in some secondary purpose or capacity or used as a scrap when removed from its initial service. While estimating the overall depreciation cost chargeable on the period of initial service, the salvage value realized by the disposal of salvage is deducted from the original capital cost.

Depreciation. It is defined as the total overall cost of a property minus its salvage value at any particular time period. It is an element of cost of production. It is chargeable to the end product. The operating cost of a truck, the expenditure for fuel, tyres and repair are the elements of cost as the cost of new truck itself less its trade-in allowance i.e., its salvage value. Until the property is retired, depreciation cost is an estimated figure that allocates a fraction of the estimated depreciation of the property to each accounting period or unit of service rendered, usually applied on some systematic bases.

Capital Recovery. It is the combined sum of return received for depreciation of capital, the ownership cost of the capital investment in the form of interest recovered annually spread over the life time of the property.

Service Life. It is the period of time extending from the date of beginning of service to the facility to the date when it is removed from the service.

Each component of the highway, grading, bridge, pavement etc. has its service life.

Motor Vehicle Running Cost. The cost incurred for the operation of a vehicle is called its *Running Cost.* It comprises of cost of fuel, tyres, lubricants, repair and replacement of parts, permit and licencing cost, depreciation cost, garage, parking, insurance etc. costs not related to milage such as interest, depreciation etc. are called *Ownership Cost.*

Value of Occupants' Time. It is the estimation of the value of time in terms of money to the owner of the vehicle, the driver and other occupants. The element of time is an important factor of cost of highway users. The saving in time by highway improvement or otherwise may be taken into account by converting it to money value.

Running Time and Speed. Running time is the period of time taken by a vehicle to cover a specified distance or it is the time period during which a vehicle remains in motion from the beginning to the end of a journey of specified distance. The running speed of a vehicle over a specified distance or section of highway. The average running speed of all types of traffic is the sum of distances covered by each vehicle and the sum of running time of all vehicles.

Benefit Ratio. If the ratio of benefits accuring to the road users per year expressed in terms of money and the annual cost of the highway. It is also expressed as a measure of desirability of constructing a specific improvement.

20.2 ECONOMIC EVALUATION OF HIGHWAYS

Economic evaluation is a national approach of estimating the future benefits of a proposed highway or its innovations or improvements for assessing the extent to which the scheme will benefit the road user. It provides for a systematic approach for the selection of alternative proposals for implementation of a plan. Following are some of the important objectives of highway evaluation :

1. To select the best possible scheme for implementation keeping in view purely the economic consideration.

2. To fix up priorities for various schemes of implementation for the meagre and scare resources.

3. To assist in phasing the construction programme.

4. To venture alternative transport projects such as railways, waterways etc.

The economic evaluation is based on the following principles :

1. Economic evaluation and financial analysis are two different aspects of planning.

2. Economic evaluation should be based on minimum interest rate and maximum return rate.

3. The economic evaluation should not look back but should give prime importance to the future flow of costs and benefits.

4. The economic evaluation should consider opportunity cost of capital and resources whenever they are important or deserve consideration.

20.3 MOTOR VEHICLE OPERATION COST

The cost of operation of a motor vehicle on the road comprises of the following:

1. The cost which is directly proportional to the distance covered. This includes cost of petrol or diesel, cost of lubricant, wear and tear of tyres, repair and maintenance of vehicle, its depreciation etc.

2. The cost which depends upon the speed and the time of travel of the vehicle including the value of time of driver and the occupants.

3. The cost which increase with time but are constant for a unit time say one year such as licensing fee, registration fee, insurance etc. If expressed in terms of costs per vehicle kilometre and are inversely proportional to the kilometre driven.

4. The accidents occurred on roads also contribute to the cost of operation. The accidents may be fatal or otherwise has its cost impact on the operating cost of the vehicle.

20.4 RUNNING COST OF A VEHICLE

The average running cost of a vehicle consists of (i) fuel consumed, (ii) Mobile or lubricating oil, (iii) Tyre cost, (iv) Maintenance and repair cost, (v) Depreciation cost and (vi) Stopping cost.

(i) Fuel Cost

The cost of fuel consumed by a vehicle depends on the following :

(a) Running speed

(b) Condition of road surface

(c) Gradient

(d) Curvatures

(e) Super elevation

(f) Type of traffic on the road and its degree of congestion

(g) Number and duration of stops

(h) Weight and horsepower of the vehicle

(i) Efficiency of engine

(j) Skill of the driver.

(ii) Mobile Oil or Lubricant Cost. The consumption of mobile oil depends on the speed of the vehicle. More mobile oil is used with increased speeds. More mobile oil is used on rough surfaces like gravel and water bound macadam surfaces. But no rational method has been devised for calculating the actual consumption of mobile oil and the surface conditions.

(iii) Tyre Cost. The wear and tear of tyres directly depends on the surface conditions, grade, speed, curvatures and type of operation, inflation pressure, wheel balancing, alignment of wheels, overload. Tyre rotation are also responsible for the wear and tear of tyres. The wear of tyres increases considerably with the increased speed. Similarly sharp gradients curves also contribute to the wear of tyres. It is estimated that for the same speed, the wear of tyres is almost double on 7–10 degree grades than on a level stretch.

(iv) Maintenance and Repair Cost. Maintenance cost of high speed vehicles directly depends upon the type of pavement. On smoother and high tech roads the maintenance cost is considerably reduced. Maintenance also depends on the age of the vehicle. Old vehicles maintenance is much more than new vehicles.

(v) Depreciation Cost. It is defined as the cost of vehicle minus the salvage value at any particular time. Depreciation of a vehicle is a march towards the junkyard. Depreciation cost depends on the milage covered, the wear and tear, length of service of the vehicle and obsolescence or out of date.

(vii) Stopping Cost. Frequent stopping and starting the engines means more fuel consumption, more wear and tear of tyres, brakes, more lubricating and brake oil. More time is consumed at stopping.

20.5 TIME OF TRAVEL AND ITS VALUE

The modern highway planning and improvement aims at achieving higher and un-obstructed speed of vehicle so as to save time of travel. Time of travel is directly related to money. Lesser the time of travel more the savings by way wages. Frankly speaking these days time is money. For quicker system of transport public pays more money just to save travel time which in turn can be usefully utilized for more productive work. But there is no yardstick for measuring time for money. A conservative basis is desirable in economic studies and justification. Estimates of time saved and the value assigned to it should therefore be conservative and should be considered only in those proposals where such time servings has a realistic value. But as the travel time decreases or in other words speed increases the running cost of the vehicle also increases as more fuel is consumed at higher speeds. The optimum economical speed of general vehicles is between 45 kmph to 55 kmph.

20.6 COSTS AND BENEFITS

During the time of travel by any vehicle will cost something and will give certain benefits to the user. The cost of travel i.e., the operating cost of a vehicle have already been discussed. The benefits can be summarised as under:

(i) *Convenience and comfort to the drivers and passengers.* For the sake of comfort and convenience some people prefer to take up longer, thorough and safer route than shorter and congested route. Longer route will naturally cost more. Sometimes they are even prepared to pay heavy toll tax for the sake of comfort and convenience, which proves that people place a money value on the comfort and convenience of travel. So highways provide benefits to the public.

(ii) *Increased land value.* The value of land adjacent to the highway appreciates considerably with the construction of a highway.

(iii) *Increased trade.* This is an indirect benefit to the public.

In economic evaluation the main objective is to compare the costs and benefits of various alternatives. It is very difficult to categorise various costs and benefits. Cost can be defined as the minus effects and benefits as the plus effects of a proposal. The cost is an out-flow whereas the benefits are the inflow. Some of the positive and negative effects of a scheme can be quantified in monetary terms whereas some cannot. The aim of the highway engineer is to quantify as many elements as can be monetarily quantified, so as to justify the proposal in terms of money also, keeping in view that everything cannot be weighed by money alone.

The savings of positive and negative effects of a highway proposal are the benefits. So in short the benefits to the road user are convenience, time saving, mental satisfaction etc. The ratio of costs and benefits will be the actual criterion for accepting or rejecting a proposal. Following example will justify the above discussion.

Example 20.1 *An existing single lane highway 20 km long is proposed to be widened to a two-lane highway at the estimated cost of Rs. 20,000,000 (Two crores). The various costs for 10 year period are as under :*

Capital recovery formula is recommended for use in economy studies of highways.

Each component of the annual cost is to be calculated. It is, therefore essential that the proposal for highway must have complete details of each and every component.

$$C_D = R \times C$$

The highway capital cost C_D can be calculated from the relation

where R = capital recovery factor

$$= \frac{i\,(Hi)^n}{(1+i)^n - 1}$$

C = construction cost of highway.

Capital Recovery Factor Table

Life Year	Interest rate 5%	6%	8%	10%
10	0.12930	0.13587	0.14901	0.16275
15	0.09634	0.10296	0.11683	0.13147
20	0.08024	0.08718	0.10185	0.11746
30	0.06505	0.07265	0.08883	0.10608
50	0.05478	0.06344	0.08174	0.16086
100	0.05038	0.06018	0.08004	0.10001

20.8 RATE OF RETURN

To justify a proposal facility, the total annual net benefits should be equal to or greater than the average annual cost of ownership and operation. Therefore in the limiting case

$$B = A + O + M + C_D$$

where

B = annual net benefit.

But

$$C_D = R \times C$$

$$B = A + O + M + R \times C$$

or

$$R = \frac{B - A - O - M}{C}$$

Sometimes the proposal is to be compared with the existing facility, then the administrative cost, maintenance cost and operation cost of both the existing and proposed facility is to be taken into consideration and can be re-written as

$$R = \frac{(U_1 - U_2) - (A_1 + O_1 + M_1) - (A_2 + M_2)}{C}$$

where

$U_1 - U_2$ = Road use cost on existing and proposed facility

A_1 and A_2 = Administrative cost

O_1 and O_2 = Operational cost

M_1 and M_2 = Maintenance cost.

The benefits cost ratio

$$= \frac{\text{Benefits from improvement}}{\text{Cost of improvement}}$$

$$= \frac{U_1 - U_2}{C_1 - C_2}$$

where

C_1 = annual cost of existing facility

C_2 = annual cost of proposed facility.

Obviously the benefit cost ratio should be greater than one to justify the proposed improvement.

Example 20.1 *Calculate the annual cost of a highway from the following data:*

1. *Cost of land Rs.12.0 lakhs*
2. *Cost of pavement Rs. 14.0 lakhs.*

Assuming estimated life of land as 100 years and that of pavement as 15 years. The rate of interest being 10%. Average cost of maintenance is Rs. 1.5 lakhs.

Solution

(i) Annual cost of land = $12 \times R$

where R is capital recovery factor which is given in table page 522 is

$= 12 \times 0.10001 = \text{Rs. } 0.72216$ lakhs

(ii) Annual cost of pavement = $14 \times R$ Here $R = 0.13147$

$= \text{Rs. } 1.84058$ lakhs.

(iii) Average maintenance cost = Rs. 1.5 lakhs

Total annual cost of highway

$$= 0.72216 + 1.84058 + 1.5 = \text{Rs. } 3.06374 \text{ lakhs.}$$

Example 20.2 *Compare the annual cost of two proposals.*

(A) W.B.M. wifh bituminous surfacing expected life 5 years and total cost of construction Rs. 2.2 lakhs @ 10% compound interest rate. The salvage value after 5 years is estimated to be Rs. 0.9 lakhs.

(B) Bituminous Macadam pavement at the estimated cost of Rs. 4.2 lakhs with 10% compound interest. The estimated life is 15 years with a salvage value of Rs. 2.0 lakhs.

The annual maintenance cost in each case is 0.35 lakhs.

Solution

(A) W.B.M. with bituminous surfacing

Annual average cost

$$= (C - V) \times R + i \times V$$

where

C = capital cost = (2.2 – 0.9)

R = capital recovery factor = 0.26180

V = salvage value = 0.9 lakhs

$=(22 - 0.9) \times 0.26180 + 0.01 \times 0.9$

= Rs. 0.78294 lakhs.

(B) Bituminous Macadam Road

Annual average cost

$$= (C_1 - V_1) \times R_1 + i \times V_1$$

Here

C_1 = 4.4 lakhs

V_1 = Rs. 2.0 lakhs

R_1 = 10%

i = 10%

$(4.4 - 2.0) \times 0.13147 + 0.01 \times 2.0$ = Rs. 0.66702 lakhs.

As is being clear from above Bituminous Macadam road should be preferred as it is economically cheaper. The maintenance cost in both the cases being the same.

Example 20.3 *The distance between two stations is 100 km. The road between these stations is almost straight and the first 25 km stretch almost level, the* *other 55 km stretch a slope of 0–2%, 4% gradient for 15 km and 6% gradient for the rest 5 km length. Fuel consumption on the level surface is 0.06 lit/km, on 0–2% grade 0.09 lit/km, on 4% grade 0.12 lit/km and on 6% grade 0.17 lit/km.*

Cost of fuel is Rs. 9.0 per litre. Cost of tyre is Rs. 1000 per tyre. Depreciation of vehicle is estimated to be Re. 0.05 per km. Number of slopes on the way are 10 without delay, 5 with 45 second delay and 4 of 60 second delay. Assume any other data not given. Use of standard graphs is permitted.

Solution Total consumption of petrol for covering 100 km distance

$$\begin{aligned} 25 \times 0.06 &= 1.5 \text{ lit} \\ 55 \times 0.09 &= 4.95 \text{ lit} \\ 15 \times 0.12 &= 1.80 \text{ lit} \\ 5 \times 0.17 &= \underline{0.85 \text{ lit}} \\ & 9.10 \text{ lit} \end{aligned}$$

Cost of petrol = 9·10 × 9 = Rs. 81.90

Assuming an average economical speed of 42 km, the average life of a tyre is 32,000 km (from standard charts).

Cost of tyres for 100 km journey

$$= \frac{4 \times 1000 \times 100}{32{,}000}$$

$$= \text{Rs. } 12.50$$

Average stopping cost of the vehicle from the standard graphs is

10 no delay stops @ Re. 0.14	= Re. 1.40
5 stops of 45 seconds delay @ Re. 0.18	= Re. 0.90
4 stops of 60 seconds delay @ Re. 0.19	= Re. 07.6
Total	Re. 3.06

(**N.B.** : Each stop costs additional fuel).

Depreciation cost

Rs. 0.05 × 100 = Rs. 5.00

Total cost for covering a distance of 100 km will be

Rs. 81.90 + Rs. 1250 + Rs. 306 + Rs. 5.0

= Rs. 102.46.

Example 20.4 *Calculate the running cost of a car to cover a distance of 120 km. The average speed of the car is 50 km/hour including stoppages. Average* *depreciation is 10 paise per kilometre. The fuel consumption is 0.06 lit/km on level road, 0.08 lit/km on 2% gradient, 0.14 lit/km on 4% grade. The road surface is 50 km level, 25 km at 2% and the remaining 45 km at 4% gradient. There had been on an average 10 stops of 60 seconds. Assume no sharp curves on the stretch. Rate of petrol is Rs. 9.0 per lit and the average of tyre is Rs. 1000.*

Assume any other data and use standard graphs as and when required.

Solution

Fuel cost

$$50 \text{ km} \times 0.06 = 3.0 \text{ lit}$$
$$25 \text{ km} \times 0.08 = 2.0 \text{ lit}$$
$$45 \text{ km} \times 0.14 = 62 \text{ lit}$$
$$11.2 \text{ lit}$$
$$11.2 \times 9 = \text{Rs. } 1.008$$

Tyre cost

For an average speed of 50 km/hour the average life of tyre is 30,000 km.

$$\text{Tyre cost} = \frac{4 \times 1000 \times 120}{32,000} = \text{Rs. } 8.0$$

Depreciation cost

$$12.0 \times 0.10 = \text{Rs. } 12.0$$

Stoppage cost.

10 stoppages @ Re. 0.24 per stop = Rs. 2.40

Total journey cost = 100.80 + 8.00 + 12.0 + 2.40

= **Rs. 123.20.**

Example 20.5 A *tourist spot from a town is 25 km. The link road is in a bad condition. It is proposed to modify the link road. There are three proposals under consideration. You are required to examine the proposals purely on economic considerations and recommend the best proposal.*

Proposal 1. The existing road should be widened and resurfaced. The total cost is worked out to be Rs. 6.20 lakh, Rs. 1.5 lakh on grading, expected life 50 years, Rs. 1.6 lakh on structures expected life 50 years, Rs. 3.10 lakh on pavement expected life 15 years. The average speed of vehicles will be 40 km/hr.

Proposal 2. The existing alignment is defective, a new alignment is suggested which will decrease the length to 20 km. The surfacing will be bituminous macadam. The average speed will increase to 60 km/hr.

The cost of this proposal is

(i) *Land Rs. 27.0 lakh expected life 100 years*

(ii) *Grading Rs. 2.90 lakh expected life 50 years*

(iii) *Structures Rs. 2.50 lakh expected life 50 years*

(iv) *Pavement Rs. 15.50 lakh expected life 15 years.*

Proposal 3. A new alignment with 18 km length is suggested. The surface will be bituminous macadam. Expected speed of traffic will be 60 km/hr (average). The cost of construction is

Land Rs. 3.10 lakh expected life 100 years

Grading Rs. 3.30 lakh expected life 50 years

Structures Rs. 2.90 lakh expected life 50 years

Pavement Rs. 14.50 lakh expected life 15 years.

Assume present day traffic volume of 400 vehicles to double in 50 years. Rate of interest is 10% and maintenance cost Rs. 5,000 per kilometre for all the three proposals.

Assume any other data and use standard tables and graphs.

Solution Proposal 1

(i) Grading = 1,50,000 × R

where R is the capital recovery factor and from the table at 10% interest rate 50 years expected life, the value of R = 0.10086.

1,50,000 × 0.10086 = Rs. 15,129.00

(ii) Structures

1,60,000 × 0.10086 = Rs. 16,13,760

(iii) Pavement

Rs. 3,10,000 × 0.13147 = Rs. 41,755.70

(v) Maintenance cost

25 × 5,000 = Rs. 1,25000

On similar ground the cost of each element of proposal 2 and 3 are worked and tabulated.

1. Highway cost

Cost element	*Propsoal 1*	*Propsoal 2*	*Proposal 3*
Land	—	Rs. 27,000.70	Rs. 21,003.3300
Grading	Rs. 15,129.00	Rs. 29,249.40	Rs. 33,373.80
Structures	Rs. 16,137.60	Rs. 29,249.40	Rs. 33,373.80
Pavement	Rs. 41,755.70	Rs. 193,778.50	Rs. 1,88,641.50
Maintenance	Rs. 1,25,000.00	Rs. 1,00,000.00	Rs. 90,000.00
Total Annual cost (H)	Rs. 1,88,022.30 H_1	Rs. 3,75,243.60 H_2	Rs. 3,81,167.70 H_3

2. Road User Cost

	1	*2*	*3*
Average traffic during design period of 50 yeras	At 40 km average speed	At 60 km Average speed	At 60 km average speed
$\frac{(400+800)}{2}$	0.55 × 600	0.50 × 600 ×	0.50 × 600 ×
	365 × 25	365 × 20	365 × 18
	= 30,11,250.00	= 21,90,000.00	= 19.71,000.00

Cost element	*1*	*2*	*3*
Time cost per vehicle km for all the vehicles for the road length	At 40 km/h Re. 0.14 speed	At 60 km/hr Re. 1.12 speed	At 60 km/hr Re. 0.12
(use standard graph)	0.14 × 600 × 365 × 25 = 7,66,500.00	0.12 × 600 × 365 × 20 = 5,25,600.00	0.12 × 600 × 365 × 18 = 4,73,040.00
Total annual road use cost R	37.77,750 R_1	27,15,600 R_2	24,44,840 R_3
Grand total of Highway Cost and road use cost (annually)	Rs. 39,65,722.00	Rs. 27,53,123.00	Rs. 28,25,207

Benefit cost ratio of Proposal 1 and Proposal 2

$$= \frac{R_1 - R_2}{H_2 - H_1}$$

$$= \frac{37,77,750 - 27,15,600}{3,75,243.60 - 1,88,022} = \frac{10.62,150}{1,87,221.60}$$

$$= 5.67$$

Benefit cost ratio of Proposal 1 and Proposal 3

$$= \frac{R_1 - R_3}{H_3 - H_1}$$

$$= \frac{37,77,750 - 24,44,840}{3,81,167, -1,88,022}$$

$$= \frac{13,32,910}{1,93,145}$$

$$= 6.90$$

Proposed number 3 is more economical and hence recommended.

PROBLEMS

20.1. Explain the purpose of highway economy studies.

20.2. Write short notes on :

Road use services. Capital cost of a highway, Maintenance cost, Capital recovery factor, Highway operating cost.

20.3. A link road is 30 km long. Its condition of surfacing and other structures are in a bad shape. The alignment is also not proper. The proposal for its reconditioning and surface dressing is enclosed with two different proposal of fresh alignment. The length of road along the new alignments is 26.5 km'and 27.8 km respectively. The present traffic volume is 600 vehicles per day likely to be doubled in 15 years. The average speed of vehicles on the reconditioned surface will be 50 km/hr whereas on new proposed alignment it will be 60 km/hr.

Assume Rs. 10,000 per kilometre as the maintenance cost for the three roads at rate of interest is 10%. The estimated details of cost are give in the following table.

Cost element	*Estimated Life years*	*Cost in lakhs rupees*		
		Proposal 1	*Proposal 2*	*Proposal 3*
Land	100	–	2.70	3.10
Grading	50	1.50	3.10	3.40
Structures	50	1.80	2.90	2.80
Pavement	15	3.40	15.50	16.20

Compare the three proposals on cost benefit basis.

20.4. Distance, between Allahabad and Lucknow is 200 km (approximately). Calculate the average operation cost of a car for an economical speed of 50 km/hr, given the following data :

Cost of petrol = Rs, 90 per lit
Cost of tyre = Rs. 1000 each
Grade — 80 km level, 40 km 2%
60 km 4% and 20 km 6%.

Pavement surface condition is good. Speed reduction to 25 km/hr and 20 km/hr at 8 congested zones.

Total 15 stops of 60 seconds (average) delay and time cost per vehicle hour = Rs. 84.50. Depreciation of vehicle is 10 paisa per km.

Use standard graphs and tables and assume missing data

20.5. State the procedure of comparing the various alignment proposals for a highway. What is cost benefit ratio.

20.6. Calculate the operating cost of passenger vehicle for 200 km length of a road with no sharp curves using the following data :

Gradients : 60 km level, 50 km 2%
50 km 4% and 40 km 6%

Pavement surface condition–good.

Petrol = Rs. 90 per litre

Tyre = Rs. 1000 each.

No. of stops 10 without delay, 5 with 45 seconds delay and 5 with 60 seconds delay.

Time cost vehicle-hour = Rs. 3 00

Depreciation = 10 paise per km

Use standard charts and graphs, given in the text.

20.7. The cost of improving an existing road 30 km long, various costs of with and without improvement, are tabulated below on a 10 year period after the completion of improvement.

(Cost in Rupees lakhs)

Year	*Road user cost*		*Accident cost*		*Maintenance cost*	
	with improvement	*without imprvement*	*with improvement*	*without improvement*	*with improvement*	*without improvement*
1st	110.5	126.4	1.1	3.1	3.6	2.5
2nd	112.6	128.6	1.2	3.2	3.6	2.6
3rd	115.6	131.2	1.2	3.2	3.8	2.6
4th	118.5	135.4	1.4	3.4	3.9	2.7
5th	120.6	138.5	1.5	3.4	4.2	2.8
6th	124.4	142.4	1.4	3.5	4.3	2.9
7th	129.8	144.2	1.6	3.8	4.6	2.9
8th	132.4	150.4	1.7	3.9	4.8	3.2
9th	140 .5	156.5	1.6	3.9	5.2	3.4
10th	144.8	162.2	1.8	4.1	5.4	3.4

Assume a discount rate of 10%. Find out whether the project of improvement is economically viable.

20.8. What is motor vehicle running cost operation cost on a highway ?

20.9. Describe the method of evaluating a highway proposal on the basis of economic considerations.

20.10. On what factors the fuel cost of a vehicle depends ?

20.11. What is cost and benefits of a highway? Explain cost benefit ratio of a highway.

20.12. Describe the method of calculating annual highway cost.

20.13. Explain what is recovery cost factor?

20.14. State the factors on which economic studies of highway depends.

20.15. What is discount rate of a highway project. Explain.

BIBLIOGRAPHY

1. Central Road Research Institute's Final Report, Vol. I–VIII on Road User Cost Study on Indian Roads.
2. Swaminathan and Kadiyali—Vehicle Operating Costs under Indian Road and Traffic Conditions, 1983.
3. Rad User Benefit Analysis for Highways Improvements — AASHO, 1960.
4. Winfrey R., Concept of Engineering Economics in Highway Field. Road Research Technical Paper, 1961.
5. Beesley M.E. and D.S. Reynolds — Economic Assessment — Road Research Technical Report, 1961.
6. Institution of Civil Engineers, Research and Road Safety, 1964.
7. Reynold D.S. — Cost of Road Accidents, 1956.
8. Dawson R.F. — Analysis of Cost of Road Improvement. Road Research Tech. Report No. 50–1961— HMSO London.
9. IRC–1969 Road Accidents Cost Evaluation (Seminar Papers).

21

Airport Engineering

In this Chapter you will study,

• Definitons • Componetns of an Airport • Location of an Airport • Type of Airports • Orientation of Airports • Taxi Ways • Sight Distance • Layouyt Pattern of Runways • Runways and Intensity of Traffic • Terminal Area • International Civil Aviation Organisation • Geometric Standards • Geometric Design • Structural Design of Runway Pavements • Airport Drainage

GENERAL

Roads act as feeders to aerodromes and airports. The airport engineering deals with the construction and maintenance of air strips, runways etc. In fact this is not all as far as airport engineering is concerned because it has now come into very much prominance and deals with planning and designing of runways, controls, signals and so many other things. But this chapter has been devoted mainly for the requirements, construction and maintenance of air-strips.

21.1 DEFINITIONS

1. **Airport :** It is the place where public facilities are provided for shelter, waiting lounge, restaurant etc. Airport also provides facilities for servicing and repair of aircrafts, refueling, receiving and discharging cargo etc.

2. Runways : It is a hard surfaced straight path, used for the movement of aeroplanes for taking off and landing within a landing area.

3. Landing Area : It is that portion of the airfield which is used primarily for taking off and landing of aircrafts. (The word aircraft is a general term used for aeroplanes, gliders, helicopter etc).

21.2 COMPONENTS OF AN AIRPORT

A large size international airport is in fact a city in itself. It has almost all the amenities of a modern life. It is normally spread over a vast area of open land. Following are the various components of a modern airport :

1. Runways
2. Lounge
3. Restaurants and shopping complex
4. Hangers
5. Fueling
6. Taxiways
7. Loading and unloading cargo yards
8. Aprons.

In this chapter only runways will be discussed as it is beyond the scope of this book to discuss about all the components of an airport.

21.3 LOCATION OF AN AIRPORT

For the location of an airport there are many a considerations such as political, geographical, aeronautical, military, economic and many other considerations. The following points should be particularly thought over :

1. It should be far away (minimum 3 kilometres) from the urban boundary or congested locality.

2. It should not have obstruction i.e., such objects which may obstruct the movement of flights.

3. It should not have restrictions for future expansions and developments.

4. It should be away from industrial hazardous, thereby limiting the operations of flights due to poor visibility etc.

5. The site of an airport should be away from the hillock, river and pond etc.

6. It should be well connected by a national highway or state highway.

21.4 TYPE OF AIRPORTS

In different countries, airports are classified in different ways. In India airports are classified as civil and military. The military airports are mainly meant for military aircrafts only. The civilian airports can be used for military aircrafts also. The civilian airports can be a minor, intermediate, major and international airports. For the development of airports and airways the Govt. of India under the Ministry of Civil Aviation, established two public sector enterprises namely Air India and Indian Airlines. The Air India International for foreign services and Indian Airlines for domestic flights. The airways and air services are fast growing world over and hence the airport engineering. The airways and airports are rendering creditable service to the mankind. There are various methods of classifying the airports. The International Air Transport Association (I ATA) has fixed certain norms for the classification of airports.

21.5 ORIENTATION OF AIRPORTS

In the planning of airports, wind direction, its intensity, number of air strips, and metrological conditions, play an important part. The landing and take-off is done in the direction of prevailing winds :

Small aeroplanes are more sensitive than big commercial jets or boeings and airbuses, because of less wing load. The number of airways or air strips depends on the intensity of prevailing winds. While orienting runways, cross components of the prevailing winds should be calculated. Normally properly oriented runways may persuit aircrafts with cross component of wind as 20 km per hour (cross component of wind is defined as wind velocity multiplied by sine of angle between runways and direction of wind). At the time of orienting runways, direction and velocity may be obtained from metrological departments.

21.6 LENGTH OF RUNWAYS

For taking off and landing, the length of a runway may be divided into :

(i) Distance from the runway end to the starting point.

(ii) Distance between the starting point and the take off speed.

(iii) Additional distance for eventuality when the engine of the aircraft fails and the engine comes to a halt.

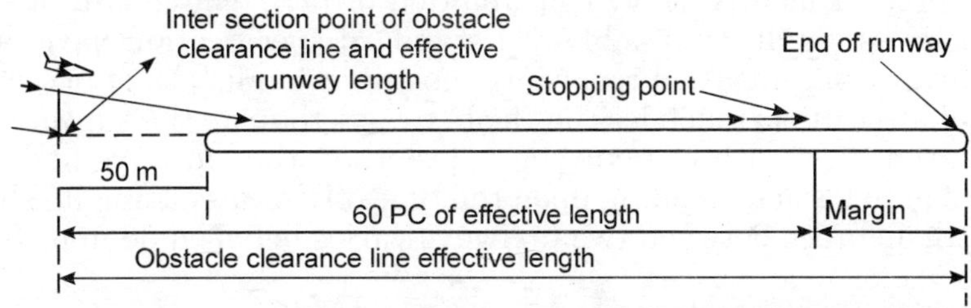

Figure 21.1 *Runway limitation for landing of planes*

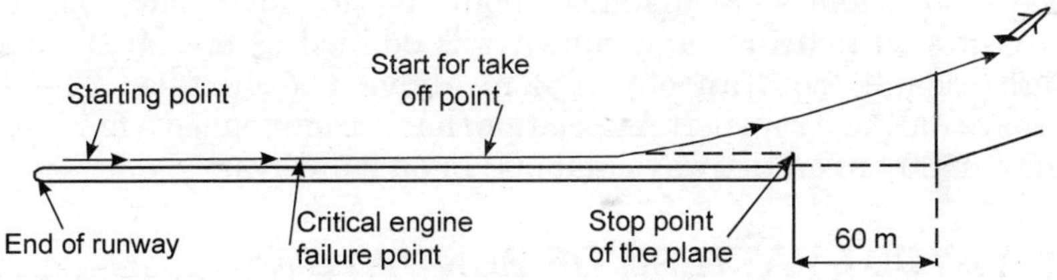

Figure 21.2 *Runway limitation for take off of planes*

The division of the length of a runway between different zones for landing and take off are shown in Fig. 21.2. The take off of the aircraft is more critical than the landing. Only 60% of the runway length is used for take off and landing and the remaining 40% is used as a margin for unforeseen or eventualities.

21.8 TAXI WAYS

It is defined as a paved path provided for the purpose of allowing aircrafts to move to and from the runway and the apron. Movement of heavy aircrafts

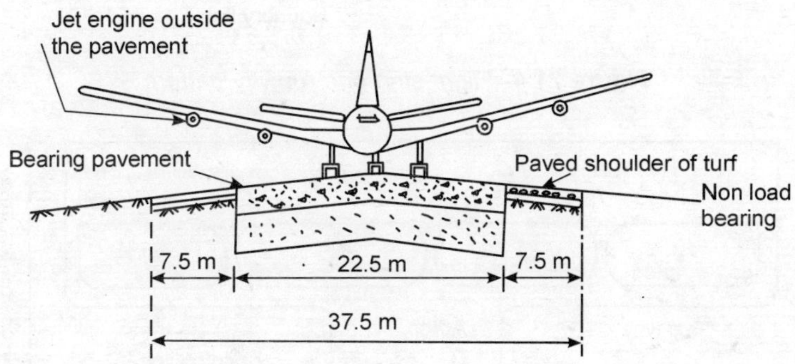

Figure 21.3 *Taxi way*

on the ground means heavy consumption of fuel. The length of taxiways should be as small as possible. No standard specification have been laid down by any organisation regarding the minimum length of taxiways. The speed of aircrafts is much less on the taxiways than on the runways. Due to this reason the width of taxiways is less than the runways and it varies from 12 m to 23 m depending upon the type of aircrafts using it. The centre to centre distance between two taxiways varies between 38 m to 100 m.

21.8 SIGHT DISTANCE

For smooth and uninterrupted movement of aircrafts on taxiways and runways, sufficient sight distance should be provided. The sight distance for aircrafts on taxiways and runways is defined as a clear distance on a straight reach from an object 3 m above the surface. The I.A.T.A (International Air Transport Association) has recommended a minimum sight distance of 300 m on taxiways and 500 in on runways.

21.9 LAYOUT PATTERN OF RUNWAYS

The layout pattern of runways may be classified as :

(i) One/way (ii) Two/way and (iii) Three way.

The various type of runways are illustrated in the following Figs. 21.5 and 21.6

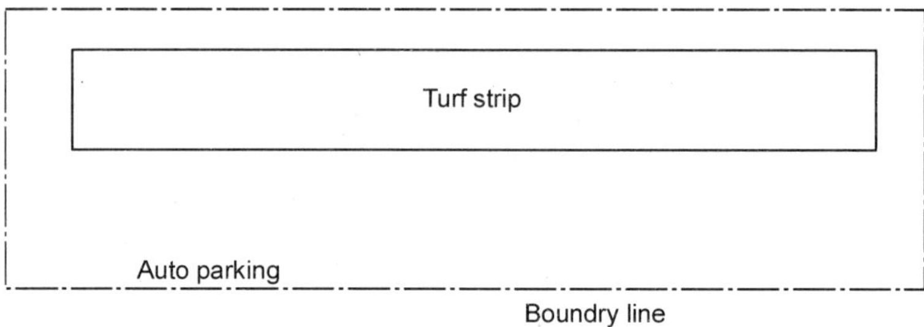

Figure 21.4 *Flight-strip one runway pattern*

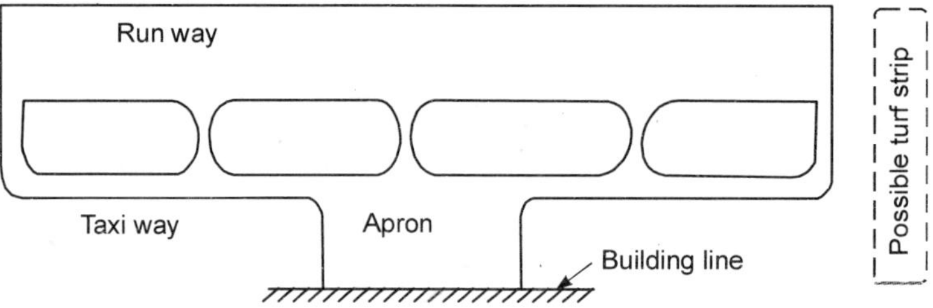

Figure 21.5 *Single runway pattern*

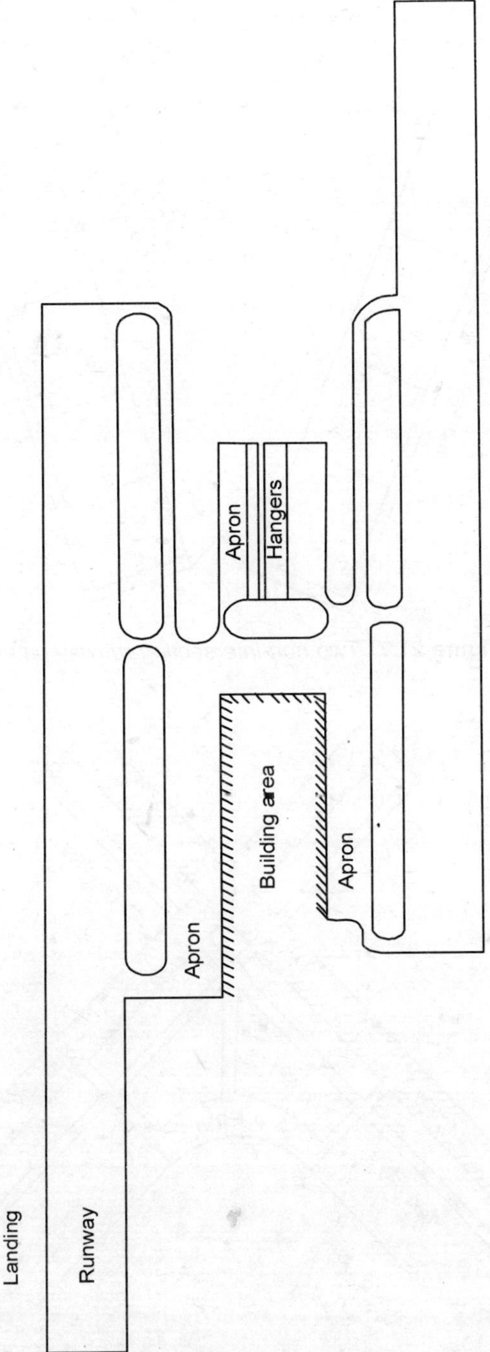

Figure 21.6 *Parallel runway*

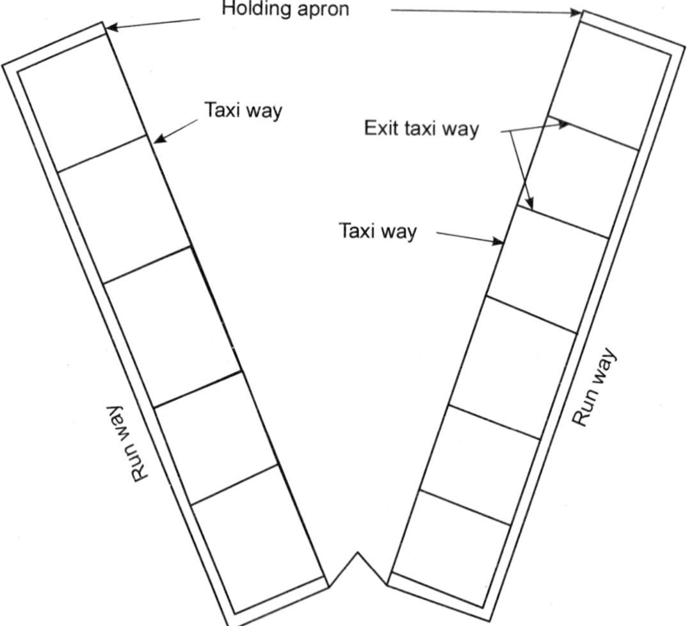

Figure 21.7 *Two non-intersecting runway pattern*

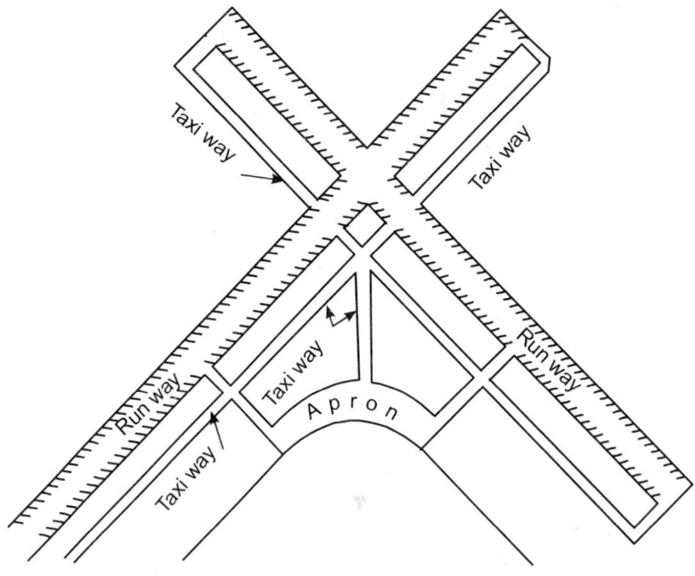

Figure 21.8 *Two non-intersecting runway pattern*

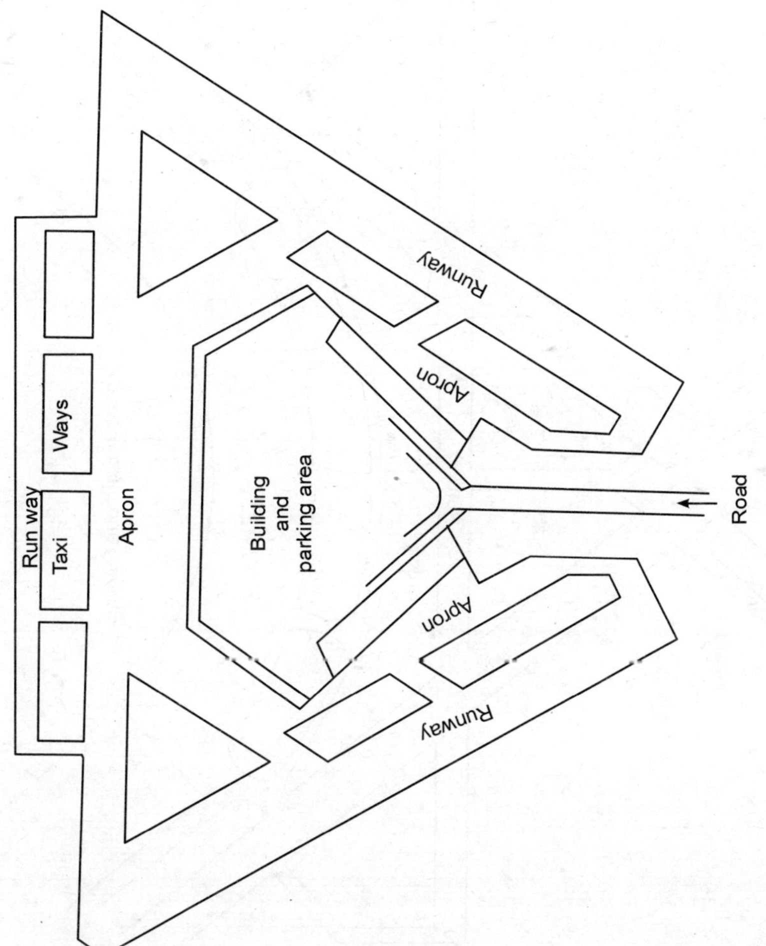

Figure 21.9 *Three non-intersecting runway pattern*

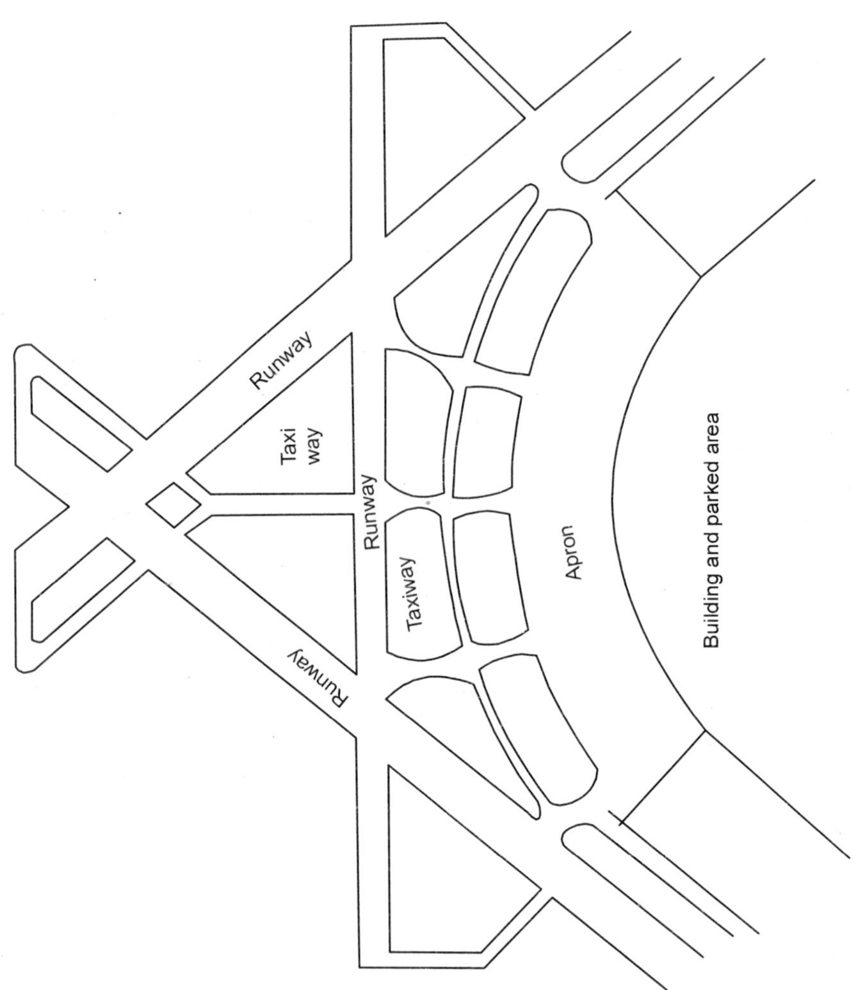

Figure 21.10 *Three intersecting runway pattern*

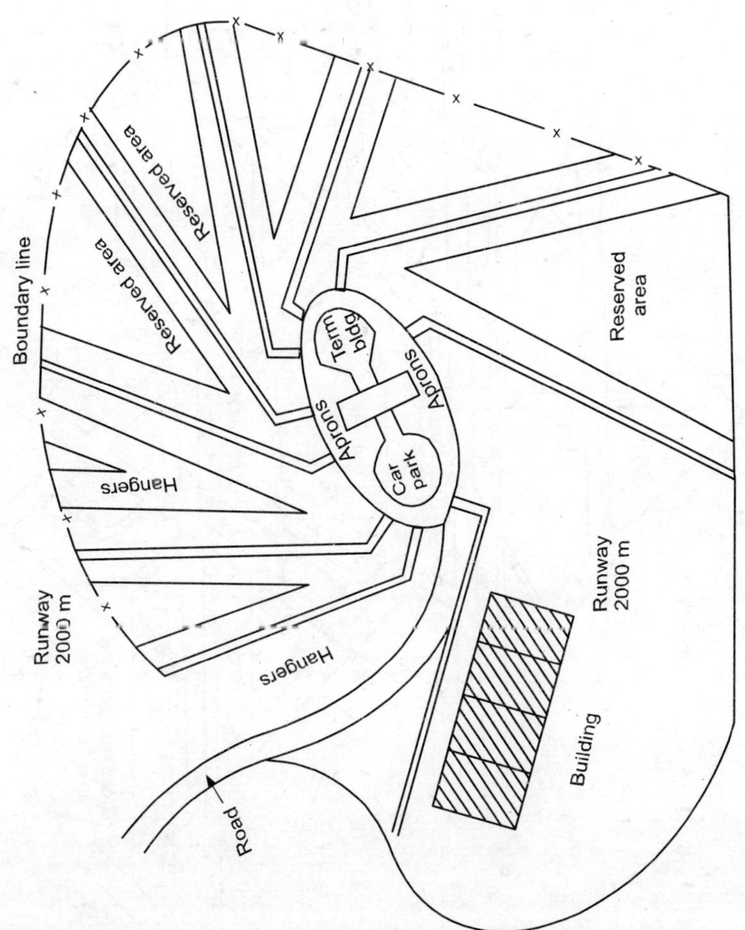

Figure 21.11 *Tangential runways pattern*

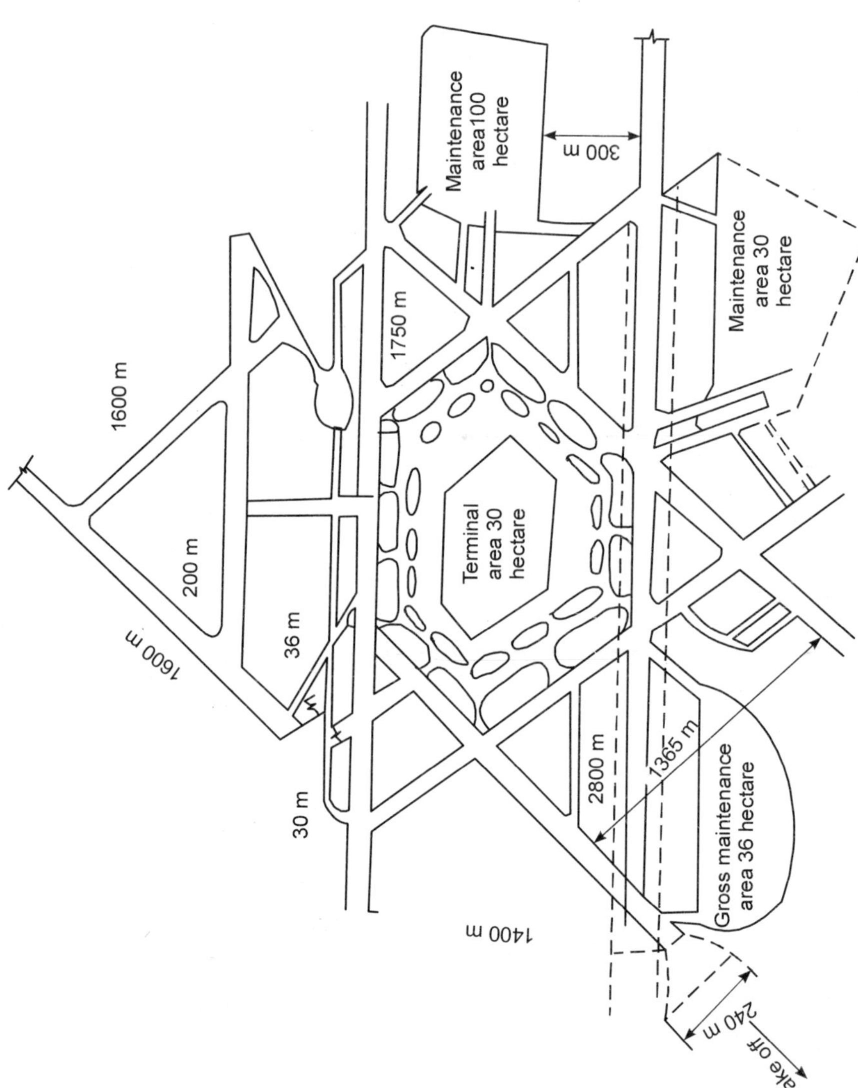

Figure 21.12 *Combination of various pattern of runways*

21.10 RUNWAYS AND INTENSITY OF TRAFFIC

The intensity of traffic and the metrological conditions decide the orientation of runways. Each runway has its capacity of accommodating aeroplanes. It will not be astonishing to note that in good weather each runway can conveniently allow 25–30 movements per hour but they will depend on the speed and direction of wind also and on the physical limitations of the site. Generally the landing and take off is done in the opposite direction of winds. The small aeroplanes are more sensitive to cross wind than bigger aeroplanes obviously. The capacity and intensity of traffic will largely depend on the cross winds and their cross component, (cross component of any wind is equal to wind velocity multiplied by sine of the angle between the runway and direction of wind). Runways should be properly oriented so that air craft could operate with the cross component of the wind at 18 kilometre per hour. For proper orientation of runway proper records of wind speed and direction should be obtained from the metrological department. Three or four directional layout are possible for runways.

21.11 TERMINAL AREA

The terminal area is meant for the transition of passengers from ground to the air. This includes space for airline operations, office for airport management and for providing facilities like booking check in, restaurant, shopping etc. Various methods are used for the transit of passengers and the cargo. The terminal area also includes parking of automobiles, servicing and storage of automobiles and aircrafts. A paved area infront of the terminal building and the landing area, called apron, is provided for parking and berthing of aircrafts. Hangers for seming, repair and storage of aircrafts are also a component of terminal area.

21.12 INTERNATIONAL CIVIL AVIATION ORGANISATION

The I.C.A.O was set up in the year 1947 for setting standards and norms for the operation and management of airports for providing comforts and adopt safety measures. With the development and introduction of supersonic and jet aircrafts the importance of I.C.A.O has increased tremendously. On the recommendations of International Civil Aviation Organisation (I.C.A.O) and the International Airport Committee an International Airport Authority of India (I.A.A.I) was set up in the year 1972 to develop, manage and control the International airports of the country.

The I.C.A.O sets standards for the design, construction and development of airports on international basis. The International Airport Authority of India (I.A.A.I) functions under the international norms. According to I.C.A.O the runway lengths are coded as A, B, C, D, E, F and G.

Code	Runway length
A	2500 m and above
B	2150 m to 2500 m
C	1800 m to 2150 m
D	1500 m to 1800 m
E	1280 m to 1500 m
F	1080 m to 1280 m
G	980 m to 1080 m

21.13 GEOMETRIC STANDARDS

I.C.A.O has laid down certain geometric standards for international airports throughout the world. Following are the international standards for the various elements of an airport.

Airport code	*Clearance between the parallel*	*Distance between runway and taxiway*	*Distance between taxiway*	*Distance between taxi-way and building*
A	210 m	165 m	70 m	39 m
B	210 m	150 m	69 m	39 m
C	210 m	150 m	52 m	30 m
D	150 m	150 m	52 m	30 m
E	150 m	150 m	45 m	24 m
F	150 m	145 m	33 m	24 m
G	150 m	145 m	26 m	18 m

The minimum distance from the centre line of runway to the building line should be 150 metres for all classes of airports.

For camber and gradient on the runways and taxiways and their width is shown in the following table :

Airport code	*Runway width*	*Taxiway wdith*	*Max. gradient*	*Effective gradient*	*Camber*
A	45 m	23 m	1 in 80	1 in 100	1 in 66
B	45 m	23 m	1 in 80	1 in 10	1 in 66.
C	45 m	23 m	1 in 66	1 in 100	1 in 66
D	45 m	18 m	1 in 66	1 in 100	1 in 66
E	45 m	15 m	1 in 66	1 in 100	1 in 66
F	30 m	12.5 m	1 in 66	1 in 100	1 in 66
G	30 m	12.5 m	1 in 66	1 in 100	1 in 66

21.14 GEOMETRIC DESIGN

The geometric design of a runway consists of the following elements :

1. Length of runway
2. Width
3. Sight distance
4. Gradients

The length of a runway as shown in the table of classification are the standard but corrections can be made depending upon the altitude, temperature and other conditions. For the design of an airport the above table should be modified in the following manner.

Basic runway length given in the table is for airports at sea level on a perfectly level ground and standard temperature of 15°C. The length of the runway must be corrected for actual reduced level and for mean temperature. The length of the runway can be increased by 7% for every 300 m rise in elevation or altitude. For temperature also the correction should be made by increasing the length by 1 % for every 1 °C rise in temperature. It is also assumed that the rise in temperature is 0.0065°C per metre rise in elevation.

Minimum width of the landing strip is 150 metres and its length is 120 metres more than the length of runway extending 60 m on both the sides or ends of the runway.

Rate of change of vertical gradient is 0.3% per 30 metre of vertical curve. Grading beyond runway ends should be 60 metres on both the sides. The transverse grade of shoulders should be 2.5% upto 75 m from the centre line of runway and for the remaining portion it may be 5%. The effective gradient is defined on the difference in elevation of the highest point and the lowest point divided by the length of the runway.

The sight distance should be such that any two points 3 metre above the surface of runway should be mutually visible within a distance equal to half the length of runway.

The manner in which an aircraft actually performs the take off and landing will decide to a large extend the length of a runway.

21.15 STRUCTURAL DESIGN OF RUNWAY PAVEMENTS

(*i*)*Flexible Pavements.* The flexible pavement of a runway consists of sub-grade having C.B.R value between 5 and 10%, the sub-base having a value of 20 to 25%, a base with C.B.R value of 80 to 90%, and a bituminous surfacing 5 – 10 cm thick. Design charts and curves are prepared and corresponding total thickness of the pavement is worked out corresponding to the C.B.R value. The total thickness of a runway pavement consists of sub-base, base and the surfacing thickness. Depending upon the C.B.R value the thickness of each component or element of the pavement can be worked out from the design charts. The theoretical methods of calculation are more or less the same as for other pavements.

(ii) Rigid pavements. Rigid pavements also consists of sub-grade of C.B.R value 5 – 10%, sub-base of CBR value 20 – 25%, base course of CBR 80–90% and 15 – 25 cm thick cement concrete laid in two layers. It has been observed that the weakest point in the rigid pavement is corner of the slab. The design procedure is the same as applicable to other highway pavements. The pioneering work in the design of rigid pavements can be attributed to Westergaad. This analysis was based on the following assumptions :

1. The rigid or concrete pavement is homogeneous and has uniform elastic properties.

2. The reaction of the sub-grade at a point is equal to the deflection of that point, where it is a constant, being modulus of subgrade reaction.

3. The reaction of the sub-grade at a point is equal to *kx* deflection at that point, where *k* is a constant, being modules of subgrade reaction.

4. The slab is of uniform thickness. Westergaad presented the following two formulae for calculating the stresses at the edge of the slab as well as at the interior :

$$S_1 = \frac{0.316p}{h^2}\,[\,4 \log_{10}(1/b) + 1.069]$$

and S_2

$$= \frac{0.572p}{h^2}\,[\,4 \log_{10}(1/b) + 0.359]$$

The equation for corner loading is

$$S_0 = \frac{3p}{3h}\;1 - \frac{a/2}{l}^{\,0.6}$$

S_1 = Maximum stress developed at the inside or interior of the slab

S_2 = Maximum stress developed at the corner of the slab

h = Thickness of slab

p = Wheel load in kg

a = Radius of wheel load distribution

b = radius of relative stiffness

b = radius of resisting thickness

Radius of relative thickness, can be calculated by the following formula.

$$l = \frac{Ed^B}{12\,(1-\mu^2)k}^{0.25}$$

where

E = Modulus of elasticity of concrete

μ = Poisson's ratio for concrete

d = Design wheel load

When the load transference devise is used the weakest point is considered to be centre of the pavement slab as the aircraft wheels seldom go near the edge of the slab. The stress in the centre of the slab is given by Westergaad formula is

$$s = 0.275\,(1+\mu)\,\frac{W}{d^2}\log_{10}\frac{Ed^2}{kb^2}\ \text{kg/cm}^2$$

where

$b = (1.6\,a^2 + d^2)^{1/2} - 0.675\,d$ if

$\frac{a}{d} < 1.724$ and $b = a$ if $\frac{a}{d} > 1.724$

21.16 AIRPORT DRAINAGE

Proper and efficient drainage system of airports is very important. Disposal of surface and sub-surface drainage is essential for proper and safe functioning of the airport. The airstrip must remain dry and no water should be allowed to collect on the surface of the airstrip. The sub-surface water may also be drained off as quickly as possible so that the subgrade does not get weakened. The weakening of the sub-grade will result in its reduced bearing capacity, resulting in the failure of pavements of runways, taxiways etc.

The surface drainage can be either through open drains or pipes. The runoff is calculated by using Rational flood formula and the capacity or size of pipe of open drains can be calculated by Manning's formula assuming minimum flow to be 0.6 metre/second.

Catch basins are provided at an interval of 30 m to 90 m along the edge of the landing area to collect water from the surface of runways. Lateral drains or pipes are provided under the catch basins to carry the collected water to the underground sewer. Natural drainage system of the airport should be followed for surface drainage.

If the water table is high, sub-surface drainage shall be provided to lower the water table so as to prevent weakening of the subgrade. Underground drains or pipes should be used as discussed in highway drainage.

PROBLEMS

21.1. Write short notes on :

Cross component of wind, sight distance, Apron.

21.2. Elucidate how taxiways are planned in an airfield and what are their requirements.

21 3. How are runways oriented. What important factors mainly govern the orientation of runways?

21.4. Draw layout plans of any two type of runway patterns.